NICKEL ALLOYS

MECHANICAL ENGINEERING

A Series of Textbooks and Reference Books

Founding Editor

L. L. Faulkner

Columbus Division, Battelle Memorial Institute
and Department of Mechanical Engineering
The Ohio State University
Columbus, Ohio

1. *Spring Designer's Handbook*, Harold Carlson
2. *Computer-Aided Graphics and Design*, Daniel L. Ryan
3. *Lubrication Fundamentals*, J. George Wills
4. *Solar Engineering for Domestic Buildings*, William A. Himmelman
5. *Applied Engineering Mechanics: Statics and Dynamics*, G. Boothroyd and C. Poli
6. *Centrifugal Pump Clinic*, Igor J. Karassik
7. *Computer-Aided Kinetics for Machine Design*, Daniel L. Ryan
8. *Plastics Products Design Handbook, Part A: Materials and Components; Part B: Processes and Design for Processes*, edited by Edward Miller
9. *Turbomachinery: Basic Theory and Applications*, Earl Logan, Jr.
10. *Vibrations of Shells and Plates*, Werner Soedel
11. *Flat and Corrugated Diaphragm Design Handbook*, Mario Di Giovanni
12. *Practical Stress Analysis in Engineering Design*, Alexander Blake
13. *An Introduction to the Design and Behavior of Bolted Joints*, John H. Bickford
14. *Optimal Engineering Design: Principles and Applications*, James N. Siddall
15. *Spring Manufacturing Handbook*, Harold Carlson
16. *Industrial Noise Control: Fundamentals and Applications*, edited by Lewis H. Bell
17. *Gears and Their Vibration: A Basic Approach to Understanding Gear Noise*, J. Derek Smith
18. *Chains for Power Transmission and Material Handling: Design and Applications Handbook*, American Chain Association
19. *Corrosion and Corrosion Protection Handbook*, edited by Philip A. Schweitzer
20. *Gear Drive Systems: Design and Application*, Peter Lynwander
21. *Controlling In-Plant Airborne Contaminants: Systems Design and Calculations*, John D. Constance
22. *CAD/CAM Systems Planning and Implementation*, Charles S. Knox
23. *Probabilistic Engineering Design: Principles and Applications*, James N. Siddall
24. *Traction Drives: Selection and Application*, Frederick W. Heilich III and Eugene E. Shube

25. *Finite Element Methods: An Introduction*, Ronald L. Huston and Chris E. Passerello
26. *Mechanical Fastening of Plastics: An Engineering Handbook*, Brayton Lincoln, Kenneth J. Gomes, and James F. Braden
27. *Lubrication in Practice: Second Edition*, edited by W. S. Robertson
28. *Principles of Automated Drafting*, Daniel L. Ryan
29. *Practical Seal Design*, edited by Leonard J. Martini
30. *Engineering Documentation for CAD/CAM Applications*, Charles S. Knox
31. *Design Dimensioning with Computer Graphics Applications*, Jerome C. Lange
32. *Mechanism Analysis: Simplified Graphical and Analytical Techniques*, Lyndon O. Barton
33. *CAD/CAM Systems: Justification, Implementation, Productivity Measurement*, Edward J. Preston, George W. Crawford, and Mark E. Coticchia
34. *Steam Plant Calculations Manual*, V. Ganapathy
35. *Design Assurance for Engineers and Managers*, John A. Burgess
36. *Heat Transfer Fluids and Systems for Process and Energy Applications*, Jasbir Singh
37. *Potential Flows: Computer Graphic Solutions*, Robert H. Kirchhoff
38. *Computer-Aided Graphics and Design: Second Edition*, Daniel L. Ryan
39. *Electronically Controlled Proportional Valves: Selection and Application*, Michael J. Tonyan, edited by Tobi Goldoftas
40. *Pressure Gauge Handbook*, AMETEK, U.S. Gauge Division, edited by Philip W. Harland
41. *Fabric Filtration for Combustion Sources: Fundamentals and Basic Technology*, R. P. Donovan
42. *Design of Mechanical Joints*, Alexander Blake
43. *CAD/CAM Dictionary*, Edward J. Preston, George W. Crawford, and Mark E. Coticchia
44. *Machinery Adhesives for Locking, Retaining, and Sealing*, Girard S. Haviland
45. *Couplings and Joints: Design, Selection, and Application*, Jon R. Mancuso
46. *Shaft Alignment Handbook*, John Piotrowski
47. *BASIC Programs for Steam Plant Engineers: Boilers, Combustion, Fluid Flow, and Heat Transfer*, V. Ganapathy
48. *Solving Mechanical Design Problems with Computer Graphics*, Jerome C. Lange
49. *Plastics Gearing: Selection and Application*, Clifford E. Adams
50. *Clutches and Brakes: Design and Selection*, William C. Orthwein
51. *Transducers in Mechanical and Electronic Design*, Harry L. Trietley
52. *Metallurgical Applications of Shock-Wave and High-Strain-Rate Phenomena*, edited by Lawrence E. Murr, Karl P. Staudhammer, and Marc A. Meyers
53. *Magnesium Products Design*, Robert S. Busk
54. *How to Integrate CAD/CAM Systems: Management and Technology*, William D. Engelke
55. *Cam Design and Manufacture: Second Edition*; with cam design software for the IBM PC and compatibles, disk included, Preben W. Jensen

56. *Solid-State AC Motor Controls: Selection and Application*, Sylvester Campbell
57. *Fundamentals of Robotics*, David D. Ardayfio
58. *Belt Selection and Application for Engineers*, edited by Wallace D. Erickson
59. *Developing Three-Dimensional CAD Software with the IBM PC*, C. Stan Wei
60. *Organizing Data for CIM Applications*, Charles S. Knox, with contributions by Thomas C. Boos, Ross S. Culverhouse, and Paul F. Muchnicki
61. *Computer-Aided Simulation in Railway Dynamics*, by Rao V. Dukkipati and Joseph R. Amyot
62. *Fiber-Reinforced Composites: Materials, Manufacturing, and Design*, P. K. Mallick
63. *Photoelectric Sensors and Controls: Selection and Application*, Scott M. Juds
64. *Finite Element Analysis with Personal Computers*, Edward R. Champion, Jr., and J. Michael Ensminger
65. *Ultrasonics: Fundamentals, Technology, Applications: Second Edition, Revised and Expanded*, Dale Ensminger
66. *Applied Finite Element Modeling: Practical Problem Solving for Engineers*, Jeffrey M. Steele
67. *Measurement and Instrumentation in Engineering: Principles and Basic Laboratory Experiments*, Francis S. Tse and Ivan E. Morse
68. *Centrifugal Pump Clinic: Second Edition, Revised and Expanded*, Igor J. Karassik
69. *Practical Stress Analysis in Engineering Design: Second Edition, Revised and Expanded*, Alexander Blake
70. *An Introduction to the Design and Behavior of Bolted Joints: Second Edition, Revised and Expanded*, John H. Bickford
71. *High Vacuum Technology: A Practical Guide*, Marsbed H. Hablanian
72. *Pressure Sensors: Selection and Application*, Duane Tandeske
73. *Zinc Handbook: Properties, Processing, and Use in Design*, Frank Porter
74. *Thermal Fatigue of Metals*, Andrzej Weronski and Tadeusz Hejwowski
75. *Classical and Modern Mechanisms for Engineers and Inventors*, Preben W. Jensen
76. *Handbook of Electronic Package Design*, edited by Michael Pecht
77. *Shock-Wave and High-Strain-Rate Phenomena in Materials*, edited by Marc A. Meyers, Lawrence E. Murr, and Karl P. Staudhammer
78. *Industrial Refrigeration: Principles, Design and Applications*, P. C. Koelet
79. *Applied Combustion*, Eugene L. Keating
80. *Engine Oils and Automotive Lubrication*, edited by Wilfried J. Bartz
81. *Mechanism Analysis: Simplified and Graphical Techniques, Second Edition, Revised and Expanded*, Lyndon O. Barton
82. *Fundamental Fluid Mechanics for the Practicing Engineer*, James W. Murdock
83. *Fiber-Reinforced Composites: Materials, Manufacturing, and Design, Second Edition, Revised and Expanded*, P. K. Mallick
84. *Numerical Methods for Engineering Applications*, Edward R. Champion, Jr.

85. *Turbomachinery: Basic Theory and Applications, Second Edition, Revised and Expanded*, Earl Logan, Jr.
86. *Vibrations of Shells and Plates: Second Edition, Revised and Expanded*, Werner Soedel
87. *Steam Plant Calculations Manual: Second Edition, Revised and Expanded*, V. Ganapathy
88. *Industrial Noise Control: Fundamentals and Applications, Second Edition, Revised and Expanded*, Lewis H. Bell and Douglas H. Bell
89. *Finite Elements: Their Design and Performance*, Richard H. MacNeal
90. *Mechanical Properties of Polymers and Composites: Second Edition, Revised and Expanded*, Lawrence E. Nielsen and Robert F. Landel
91. *Mechanical Wear Prediction and Prevention*, Raymond G. Bayer
92. *Mechanical Power Transmission Components*, edited by David W. South and Jon R. Mancuso
93. *Handbook of Turbomachinery*, edited by Earl Logan, Jr.
94. *Engineering Documentation Control Practices and Procedures*, Ray E. Monahan
95. *Refractory Linings Thermomechanical Design and Applications*, Charles A. Schacht
96. *Geometric Dimensioning and Tolerancing: Applications and Techniques for Use in Design, Manufacturing, and Inspection*, James D. Meadows
97. *An Introduction to the Design and Behavior of Bolted Joints: Third Edition, Revised and Expanded*, John H. Bickford
98. *Shaft Alignment Handbook: Second Edition, Revised and Expanded*, John Piotrowski
99. *Computer-Aided Design of Polymer-Matrix Composite Structures*, edited by Suong Van Hoa
100. *Friction Science and Technology*, Peter J. Blau
101. *Introduction to Plastics and Composites: Mechanical Properties and Engineering Applications*, Edward Miller
102. *Practical Fracture Mechanics in Design*, Alexander Blake
103. *Pump Characteristics and Applications*, Michael W. Volk
104. *Optical Principles and Technology for Engineers*, James E. Stewart
105. *Optimizing the Shape of Mechanical Elements and Structures*, A. A. Seireg and Jorge Rodriguez
106. *Kinematics and Dynamics of Machinery*, Vladimír Stejskal and Michael Valášek
107. *Shaft Seals for Dynamic Applications*, Les Horve
108. *Reliability-Based Mechanical Design*, edited by Thomas A. Cruse
109. *Mechanical Fastening, Joining, and Assembly*, James A. Speck
110. *Turbomachinery Fluid Dynamics and Heat Transfer*, edited by Chunill Hah
111. *High-Vacuum Technology: A Practical Guide, Second Edition, Revised and Expanded*, Marsbed H. Hablanian
112. *Geometric Dimensioning and Tolerancing: Workbook and Answerbook*, James D. Meadows
113. *Handbook of Materials Selection for Engineering Applications*, edited by G. T. Murray

114. *Handbook of Thermoplastic Piping System Design*, Thomas Sixsmith and Reinhard Hanselka
115. *Practical Guide to Finite Elements: A Solid Mechanics Approach*, Steven M. Lepi
116. *Applied Computational Fluid Dynamics*, edited by Vijay K. Garg
117. *Fluid Sealing Technology*, Heinz K. Muller and Bernard S. Nau
118. *Friction and Lubrication in Mechanical Design*, A. A. Seireg
119. *Influence Functions and Matrices*, Yuri A. Melnikov
120. *Mechanical Analysis of Electronic Packaging Systems*, Stephen A. McKeown
121. *Couplings and Joints: Design, Selection, and Application, Second Edition, Revised and Expanded,* Jon R. Mancuso
122. *Thermodynamics: Processes and Applications*, Earl Logan, Jr.
123. *Gear Noise and Vibration*, J. Derek Smith
124. *Practical Fluid Mechanics for Engineering Applictions,* John J. Bloomer
125. *Handbook of Hydraulic Fluid Technology*, edited by George E. Totten
126. *Heat Exchanger Design Handbook*, T. Kuppan
127. *Designing for Product Sound Quality*, Richard H. Lyon
128. *Probability Applications in Mechanical Design*, Franklin E. Fisher and Joy R. Fisher
129. *Nickel Alloys,* edited by Ulrich Heubner

Additional Volumes in Preparation

Rotating Machinery Vibration: Problem Analysis and Troubleshooting, Maurice L. Adams

Handbook of Machinery Dynamics, Lynn Faulkner and Earl Logan, Jr.

Rapid Prototyping Technology: Selection and Application, Ken Cooper

Reliability Verification, Testing, and Analysis of Engineering Design, Gary S. Wasserman

Maintenance Excellence: Optimizing Equipment Life Cycle Decisions, edited by John D. Campbell

Formulas for Dynamic Analysis, Ronald Huston and C. Q. Liu

Mechanical Engineering Software

Spring Design with an IBM PC, Al Dietrich

Mechanical Design Failure Analysis: With Failure Analysis System Software for the IBM PC, David G. Ullman

NICKEL ALLOYS

edited by

ULRICH HEUBNER

Consultant
Werdohl, Germany

contributors

Ulrich Brill
Theo Hoffmann
Rolf Kirchheiner
Jutta Klöwer
Reiner Köcher
Manfred Rockel
Frederick White

MARCEL DEKKER, INC. NEW YORK • BASEL

Although every effort has been made to provide dependable information, the publisher and authors cannot be held responsible for any errors or omissions.

ISBN: 0-8247-0440-1

This book is printed on acid-free paper.

Distributed by Marcel Dekker, Inc.

Headquarters
Marcel Dekker, Inc.
270 Madison Avenue, New York, NY 10016
tel: 212-696-9000; fax: 212-685-4540

Eastern Hemisphere Distribution
Marcel Dekker AG
Hutgasse 4, Postfach 812, CH-4001 Basel, Switzerland
tel: 41-61-261-8482; fax: 41-61-261-8896

World Wide Web
http://www.dekker.com

The publisher offers discounts on this book when ordered in bulk quantities. For more information, write to Special Sales/Professional Marketing at the headquarters address above.

Current printing (last digit):
10 9 8 7 6 5 4 3 2 1

PRINTED IN THE UNITED STATES OF AMERICA

Preface of the editor

For their use in chemical plant construction and energy engineering, nickel alloys and high-alloy special stainless steels are characterised primarily by their material number, price, wear rate in various corrosive media, and their most important mechanical and metallurgical characteristics. However, other special aspects not infrequently require consideration. For instance, the question of weldability arises again and again when a new application is under consideration. In the field of corrosion behaviour, resistance to pitting and crevice corrosion has gained in significance in recent years. New applications demand new solutions in the field of materials, and use must be made of basic knowledge in the field of materials science where no long-term practical experience is available.

This book, then, is intended to improve the users' level of knowledge about this group of materials, to draw attention to new developments and know-how, and to provide general assistance in the search for the most economical solutions to problems of materials in chemical equipment construction, power-station engineering and high-temperature technology.

Table of contents

1 Nickel alloys and high-alloy special stainless steels – materials summary and metallurgical principles

U. Heubner

1.1 Introduction

Nickel materials and high-alloy special stainless steels are becoming increasingly important for chemical plant construction, power- and offshore engineering. Nickel alloys and high-alloy special stainless steels represent a group of materials which has largely become established between the widespread stainless steels [1,2] and the so-called superalloys [3-5] and which extends into both. In many cases, the definition as steel or nickel alloy is of a purely formal nature. For this reason it is appropriate and useful, from both the application and materials point of view, to treat nickel alloys and high-alloy special stainless steels as a single materials group. Since the materials involved have been developed in response to specific applications, it is appropriate to subdivide the materials summary to be presented here according to principal applications.

1.2 Materials summary according to material composition and areas of application

1.2.1 Corrosive attack by reducing acids: sulphuric acid, phosphoric acid, hydrochloric acid and organic acids

Sulphuric acid in concentrations below approximately 85 % is the most important representative of what are known in industry as 'reducing acids'. Although scientifically incorrect, this term has become established for those acids in which, in the course of metal dissolution and thus in the case of corrosion, hydrogen ions are reduced to molecular hydrogen, thereby becoming the sole or principal oxidation agent. As soon as these acids become aerated, oxygen from the air becomes an additional oxidation agent. Furthermore, dissolved oxidizing impurities such as N_2O_3 or As_2O_3 as well as dissolved higher-valence heavy metal ions such as Fe^{3+} and Cu^{2+} lend these so-called reducing media an increasingly oxidizing character, the potential being shifted to more noble values than could have been achieved by hydrogen reduction alone. For high-alloy stainless steels it may be only by means of such oxidizing additions that passivation and, consequently, corrosion resistance in high-temperature sulphuric acid in the concentration range below about 85 % is achieved [6, 7]. It is for this reason that a high-alloy stainless steel may be corrosion resistant in, e.g. „chemical grade" sulphuric acid of 80 % at 80° C but may corrode actively in „analysis grade" sulphuric acid under the same conditions [7]. Though chemical composition and microstructure of nickel-alloys and high-alloy special stainless steels are a necessary prerequisite for their resistance to corrosion, as will be pointed out below, corrosion resistance is not a materials property but a mode of behaviour, i.e., it is always dependent on an interaction between the material and the surrounding corrosive medium. An increase in temperature also brings about a shift in potential to more noble values until transpassive corrosion occurs and the materials lose corrosion resistance, as can be seen in the isocorrosion diagrams (cf. Chapter 8, Fig. 8.5-8.8). These isocorrosion diagrams are valid for the materials' interaction with „chemical grade" sulphuric acid.

Table 1.1: Nickel alloys and special stainless steels for corrosive attack by reducing acids: sulphuric acid, phosphoric acid, hydrochloric acid, organic acids

Krupp VDM trade name	**Material No. DIN**	**Main alloying elements (typical values in %)**					
		Ni	**Cr**	**Mo**	**Cu**	**Fe**	**others**
Cronifer							
2205 LCN - alloy 318 LN	1.4462	6	22	2.8		66	0.14 N
1810 LC - alloy 316 L	1.4404	12	17	2.3		66	
1810 Ti	1.4571	12	17	2.3		66	0,4 Ti
1812 LC	1.4435	13	17	2.6		65	
1713 LCN (- alloy 317 LN)	1.4439	13	18	4.5		62	0.16 N
1925 LC - alloy 904 L	1.4539	25	21	4.8	1.5	46	
1925 hMo - alloy 926	1.4529	25	21	6.5	0.9	45	0.20 N
2328	1.4503	27	23	2.8	2.8	43	0.6 Ti
Nicrofer							
3127 LC - alloy 28	1.4563	31	27	3.5	1.3	35	
3127 hMo - alloy 31	1.4562	31	27	6.5	1.3	31	0.20 N
3033 - alloy 33	1.4591	31	33	1.6	0.6	32	0.40 N
3620 Nb - alloy 20	2.4660	38	20	2.4	3.4	34	0.2 Nb
4221 - alloy 825	2.4858	40	23	3.2	2.2	31	0.8 Ti
6020 hMo - alloy 625	2.4856	62	22	9		3	3.4 Nb
5716 hMoW - alloy C-276	2.4819	57	16	16		6	3.5 W
5923 hMo - alloy 59	2.4605	59	23	16		1	
6616 hMo - alloy C-4	2.4610	66	16	16		1	
Nimofer							
6928 - alloy B-2	2.4617	69	0.7	28		1.7	
6629 - alloy B-4	2.4600	68	1.2	27		3	
6224 - alloy B-10	2.4604	62	7.5	24		6	

*) Cronifer, Nicrofer, Nimofer, Nicorros, Cunifer and Cronix are registered trade names of Krupp VDM GmbH.

The range of materials shown in Table 1.1 can be considered for corrosive attack by sulphuric acid and other reducing acids as defined earlier. The pairs of figures attached to each designation Cronifer and Nicrofer*) indicate the percentage contents of the two main alloy components, chromium and nickel (or vice versa), in accordance with the order of the syllables. However, developments have taken place without the name being changed. The suffix LC signifies an alloy with particularly low carbon content (Low Carbon), and N, Nb and Ti indicate the presence of the alloying elements nitrogen, niobium and titanium. The designation hMo indicates high molybdenum; W indicates the addition of tungsten and Si of silicon. Some minor alloying additions made for manufacturing requirements, particularly those of manganese and silicon, and a few other elements, are not included. Examining Table 1.1 from the top downwards, i.e. from low to high nickel contents, it is apparent that, with exception of the nickel-molybdenum alloys listed below, chromium contents are always within the range 16 to 33 %. In fact a chromium content of at least 12 % is necessary for the formation of stable passive surface layers, upon which these materials' corrosion resistance depends. *As a rule, the more oxidizing agents are present besides the hydrogen ions, the higher the chromium content has to be. This is the first basic principle for the selection of a material. The very high chromium content of the new alloy Nicrofer 3033 is a striking example of this.*

Examining Table 1.1 from top to bottom, that is towards higher nickel contents, general resistance to corrosion increases, since it is only due to their nickel contents that stainless iron-chromium alloys gain their resistance to acids [8]. *The increase in corrosion resistance with increasing nickel content should, in this context, be regarded as the second basic principle for the selection of a material.* However, there may be individual deviations from this general principle, since other alloy components and, most specifically, chromium, molybdenum and copper, also play a role. For example, at lower acid concentrations and lower temperatures, the ferritic-austenitic steel at the very top is somewhat superior to the subsequent, low-alloy austenitic steels, and Cronifer 2328 (1.4503), with only 27 % nickel, demonstrates lower general corrosion in sulphuric acid than Nicrofer 4221 - alloy 825 (2.4858), which has a higher nickel content (Figure 1.1). Here the differences in chromium content, as well as in the alloying elements molybdenum and, above all, copper (described later) have a greater effect than the differences in the nickel content.

Figure 1.2 clearly shows, that the basic principle of increasing corrosion resistance with increasing nickel content, also applies to the resistance to stress-corrosion cracking. The ferritic-austenitic steels such as Cronifer 2205 LCN - alloy 318 LN (1.4462), indicated in Table 1.1 also respect this basic principle [9, 10]. They present themselves as more resistant than the low-alloy austenitic steels below, only at low stress levels , thus offering no advantage at all from their considerably higher 0.2 % proof stress.

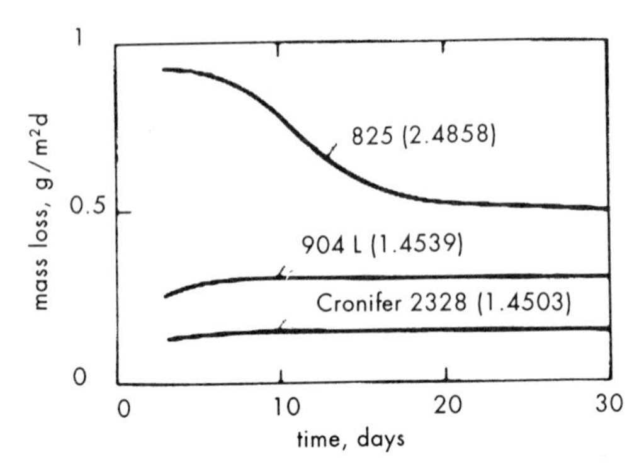

Figure 1.1: Corrosion rates of Cronifer 2328 (1.4503), Cronifer 1925 LC - alloy 904 L (1.4539), and Nicrofer 4221 - alloy 825 (2.4858) in a technical sulphuric acid solution of a zinc electrolysis (140g/l H_2SO_4, 1g/l Cl^-, 60g/l Zn, 30g/l Fe^{3+}, 8g/l Mn^{2+} at 80 °C) according to M.B. Rockel, unpublished

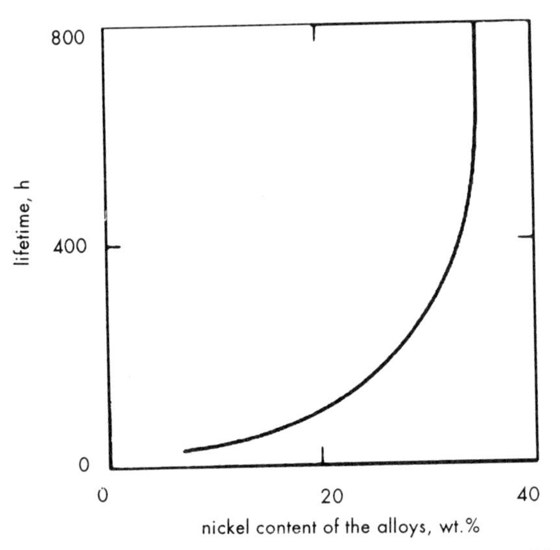

Figure 1.2: Resistance to stress-corrosion cracking (SCC) of high-alloy stainless steels and nickel-base alloys as a function of their nickel content, tested in boiling 62% $CaCl_2$ solution and at a load of 120% $Rp_{0.2}$ [10]

Copper, cited as a main alloying element in the range of medium nickel contents and which can be present in mass proportions of up to approximately 4%, *facilitates passivation in the reducing medium sulphuric acid* and thus reduces corrosion weight loss. This, probably, is a consequence of the copper coating [11] which forms on the corroding material.

Molybdenum increases resistance to general corrosion attack in reducing media and is thus of considerable significance for acid resistance. *Molybdenum also enhances resistance to stress-corrosion cracking* [10]. *Above all, however, it is, together with chromium, of paramount importance for resistance to pitting and crevice corrosion.* The mechanism of its effectiveness is not yet clear [12]. The more highly molybdenum-alloyed materials are thus used to combat corrosive attack from sulphuric acid contaminated by chlorides and fluorides, as for instance in flue gas scrubbers. In this context, the material must generally also have a higher nickel content the more the halogenide concentration increases and the pH value decreases. Molybdenum, after nickel and chromium, is the most important alloying element of this group of alloys, and it can be seen from Table 1.1, that it may be present in proportions ranging from about 1.6 to 28%. *For resistance to pitting and crevice corrosion the total sum of the chromium content plus approximately 3.3 times the molybdenum content, the so-called pitting resistance equivalent number (PREN), is the decisive factor [13]. In many cases, 30 times the nitrogen content, may also be added [14]. This is a further, important basic principle for the selection of a material for use with chloride-containing acid media.*

The element *tungsten* which is included in the column 'others' on the right-hand side of Table 1.1, works in corrosion-chemical terms in a way similar to molybdenum and should therefore be assimilated to it when selecting a material [15].

Although the user may find the wealth of material data in Table 1.1 confusing, there is in reality an even greater number of eligible materials for this range of applications. Selection is based on the specific corrosive attack, usually with the aim of achieving the most cost-effective way of use. Examples are demonstrated in the following sections (cf. e.g. Fig. 8.11).

The challenges presented by new applications such as flue-gas desulphurization, and the competition between materials producers, are generating still further refinements to existing materials, and even the development of new alloys. Thus the average molybdenum content of the stainless steel Cronifer 1925 hMo - alloy 926 (1.4529) has been raised to 6.5%, and the nitrogen content from 0.14 to approximately 0.2%. The molybdenum content is consequently 0.5% higher than it was originally. Since molybdenum is included in the pitting resistance equivalent number with a factor of 3.3 and nitrogen with a factor of 30, this has raised the pitting resistance equivalent number by approximately 3 points, a substantial increase in resistance to pitting and crevice corrosion. The chromium and molybdenum contents in the case of the classical material Nicrofer 4221 - alloy 825 (2.4858) also experienced an increase in the same sense. The materials Nicrofer 3127 hMo - alloy 31 (1.4562) and Nicrofer 5923 hMo - alloy 59 (2.4605), also included in Table 1.1, were the result of development work in the 1980's. One of the features of Nicrofer 3127 hMo - alloy 31 (1.4562) is its outstanding resistance to hot chemical grade sulphuric acid over a very wide range of concentrations [16], as shown in the isocorrosion diagram in Figure 8.6. Accordingly, the material has to be regarded corrosion-resistant in 20 to 60% sulphuric acid up to 100°C and in 80% sulphuric acid up to 80°C. Nicrofer 5923 hMo - alloy 59 (2.4605) is clearly superior to other high-alloy metals under many extremely corrosive conditions prevailing in today's processing and environmental-protection technology, such as flue-gas desulphurization and including reprocessing of flue-gas desulphurization waste water [17-21]. Therefore, older materials, such as Nicrofer 4823 hMo - alloy G-3 (2.4619), are no longer included in Table 1.1, as they have lost their practical significance.

The range of materials shown in Table 1.1 for handling sulphuric acid also applies to corrosive attack by phosphoric acid. Here Nicrofer 3127 LC - alloy 28 (1.4563) is steadily gaining in significance [22]. The new Krupp VDM alloy Nicrofer 3127 hMo - alloy 31 (1.4562) has also proved excellent in initial field tests for phosphate digestion and Nicrofer 3033 - alloy 33 (1.4591) seems to be a promising material for handling of phosphoric acid too [23]. *One basic principle for the selection of materials to handle phosphoric acid therefore is a high chromium content.*

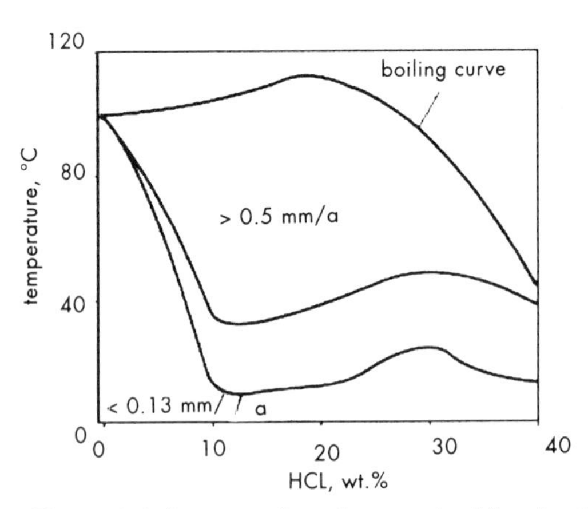

Figure 1.3: Isocorrosion diagram for Nicrofer 3127 hMo - alloy 31 (1.4562) in hydrochloric acid, determined by static immersion tests [16]

In contrast, only a few of the materials shown in Table 1.1 can be used for hydrochloric acid, the third reducing acid of technological importance. As shown in the isocorrosion diagram in Figure 1.3, Nicrofer 3127 hMo - alloy 31 (1.4562) can be used, provided that HCl concentration at room temperature does not exceed 8%. This alloy can, for example, be considered for chemical-process applications where small quantities of hydrochloric acid are present. The recently developed alloy Nicrofer 5923 hMo - alloy 59 (2.4605) is the most resistant of the nickel-chromium-molybdenum alloys to hydrochloric acid. The isocorrosion diagram in Figure 1.4 shows the 0.13 mm/year isocorrosion curve, below which the alloy can be regarded as corrosion-resistant across the entire concentration range of up to 40 % HCl and at temperatures above 40°C [18,19]. In all other cases, in particular in the elevated temperature range, the nickel-molybdenum alloy Nimofer 6928 - alloy B-2 (2.4617), listed in the lower part of Table 1.1, should be used. Nimofer 6629 - alloy B-4 (2.4600) is an advanced version of alloy B-2 offering an improved resistance to stress corrosion cracking [24]. Its use may be advisable if during manufacturing a very high heat-input, caused e.g. by extended welding operations, is unavoidable. Nimofer 6224 - alloy B-10 (2.4604) is a transitional alloy between the nickel-chromium-molybdenum and the nickel-molybdenum alloys. It is intended to offer sufficient corrosion resistance to acid solutions which are exempt of oxidizing media, such as may occur under incrustations of gypsum in flue gas desulphurization plants, or during reprocessing of spent sulphuric acid under very reducing conditions.

Similarly, organic acids also belong to the range of reducing acids and materials suitable for applications with organic acids should be selected from Table 1.1, depending on the type of attack.

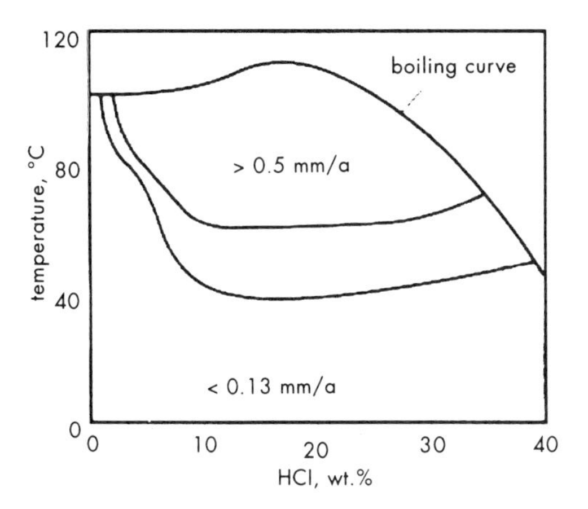

Figure 1.4: Isocorrosion diagram for Nicrofer 5923 hMo - alloy 59 (2.4605) in hydrochloric acid, determined by static immersion tests [18]

Table 1.2: Nickel alloys and special stainless steels for corrosive attack by oxidizing acids: nitric acid and other oxidizing media, and for urea production

	Krupp VDM trade name	Material No. DIN	Main alloying elements (typical values in %)				
			Ni	Cr	Mo	Fe	others
Nitric acid	Cronifer						
	1809 LC - alloy 304 L	1.4306	10	18		68	
	1809 LCLSi	1.4306	13	20		66	
	2521 LC	1.4335	21	25		53	
	1815 LCSi	1.4361	15	18		61	4 Si
	Nicrofer						
	3033 - alloy 33	1.4591	31	33	1.6	32	0.40 N 0.6 Cu
	6030 - alloy 690	2.4642	61	29		9	0.25 Ti
Urea production	Cronifer						
	1812 LC	1.4435	13	17	2.6	65	
	1812 LCN	1.4429	13	17	2.6	65	0.17 N
	2522 LCN	1.4466	22	25	2.1	48	0.13 N
	2525 Ti	1.4577	25	25	2.1	46	0.25 Ti
Highly-concentrated sulphuric acid	Cronifer 2803 Mo (superferrite)	1.4575	3.7	29	2.3	64	0.35 Nb
	Nicrofer						
	2509 Si 7 - alloy 700 Si	1.4390	25	9		57	7 Si
	3033 - alloy 33	1.4591	31	33	1.6	32	0.40 N 0.6 Cu

1.2.2 Corrosive attack by oxidizing acids: nitric acid and other oxidizing media, and in urea production

In industry, oxidizing acids are understood to be media in which substances other than the hydrogen ion act as oxidizing agent during metal dissolution or corrosion, the nitrate ion as the most important representative in the case of nitric acid, the chromate ion in the case of chromic acid, as well as higher valency, heavy-metal ions, Fe^{3+} and Cu^{2+} ions, or dissolved oxygen.

As shown in Table 1.2, the materials which can be considered for handling nitric acid fall in the range of between approximately 18 and 33 % for the main alloy component, chromium. Particularly with regard to corrosion resistance in a maximum 67% nitric acid concentration, relatively low alloyed steels can be used, provided however that, at temperatures up to the boiling point, only limited silicon, phosphorus and sulphur contents are present. At boiling point, their use is at the expense of a relatively high corrosion rate. A corresponding material is Cronifer 1809 LCLSi (1.4306), the suffix LSi indicating limited silicon content (Low Silicon). In severely oxidizing nitric-acid solutions, not limiting the silicon and phosphorus contents results in intercrystalline corrosion in the transition area from the passive to the transpassive state, as well as in the transpassive state [8]. This intercrystalline attack is particularly severe when the nitric acid contains specific higher valency metal ions, for example Cr^{6+}. For this reason the phosphorus content in Cronifer 1809 LCLSi (1.4306) is restricted to a maximum of 0.02%, while the silicon content is targeted to be less than 0.1%. However, it is generally true, as shown in Figure 1.5, that *the corrosion resistance of iron-nickel-chromium alloys to nitric acid increases with their chromium content. This is the overriding basic principle of material selection.* By using the alloy Cronifer

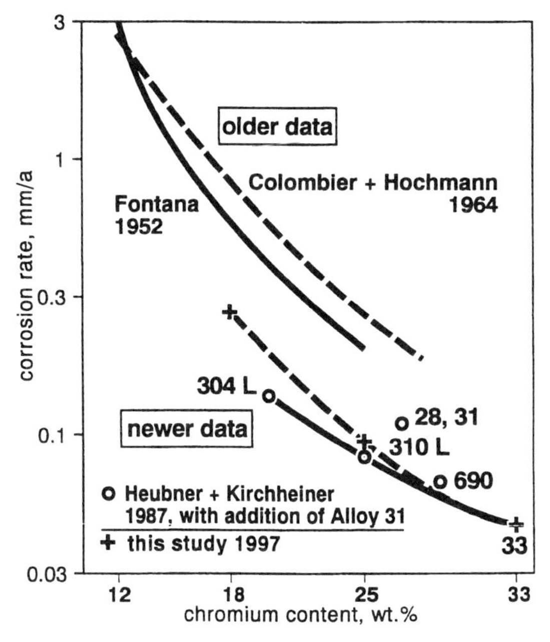

Figure 1.5: Corrosion of iron-chromium-nickel alloys in boiling azeotropic (67 %) nitric acid [23]. The newer data are clearly below the older data, this shift to improved corrosion resistance reflects the progress in metallurgy towards cleaner and more uniform materials [25]

2521 LC (1.4335), featuring a substantially higher chromium content of approximately 25%, it is possible to increase corrosion resistance to an extent sufficient for the majority of applications. When testing in 67% boiling nitric acid (Huey Test), one can expect corrosion rates of approximately 0.08 mm/year compared with an average rate of approximately 0.13 mm/year of the stainless steel 1.4306 [25].
In both cases, however, the precondition is an extremely clean and uniform structure, which means *the alloys have to be ordered and manufactured in their special nitric-acid grades [26]. This is a second major principle of material selection.* Otherwise long-term corrosion rates may be several times higher.

High-nickel alloys such as Nicrofer 6030 - alloy 690 (2.4642), also listed in Table 1.2, provide advantages where halogen compounds are present, or where *nitric-acid/hydrofluoric-acid mixtures* are used as, for example, in the reprocessing of nuclear-reactor fuel elements. It should be noted that, apart from the alloys shown in Table 1.2, Nicrofer 4221 - alloy 825 (2.4858) is also of specific interest in certain cases despite the fact that the alloy is alloyed with molybdenum and copper which, in many materials, reduce corrosion resistance to nitric acid. The alloy may be considered for applications requiring greater versatility in corrosion resistance. It has passed the Huey Test many times [27]. However the new alloy *Nicrofer 3033 - alloy 33 (1.4591) is* expected to become *a better alternative* to nickel-base alloys *in this application, due to its high resistance to corrosion by nitric-acid/hydrofluoric-acid mixtures and ease of fabrication* [23].

Also not included in Table 1.2 is Nicrofer 3127 hMo - alloy 31 (1.4562). Its use is to be considered in selected cases and at locations where high resistance to pitting and crevice corrosion is of great importance, in addition to a relatively high resistance to nitric acid, i.e. in situations where deposits and incrustations are likely to occur. A typical corrosion rate for this material in boiling azeotropic nitric acid (Huey Test) is approximately 0.11 mm/year. The molybdenum content of this material does in fact have no detrimental effect on the material's corrosion resistance to nitric acid [28].

Cronifer 1815 LCSi (1.4361), with approximately 4% silicon, has excellent resistance to nitric acid up to the boiling point, at concentrations greater than 67% and for other extreme, severely oxidizing conditions [29]. Under such conditions, the addition of 4% silicon results in the spontaneous formation of protective silicon- and chromium-oxide layers. Also with this alloy, much progress has been made in recent years with regard to its manufacturing technology. The alloy's carbon content has to be kept extremely low, otherwise the high silicon content causes carbide and carbonitride precipitation which would affect the corrosion resistance. The processing procedure, too, has to be adjusted to achieve a precipitation-free structure. Provided everything has been done correctly, corrosion rates of around 0.25 mm/year and below are possible to achieve, even in hot, highly concentrated nitric acid [30].

Materials which can be considered *for urea production* are in a similar range to those stainless steels particularly resistant to nitric acid, as long as one excludes the material combination Cronifer 2522 LCN/zirconium for the most severe levels of attack [31]. The corrosion resistance of Cronifer 2522 LCN (1.4466) in boiling azeotropic nitric acid equals approximately that of Cronifer 1809 LCLSi (1.4306).

The stainless steel Nicrofer 2509 Si7 - alloy 700 Si (1.4390), alloyed with 7% silicon, has only recently been developed for working with *hot, highly concentrated sulphuric acid* [32]. Design of plants and equipment has to take into consideration the restricted weldability of this material.

Also the superferrite Cronifer 2803 Mo (1.4575) is of special interest for this application. However, because of their restricted fabrication properties, superferrites should only be considered for wall thicknesses of about 2 mm and below, and preferably not exceeding 6 mm. This is why the new alloy *Nicrofer 3033 - alloy 33 (1.4591),* due to its excellent corrosion behaviour combined with ease of fabrication including welding [23] *has a paramount potential as a material for handling of hot, highly concentrated sulphuric acid.*

1.2.3 Corrosive attack by hot concentrated alkaline solutions (KOH, NaOH) and salt solutions (KCl, NaCl)

Resistance to corrosion by alkaline solutions increases with the nickel content of the alloys [33, 34, 35]. *This is the basic principle for the selection of materials for applications with alkaline solutions. Increasing molybdenum contents also reduce the degree of corrosion* [33]. Low-alloy austenitic steels can thus only be used under relatively mild corrosive attack, that is at concentrations of less than 50% NaOH and at temperatures below 100°C. By contrast, unalloyed nickel is the material most worthy of consideration for these applications., A quality with particularly low carbon content, i.e a LC-grade must be used at working temperatures greater than 300°C, in order to avoid loss of ductility due to precipitation of graphite at the grain boundaries [33, 34]. Of the materials additionally presented in Table 1.3, Nicrofer 4221 - alloy 825 (2.4858) or even better [23] Nicrofer 3033 - alloy 33 (1.4591) should be considered for applications with particularly wide-ranging demands on resistance to corrosion attack, that is when, besides an alkaline attack, some other kind of corrosive attack is expected. If, for example, alkaline solutions contain oxidizing agents, chromium-containing alloys may show a better corrosion resistance than unalloyed nickel. Also the risk of pitting corrosion, caused by the presence of chlorides and hypochlorites, may make the use of nickel-chromium-molybdenum-iron alloys necessary. Nicrofer 7216 LC - alloy 600 L (2.4817) should be used for applications requiring higher strength at higher temperatures or when the alkaline solutions or the external heating atmosphere contain sulphur.

Table 1.3: Nickel alloys and special stainless steels for corrosive attack by hot concentrated alkaline solutions (KOH, NaOH) and salt solutions (KCl/NaCl)

	Krupp VDM trade name	Material No. DIN	Main alloying elements (typical values in %)					
			Ni	Cr	Mo	Cu	Fe	others
Alkaline solutions	Nickel							
	VDM Nickel 99.2 - alloy 200	2.4066	>99					
	VDM LC-Nickel 99.2 - alloy 201	2.4068	>99					
	Nicrofer							
	3033 - alloy 33	1.4591	31	33	1.6	0.6	32	0.40 N
	4221 - alloy 825	2.4858	40	23	3.2	2.2	31	0.8 Ti
	7216 LC - alloy 600 L	2.4817	73	16			9	0.25 Ti
	6030 - alloy 690	2.4642	61	29			9	0.25 Ti
	Nicorros - alloy 400	2.4360	64			32	1.8	1.0 Mn
Salt solutions	Cronifer							
	1713 LCN	1.4439	13	18	4.5		62	0.16 N
	1925 LC - alloy 904 L	1.4539	25	21	4.8	1.5	46	
	1925 hMo - alloy 926	1.4529	25	21	6.5	0.9	45	0.20 N
	Nicorros - alloy 400	2.4360	64			32	1.8	1.0 Mn
	Cunifer 30 - alloy CuNi 70/30	2.0882	30			68	0.6	0.6 Mn

Nicrofer 6030 - alloy 690 (2.4642) has been developed as a material with particularly high resistance to *caustic-induced stress-corrosion cracking.* In fact, even unalloyed nickel may show caustic-induced stress-corrosion cracking. In material clad with nickel, the development of residual tensile stresses must therefore be avoided [36]. In view of its high nickel content, Nicorros - alloy 400 (2.4360) can also be considered for applications involving alkaline solutions, its corrosion resistance here not being significantly below that of unalloyed nickel, but only in cases where absorption of copper has no effect on the product manufactured or treated. *Materials* which can be considered *for concentrated salt solutions* are also shown in Table 1.3. Here the Cronifers are supplemented by the nickel-copper and copper-nickel alloys Nicorros - alloy 400 (2.4360) and Cunifer 30 (2.0882).

Table 1.4: Copper-nickel- and nickel-copper alloys for corrosive attack by seawater

Krupp VDM trade name	**Material No.**	**Main alloying elements (typical values in %)**				
	DIN	**Ni**	**Cu**	**Fe**	**Mn**	**others**
Cunifer						
10 - alloy CuNi 90/10	2.0872	10	88	1.5	0.7	
30 - alloy CuNi 70/30	2.0882	30	68	0.6	0.7	
302	2.0883	30	66	1.8	1.8	
Nicorros - alloy 400	2.4360	64	32	1.8	1.0	
Nicorros LC	2.4361	65	32	1.5	0.7	
Nicorros Al - alloy K-500	2.4375	65	30	1.0	0.6	2.8 Al; 0.45 Ti

1.2.4 Corrosive attack by seawater and chloride-contaminated cooling water

The *alloys of nickel with copper* should be given first consideration in cases of corrosive attack by seawater [37]. Typical representatives of such materials are shown in Table 1.4. It can be seen that all of these alloys also contain amounts of iron and manganese, which are of considerable significance for their resistance to corrosion and erosion. Alloys with 10% nickel, equivalent to Cunifer 10 - alloy CuNi 90/10 (2.0872), provide more protection against corrosion, erosion and corrosion fatigue than brasses of the CuZn20Al2 (aluminium brass) type, even under more severe operating conditions. They are not susceptible to stress corrosion cracking and are immune to incrustation by marine organisms. For this reason, they have been a preferred material for seawater lines on ships and for the construction of seawater desalination plants. For even more severe attack, arising, for example, from increased flow velocity, alloys of the type Cunifer 30 - alloy CuNi 70/30 (2.0882) are selected and, in the case of extremely high erosive attack due to sand-bearing cooling water, alloys with higher iron and manganese content of the type Cunifer 302 (2.0883) are employed. These copper alloys, with 10 or 30% nickel, are standard materials today for seawater-cooled condenser tubes.
Nicorros - alloy 400 (2.4360), with about 65% nickel, is used where high resistance to corrosion and erosion is required in *seawater splash and tidal zones,* or

where enhanced strength is demanded in combination with resistance to seawater. It is even possible, with the age-hardenable variant Nicorros Al - alloy K-500 (2.4375), to obtain a material of extremely high strength, for instance for shafts and rods. Limitations for the application of copper-nickel and nickel-copper alloys in seawater depend on the relevant flow velocities of the seawater [38]. In order to safely avoid pitting corrosion, the seawater should not be stagnant, and to avoid erosion-corrosion, specific flow velocities should not be exceeded for copper-nickel alloys. Furthermore, attention should be paid to avoiding the formation of galvanic elements, since a galvanic coupling with less noble elements such as carbon steel destroys the growth-inhibiting effect of copper-nickel alloys because this requires a minimum amount of copper ions passing into the water. A galvanic coupling with more noble elements such as stainless steels or titanium results in a corrosion of the copper material, if no cathodic protection is provided. When in continuous or repetitive contact with high-sulphide containing seawater, other materials than the ones mentioned above have to be used.

As alternative to copper-nickel and nickel-copper alloys, *standard molybdenum-free austenitic stainless steels* may only be used if a shorter service life is accepted or cathodic protection is provided, using aluminium or carbon steel. Under closer examination cathodic protection also appears to be the reason for positive experiences in seawater with the steels Cronifer 1810 Ti (1.4571) and Cronifer 1810 LC (1.4404) mentioned in Table 1.5, unless a certain amount of crevice corrosion is simply accepted or special measures are adopted, such as the complete emptying of all horizontal tubes and evaporation chambers in seawater desalination plants during downtime [39]. The latter also applies to *ferritic-austenitic duplex steels* of the type Cronifer 2205 LCN - alloy 318 LN (1.4462). Better pitting- and crevice-corrosion resistance, together with an even higher 0.2% proof stress, are shown by the so-called 'superduplex' steel Cronifer 2507. The alloy compares with Cronifer 2419 MoN - alloy 24 (1.4566), which, being austenitic, exhibits a slightly lower 0.2% proof stress, but with better ductility in cold working, a significantly smaller drop in the notch-impact toughness with lower temperatures and a reduced tendency to hydrogen embrittlement [40].

Austenitic 6% molybdenum steels experienced a significant breakthrough in recent years as materials for use in seawater [41, 42]. Cronifer 1925 hMo - alloy 926 (1.4529), also included in Table 1.5, is a typical and technically advanced [43] representative of this group. The most popular copper-nickel material Cunifer 10 - alloy CuNi 90/10 (2.0872) and the 6%-molybdenum steels exhibit the same number of positive factors. This, more or less, also applies to the cost of materials [41]. The 6%-molybdenum steels, however, have a 0.2% proof stress about 3.3 times higher and an E modulus about 1.5 times higher. At the same time there is no limitation in flow velocity. This allows a lighter construction of off-shore platforms. The copper-nickel materials on the other hand are distinguished by easier working characteristics and their natural marine-growth inhibitions.

VDM Nicrofer 3127 hMo - alloy 31 (1.4562) is the most recent material development within the group of 6%-molybdenum steels. As shown in Table 1.5 the alloy differs from Cronifer 1925 hMo - alloy 926 (1.4529) mainly by its nickel and chromium contents, each being 6% higher, which has a considerable influence on the material's resistance to pitting and crevice corrosion [28]. Furthermore, the alloy's high resistance to reducing and oxidizing acidic media [16, 28] should be noted. The combination of these features makes the new *alloy Nicrofer 3127 hMo - alloy 31 (1.4562) an ideal heat-exchanger material for hot process coolers in the chemical industry, including when cooling with brackish or seawater.* Also in case of chlorinated seawater Nicrofer 3127 hMo - alloy 31 (1.4562) exhibits superior characteristics compared to other seawater-resistant stainless steels [44].

The material *Nicrofer 5923 hMo - alloy 59 (2.4605) has proved to be absolutely resistant even to filtered seawater under extreme crevice-corrosion conditions.*

Corrosive attack due to *chloride-contaminated cooling water* generally necessitates materials equivalent to Cronifer 1713 LCN (1.4439), or Cronifer 1925 LC (1.4539), or again to Cronifer 1925 hMo - alloy 926 (1.4529). When using low-alloyed stainless steels such as Cronifer 1810 Ti - alloy 316 Ti (1.4571), *micro-biologically-induced corrosion* [45] may become a nuisance [46] should chlorination not be allowed due to environmental legislation.

Table 1.5: Nickel alloys and special stainless steels for corrosive attack by seawater and contaminated cooling water

Krupp VDM trade name	**Material No.**	**Main alloying elements (typical values in %)**					
	DIN	**Ni**	**Cr**	**Mo**	**Cu**	**Fe**	**others**
Cronifer							
2205 LCN - alloy 318 LN	1.4462	6	22	2.8		66	0.14 N
2507	1.4469	7	25	3.8	0.6	61	0.27 N 0.6 W
1810 LC - alloy 316 L	1.4404	12	17	2.3		66	
1810 Ti	1.4571	12	17	2.3		66	0.4 Ti
1713 LCN - alloy 317 LN	1.4439	13	18	4.5		62	0.16 N
2419 MoN- alloy 24	1.4566	18	24	4.3	0.6	46	0.45 N 6.2 Mn
1925 LC - alloy 904 L	1.4539	25	21	4.8	1.5	46	
1925 hMo - alloy 926	1.4529	25	21	6.5	0.9	45	0.20 N
Nicrofer							
3127 hMo - alloy 31	1.4562	31	27	6.5	1.3	31	0.20 N
4221 - alloy 825	2.4858	40	23	3.2	2.2	31	0.8 Ti
6020 hMo - alloy 625	2.4856	62	22	9		3	3.4 Nb
5923 hMo - alloy 59	2.4605	59	23	16		1	

1.2.5 Corrosive attack by special aqueous media

Special aqueous media exist, ranging from weak acid to weak alkaline, which are very corrosive and which, therefore, need special consideration. Examples include *chlorine dioxide bleaching solutions in the pulp and paper industry.* High-alloy special stainless steels such as Nicrofer 3127 hMo - alloy 31 (1.4562) are corrosion resistant versus those media, whereas nickel-chromium-molybdenum alloys may show an increased rate of corrosion [47].

1.2.6 Corrosive attack by hot gases and combustion products: heat-resisting materials

Heat-resisting materials are characterized by their specific resistance to the attack of hot gases and combustion products above 550°C. *High-temperature materials* have especially high mechanical properties at high temperatures under long-term loading; this implies high resistance to creep and high creep-rupture strength, even at temperatures above about 550°C. In many applications, both characteristics are required simultaneously.

Table 1.6: Heat-resistant nickel alloys and special stainless steels

Krupp VDM trade name	**Material No.**	**Main alloying elements (typical values in %)**							
	DIN	**Ni**	**Cr**	**Si**	**RE**	**Al**	**Ti**	**Fe**	**others**
Cronifer									
2504	1.4821	5	25	1.2				67	
1809TiH	1.4878	9	18				0.4	70	
2111 N	14835	11	21	1.8	0.05			65	0.16 N
2012	1.4828	12	20	1.9	0.05			65	
2413 - alloy 309	1.4833	13	22		0.05			61	
2520 - alloy 310 S	1.4845	20	25		0.05			54	
2520 - alloy 310/314	1.4841	20	25	1.9	0.05			50	
Nicrofer									
2020 - alloy 840	1.4847	20	20			0.3	0.4	58	
3220 - alloy 800	1.4876	31	20			0.25	0.4	47	
3718 - alloy DS*	1.4862	36	18	2.2		0.15	0.15	42	
45 TM - alloy 45 TM	2.4889	47	27	2.7	0.10			23	
6023 - alloy 601	2.4851	60	23			1.4	0.4	14	
6030 - alloy 690	2.4642	61	29					9	
7216 - alloy 600	2.4816	73	16			0.2	0.2	9	
7520 - alloy 75	2.4951	75	20			0.2	0.4	3	
6009 Al 10 - alloy 10 Al		57	8			10		26	0.1 Hf
Cronix									
70 - NiCr 70/30	2.4658	68	30	1.3	0.03				
80 - NiCr 80/20	2.4869	78	20	1.3	0.03				
*Material 1.4864 has somewhat lower chromium and silicon contents than Alloy DS									

Heat-resisting materials based on the iron-nickel-chromium alloy system and suitable for applications without very strict requirements on strength at high temperatures are shown in Table 1.6. Their chromium contents lie between approx. 8 to 30%. Of the three elements aluminium, silicon and chromium which can be used for the formation of protective oxide layers, chromium can be used most universally, since it is very soluble in nickel, iron and even in cobalt, and has the lowest tendency to form intermetallic phases. *Chromium contents of between 20 and 30% provide the best protection against corrosion and oxidation at elevated temperatures of up to approximately 1000°C. This is the first basic principle for selecting heat-resisting materials* with the result, that a material such as Cronifer 2520 - alloy 310 (1.4841) can be used in furnace construction for a variety of applications. Although this material may be sufficient for many requirements, alloys possessing even better resistance to scaling can be produced by means of further increases in nickel content. Further down in Table 1.6 these are Nicrofer 3220 - alloy 800 (1.4876), 45TM - alloy 45TM (2.4889), 6023 - alloy 601 (2.4851), 6030 - alloy 690 (2.4642), 7520 - alloy 75 (2.4951) and finally Cronix 80 - NiCr 80/20 (2.4869).

This improved resistance to scaling is particularly useful under fluctuating temperatures. For this reason, *an important principle for material selection is that, in most cases, nickel-chromium alloys are more resistant to oxidation under cyclic temperatures than iron-based alloys with chromium* [34].

Rare earth metals stated in Table 1.6 as alloying elements improve the adhesive strength of the protective oxide surface layer to the metal and thereby the resistance to oxidation. A further differentiation in the oxidation resistance of these alloys is based therefore on their content of those elements which increase the adhesive strength of the protective oxide surface layer. Apart from rare earth metals, these are the elements *zirconium, calcium and magnesium* [48]. Other alloying elements which are present in smaller quantities also have their influence. *Silicon* is effective, mainly in the initial stages of oxidation, by very rapidly forming protective oxide films.

Aluminium primarily raises the maximum working temperatures of high-temperature alloys. Since chromium oxide noticably vaporizes at temperatures above 1000°C, oxidation resistance, particularly in flowing atmospheres, is affected as a result of continuous loss of chromium. To a certain extent, this can be compensated by increasing the chromium content to approximately 30%. *A basic principle of material selection for application above 1000° C in oxidizing gas atmospheres is, however, to work with protective layers of aluminium oxide instead of chromium oxide.*

Thus, for equipment manufacture there is, according to Table 1.6, Nicrofer 6023 - alloy 601 (2.4851), an alloy with an aluminium content of approximately 1.4%. Despite its relatively low aluminium content, the alloy, in oxidizing media, exhibits a markedly improved resistance under fluctuating temperatures above 1000°C, when compared to other materials listed in Table 1.6. Clearly superior to this material with regard to oxidation resistance above 1000°C, is the high-temperature alloy *Nicrofer 6025 HT - alloy 602 CA (2.4633),* mentioned in Table 1.7, which, despite its aluminium content of only about 2.3 %, forms continuous Al_2O_3 protective oxide layers, having its application potential in the temperature range up to approximately 1200°C [49, 50].

Where resistance to scaling in air is not of significance, but rather resistance to other atmospheres, *the basic principle for material selection is that increasing nickel content produces increasing resistance to uniformly carburizing gases, but has a more detrimental effect with regard to resistance to sulphur-bearing atmospheres.* For this reason Nicrofer 3220 - alloy 800 (1.4876), placed in the middle section of Table 1.6, has proved to be a good universal solution for appli-

Table 1.7: High-strength high-temperature nickel alloys and special stainless steels

Krupp VDM trade name	Material No.	Main alloying elements (typical values in %)							
	DIN	Ni	Cr	Mo	C	Al	Ti	Fe	others
Cronifer									
1809 H	1.4948	10	18		0.06			69	
1812 H	1.4919	12	18	2.1	0.05			66	
1613 Nb	1.4961	13	16		0.07			68	0.9 Nb
1613 NbN	1.4988	13	16	1.3	0.07			65	0.9 Nb; 0.1 N; 0.8 V
1613 N	1.4910	13	17	2.4	0.03			65	0.14 N
1616 Nb	1.4981	16	16	1.8	0.07			63	0.9 Nb
Nicrofer									
3220 H - alloy 800 H	1.4958	31	21		0.07	0.25	0.35	47	
3220 HT - alloy 800 HT	1.4959	30	21		0.09	0.5	0.5	47	
3228 NbCe - alloy AC 66	1.4877	32	28		0.05			39	0.8 Nb; 0.07 Ce
6023 H - alloy 601 H	2.4851	60	23		0.06	1.4	0.5	14	
6025 HT - alloy 602 CA	2.4633	62	25		0.18	2.3	0.2	9.5	0.1 Y, 0.1 Zr
6125 Gt - alloy 603 GT	2.4647	62	25		0.22	2.8	0.2	9	0.1 Y; 0.1 Zr
7216 H - alloy 600 H	2.4816	74	16		0.07	0.2	0.2	9	
4626 MoW - alloy 333	2.4608	46	25	3	0.05	0.1		17	3 W; 3 Co; 1 Si
4722 Co - alloy X	2.4665	48	22	9	0.07			18	1 Co
5520 Co - alloy 617	2.4663	54	22	9	0.06	1	0.5	1	12 Co
5120 CoTi - alloy C-263	2.4650	51	20	6	0.06	0.5	2.1	0.5	20 Co
5219 Nb - alloy 718	2.4668	53	19	3	0.05	0.6	0.9	18	5.2 Nb
7016 TiNb - alloy X-750	2.4669	72	16		0.05	0.6	2.7	7.5	1Nb
7520 Ti - alloy 80 A	2.4952	75	20		0.06	1.4	2.3		
Cronix 70 NS		67	30		0.08	0.15			2.3 Si, 0.4 N, 0.2 Y

cations in alternately oxidizing and carburizing, and simultaneously sulphur-bearing media, as in coal processing and crude-oil processing. Also placed in this middle section of Table 1.6 is Nicrofer 45 TM - alloy 45 TM (2.4889), a material with outstanding resistance to typical waste incineration gases in the temperature range between 550 and 850°C [51]. These are oxygen-bearing, chlorinating, and simultaneously sulphidizing media. In addition, the alloy exhibits good resistance to carburization, making it attractive for applications such as coal dust burners and heat-exchanger tubes in coal gasification processes [52]. *The lower chromium-content materials*, such as Nicrofer 3718 - alloy DS (1.4864) and 7216 - alloy 600 (2.4816), *are particularly resistant to nitrogen pick-up in low-oxygen, nitrogen-bearing atmospheres, this being a further basic principle of material selection.*

Finally, it must be pointed out that, just as with aqueous corrosion, *localized corrosive attack in high temperature corrosion can be significant* and may become decisive for the life of equipment and plant [53]. Most important in this context is deposit-induced corrosion, exemplified by sodium sulphate as the deposit condensing from the gas phase. This is also known as *hot gas corrosion.* This type of corrosion proceeds most rapidly in the temperature range around 900°C (Type I), and in the temperature range around 750°C (Type II) with lower chromium-containing alloys and higher SO_3 contents in the atmosphere. In such a case, corrosion can quickly ruin equipment components. It is therefore necessary to change the design or operating conditions of the plant to prevent condensation of sodium sulphate from the gaseous phase. Similarly local corrosion may take place when burning alkali-rich low grade coals in the temperature range of around 650°C, with the formation of low-melting, i.e., *liquid alkali-iron-sulphate slag.* So-called 'oil ash corrosion' can also be a significant form of deposit-induced corrosion [35]. High-silicon alloyed materials such as Nicrofer 45 TM (2.4889) are best suited to resist the attack of such salts and slags [54].

A completely new material design is offered with *Nicrofer 6009 Al 10 - alloy 10 Al,* mentioned in the lower part of Table 1.6. This unique nickel-base alloy features an aluminium content of 10 %, in combination with a chromium content of 8 %; the microstructure is duplex γ/β (NiAl). This new alloy provides an excellent resistance to oxidizing gases, including air, up to 1300° C, as well as to hot gases with a high activity of sulphur and carbon [55].

1.2.7 Mechanical loading at very high temperatures: high-temperature, high-strength materials

Close study of current conceptions on the state of technology leads one to define the difference between heat-resisting and high-temperature, high-strength austenitic materials in terms of the latter being required to have a creep rupture strength $Rm/10^5$ h of above approximately 90 N/mm^2 at 600°C, 30 N/mm^2 at

700°C, 10 N/mm² at 800°C and 3.5 N/mm² at 900°C, at least in one partial temperature range. Likewise a similar, somewhat optional limit can be established for Rm/10⁴ h with 140 N/mm² at 600°C, 70 N/mm² at 700°C, 35 N/mm² at 800°C, 18 N/mm² at 900°C and 9 N/mm² at 1000°C. The areas marked in Figure 1.6 are thus obtained for high-temperature, high-strength materials. With the material examples entered, one can see the broad band of very diverse creep-rupture strengths, for the group of high-temperature, high-strength alloys. This allows a large number of high-temperature materials to be included, as shown in Table 1.7.

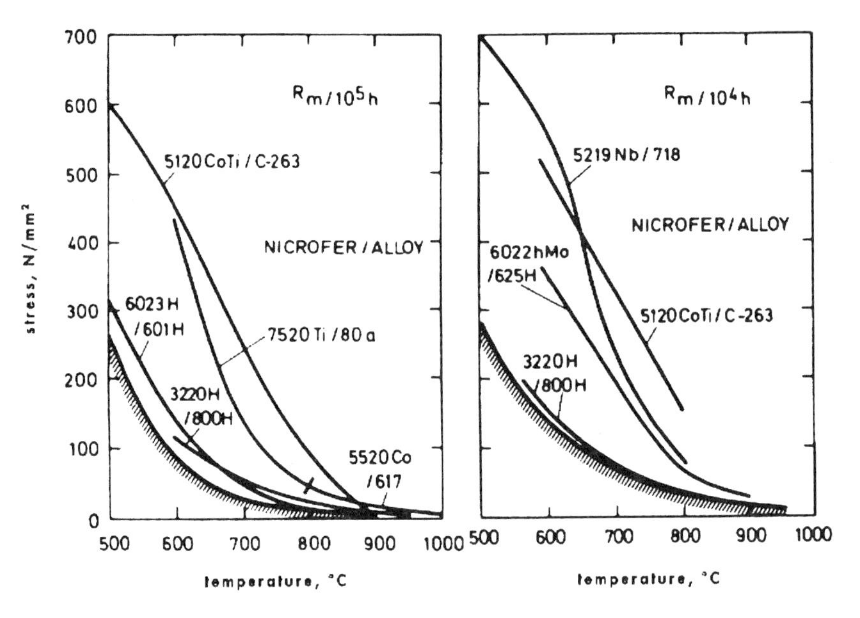

Figure 1.6: Creep-rupture strengths $Rm/10^4$ and $Rm/10^5$ of high-strength, high-temperature materials; curves based on the data of [56]

In this table, proceeding from top to bottom, Cronifer 1809 H (1.4948) falls, with its long-term high-temperature strength, below the somewhat arbitrary property levels defined above, while all the other materials lie above, especially high values being obtained for Cronifer 1613 NbN (1.4988) in the temperature range between 600 and 650°C. In order to permit welding of larger cross-sections without hot cracking, non-stabilized material variants have been developed in recent years, such as Cronifer 1613 N (1.4910), having no niobium content, reduced carbon and increased nitrogen contents. Above approximately 600°C, *Nicrofer 3220 H - alloy 800 H (1.4958),* which has a higher nickel content, is superior to these special stainless steels and can be used *without danger of sigma-phase embrittlement.* Like Nicrofer 3220 - alloy 800 (1.4876), it also features good resistance to corrosive attack by a variety of hot gases.

The separation of the former alloy Nicrofer 3220 H - alloy 800 H into the two variants Nicrofer 3220 H - alloy 800 H and 3220 HT - alloy 800 HT with their corresponding new German material Nos. 1.4958 and 1.4959 is a more recent development. The previous material no. 1.4876 has been given to the heat-resistant variant of this material. Under United States *ASME* regulations the variants H and HT have different maximum allowable design stresses. In accordance with *VdTÜV*-Blatt 434 the HT material must not be used for service temperatures below 700°C.

Table 1.8: Creep-rupture strength values of Nicrofer 3220 H/HT - alloy 800 H/HT (1.4958/1.4959)

Temperature in °C	**Creep-rupture strength Rm/10^5h in N/mm²**		
	Nicrofer 3220 H - alloy 800 H (1.4958)		**Nicrofer 3220 HT - alloy 800 HT (1.4959)**
	1.4876 (previous) 1.4958 (new) VdTÜV 434	UNS N 08810 ASME calculated*	UNS N 08811 ASME calculated*
600	114	114	126
700	53	50	57.3
800	24	22.2	25.8
900	10.5	10.2	11.2
950	7	-	-

*) Maximum permitted stress according to ASME Sec. VIII, Div. 1 Code Case 1987, multiplied by 1.5 as usual safety factor.

For a comparison with the creep-rupture strength data to be used according to VdTÜV-Blatt 434, the maximum permitted design stresses as per ASME Code were multiplied by the safety factor 1.5 and entered into Table 1.8. Creep rupture strength data Rm/10^5 h as per VdTÜV-Blatt 434 are given on the left side, on the right are the calculated values for the same 3220 H material as per ASME Code Case. Above 600°C, they are below the values of the VdTÜV-Blatt. However, the creep-rupture strength values for the HT variant as calculated from the ASME Code Case data are clearly above those taken from the VdTÜV-Blatt 434.

Nicrofer 6023 H - alloy 601 H (2.4851) and Nicrofer 7216 H - alloy 600 H (2.4816) fall, with comparable or somewhat lower creep-rupture strengths than Nicrofer 3220 H - alloy 800 H (1.4958), in the temperature range 600 - 900°C, in the lower regions of the high-temperature, high-strength materials. Their corrosion resistance in ammonia cracking, nitriding and carburization plants (Nicrofer 7216 H - alloy 600 H, 2.4816), and high resistance to scaling up to 1200°C (Nicrofer 6023 H - alloy 601 H, 2.4851) are their dominant characteristics in conjunction with their strength properties, and form the basis for their selection.

Nicrofer 6025 HT - alloy 602 CA (2.4633) is a further development of Nicrofer 6023 H - alloy 601 H (2.4851). Its mechanical strength in the high-temperature range of 1000°C and above is substantially increased due to the high carbon content and the addition of zirconium. This strength is assured with excellent resistance to cyclic oxidation and carburization. Typical applications include *furnace rolls, radiant tubes, muffles for bright annealing furnaces, other furnace installations and pig-tails in ethylene cracking plants* [57, 58]. *Nicrofer 6125 GT - alloy 603 GT (2.4647)* is an advanced gas-turbine version of alloy 602 CA. Due to further increased contents of aluminium and carbon, its creep strength and oxidation resistance up to 1200° C are exceptionally high.

Nicrofer 4626 MoW - alloy 333 (2.4608), Nicrofer 4722 Co - alloy X (2.4665) and Nicrofer 5520 Co - alloy 617 (2.4663) can be regarded, with respect to their combination of oxidation resistance and creep-rupture strength, as superalloys in the broader sense. These alloys, solid-solution-hardened high-temperature materials closely related to the creep-resistant austenitic steels, obtain their *high degree of creep-rupture* strength largely from additions of *molybdenum* and from *carbide precipitation.* Table 1.9 places these alloys into three groups, in accordance with their increasing molybdenum content, whereby for Nicrofer 4626 MoW - alloy 333 (2.4608) the sum of its molybdenum content plus half of its tungsten content was chosen, on the basis that the atomic weight of tungsten is almost twice that of molybdenum. It is evident that the 10,000 h creep-rupture strength at 800 and 900°C for Group 3 with an average of 9% molybdenum is more than double that of Group 1 with no addition of molybdenum. In such a comparison it should be noted that the *scatter-band of creep-rupture values* is generally assumed to be ± 20% [59, 60]. Nevertheless, the excellent performance of alloy Nicrofer 3220 H - alloy 800 H (1.4958) stands out in Group 1, making the somewhat limited applications of Nicrofer 4626 MoW - alloy 333 (2.4608) in Group 2 more understandable. The criterion for using this material must have less to do with creep-rupture strength, therefore, than with its better resistance to oxidation, due to the higher chromium content. *Nicrofer 5520 Co - alloy 617 (2.4663)* outperforms all other materials mentioned in Table 1.9 by virtue of its high creep strength. In addition this alloy is qualified as a material of construction for *high-temperature pressure vessels* up to 1050° C, according to VdTÜV-Blatt 485. But also in this field the potential of nickel-based alloys is not limited as shown above in referring to alloy 603 GT. Another development is *Cronix 70 NS,* aiming at excellent creep strength, in combination with elevated low-cycle fatigue strength (LCF) and high oxidation resistance, by alloying nickel with chromium, silicon, nitrogen and yttrium [61].

Nicrofer 6020 hMo - alloy 625 (2.4856), sometimes used as a high-temperature material, undergoes a strong ductility loss at temperatures between 500 and 1000° C [62] and is, for this reason, not mentioned any more in Tables 1.7 and 1.9.

As a *basic principle of material selection* it should be maintained, however, that such *solid-solution- and carbide-precipitation-hardened high-temperature nickel-based alloys have their use in the temperature range above 650 to 700°C. Below that, either the high-temperature austenitic steels or, where higher strength is required, the precipitation-hardenable, nickel-based superalloys should be used,* a representative selection of which is presented in the following paragraph.

Table 1.9: Creep-rupture strength in N/mm² of various materials as a function of the weighted sum (Mo + ½ W) of their molybdenum and tungsten contents

Group	**Mo + ½ W**	**Material** Nicrofer	**Rm/10⁴h**		**Rm/10⁵h**	
			800°C	**900°C**	**800°C**	**900°C**
1	0	3220 H - alloy 800 H 6023 H - alloy 601 H 7216 H - alloy 600 H	37 26 25	17 12 9	24 14 15	10.5 6 5
2	4.5	4626 MoW - alloy 333	40	18	30	9
3	9	4722 Co - alloy X 5520 Co - alloy 617	59 65	31 30	- 43	- 16

Nicrofer 5120 CoTi - alloy C-263 (2.4650), Nicrofer 5219 Nb - alloy 718 (2.4668), Nicrofer 7016TiNb - alloy X-750 (2.4807) and Nicrofer 7520Ti - alloy 80 A (2.4952) belong to the group of classic nickel-based superalloys, which can be worked to semi-finished components and are precipitation hardenable. They are of no a great importance in chemical and power engineering. Their main applications are where particularly high strength at reasonably high temperatures is required, as in stationary and rotating components of aircraft gas turbines, for fasteners and for springs.

1.2.8 Predicting the behaviour of materials and their possible applications from alloy composition and structure

The preceding explanations have made it evident that the behaviour and potential application of materials can be predicted to some extent on the basis of their composition. Less has been said about the material structure, although there are some basic principles, for instance, with regard to the increase of the 0.2% proof stress with decreasing grain size, the increase in creep-rupture strength with increasing grain size and the advantages of a largely homogeneous structure, i.e., low in inclusions and carbide bands and of mainly uniform grain size, for many processing and service requirements in chemical, offshore and power engineering.

Difficulties with prediction in this particular case *result from the fact that aspects of corrosion resistance in the wet and high-temperature range are not a question of material properties, but are always modes of behaviour,* arising from the interplay between the properties of a material and its surrounding media. Also material surface layers, with their individual properties, are often decisive for behaviour, and which, again, are formed only in the course of such interaction and are a product thereof. A detailed prediction, therefore, is only possible on the basis of comprehensive experience, whereby the combination of many factors and influences have to be taken into account.

1.3 Phase diagrams

As is clear from the preceding materials summary, *the phase diagram of principal interest is that of the ternary system nickel-iron-chromium* with its relevant peripheral systems. Pure nickel has a face-centered cubic structure up to its melting point at around 1453°C, and is, to use the terminology of the steel industry, austenitic. In this sense it has complete miscibility with the austenitic form of iron known as the gamma phase, which can be retained even at room temperature and below, given an adequately high nickel content.

By contrast, pure chromium is of body-centered cubic structure, like the ferritic form of iron, and has total miscibility with this ferritic form. However, as can be seen from Figure 1.7, this only applies at elevated temperatures. Below 820°C, an intermetallic iron-chromium compound known as *sigma phase,* which is undesirable in engineering materials, is formed. Its occurrence leads to loss of mechanical ductility and of corrosion resistance to oxidizing media. Chromium is soluble in the face-centered cubic structure of nickel to a larger extent than in the body-centered cubic gamma structure of iron, and proportions of up to approximately 30% chromium by weight are employed for various industrial applications. As the interrupted lines plotted in Figure 1.8 illustrate, the potentially disruptive formation or precipitation of other phases takes place so sluggishly in this range that they may well not appear at all. As one would expect from the peripheral systems of Figure 1.9, the range of existence of the gamma phase is reduced by increasing the iron plus chromium content in the nickel-iron-chromium ternary system (Figure 1.9a), and below 950°C the sigma phase also occurs (Figures 1.9b and 1.9c). Its existence range is accordingly extended towards higher temperatures by nickel. *After nickel, iron and chromium, molybdenum is in many cases an important component of the materials under consideration.* It narrows the gamma range and thus makes it easier for the sigma phase to occur. In high molybdenum-bearing materials, sigma phase can only be redissolved with difficulty.

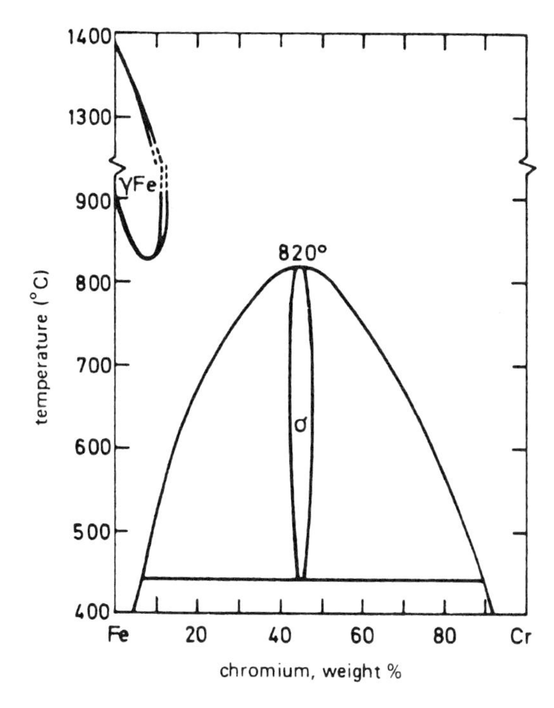

Figure 1.7: Iron-chromium phase diagram [63]

Annealing temperatures of more than 1200°C may be necessary for this. Moreover, molybdenum additions may also result in the formation of other undesirable intermetallic phases. Silicon, tungsten, niobium and titanium also restrict the gamma range. Carbon and nitrogen expand the gamma-phase range. Furthermore, nitrogen makes transformation of the gamma phase sluggish. The solubility of these elements is, however, limited and decreases with temperature, *carbides* and *nitrides* being precipitated. The carbide formers are chromium and molybdenum. The most important carbide has the composition $M_{23}C_6$, chromium being the main metallic component. The occurrence of the M_6C carbide is greatly promoted by molybdenum, tungsten or niobium as alloy components. *Carbide*

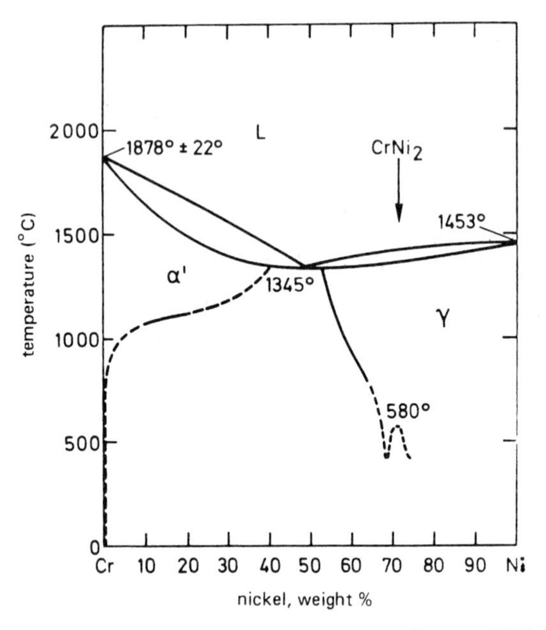

Figure 1.8: Chromium-nickel phase diagram [63]

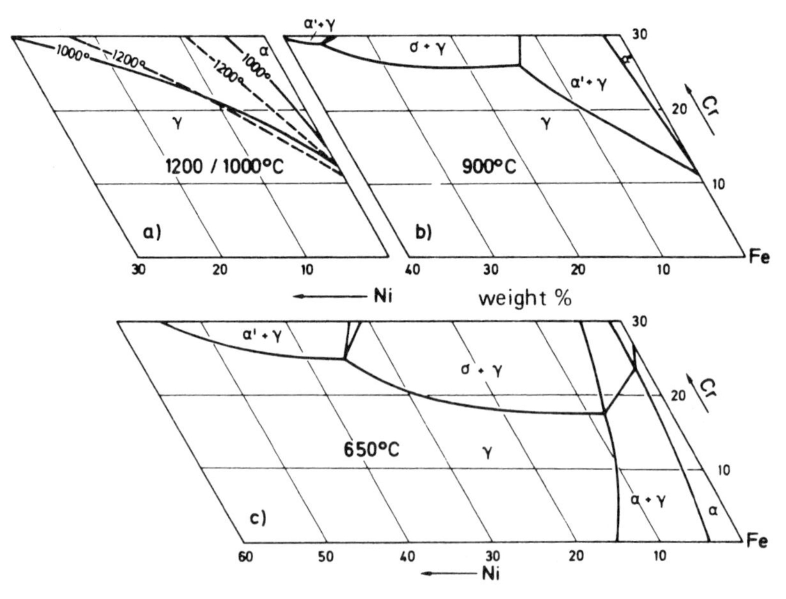

Figure 1.9: Sections from the nickel-iron-chromium ternary system [63]

precipitation can be suppressed both by extremely rapid cooling from the high solubility range and by keeping the carbon content extremely low from the beginning. This is the case with the so-called LC (low-carbon) grades. *Alternatively, the carbon can be fixed in the form of carbides by other elements, known as stabilizers,* and specifically added for this purpose. Titanium and niobium are particularly suitable for this purpose, though they also form nitrides with any nitrogen present.

Figure 1.10 shows, using the example of an austenite with 32% nickel and 20% chromium, the way in which one should conceive the solubility ratios at 980°C where titanium is used for stabilization. This diagram also facilitates estimation of the amounts of titanium carbide precipitated. Where titanium and niobium are added in larger amounts than is necessary to fix the carbon, the added elements then form intermetallic phases with the nickel, which have a temperature-dependent solubility in the gamma phase of the alloy matrix. Considerable use is made of this phenomenon in industry for the purpose of *age-hardening nickel materials,* though aluminium is a more significant alloying additive than titanium and niobium for this purpose. In this context, some of the customary precipitation-hardenable nickel-based alloys behave in a very similar manner to one another. In the corresponding alloys of such systems as Ni-Al, Ni-Cu-Al and Ni-Cr-Al, aluminium is present in the precipitation-hardened condition in the form of an A_3B phase, A signifying nickel, (nickel + copper) or (nickel + chromium), and B aluminium, to which the titanium is added, where it is present.

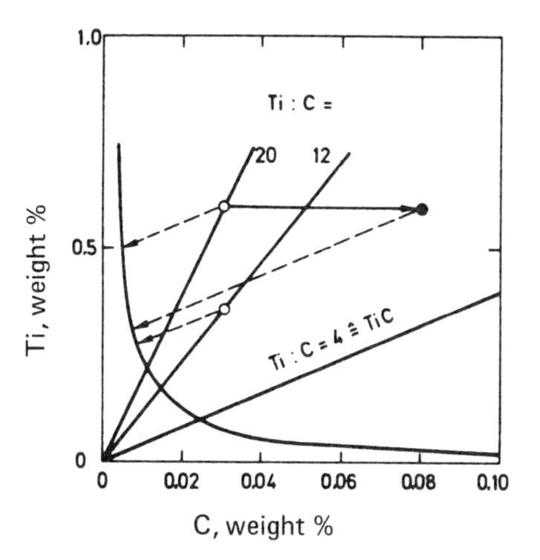

Figure 1.10: Saturation solubility of TiC in Nicrofer 3220 - alloy 800 austenite, at 980°C according to J. Orr [64]

Figure 1.11a shows, for the nickel-aluminium phase diagram, and Figure 1.11b for the nickel-chromium-aluminium-titanium quaternary system, what can be expected in the equilibrium case. Figure1.11b shows that titanium restricts the solubility of the $Ni_3(Al,Ti)$ phase in the gamma alloy matrix. Chromium, too, reduces the solubility of aluminium and titanium, iron having only a limited influence, with the result that precipitations of the $Ni_3(Al,Ti)$ phase can be formed even in the lower temperature range in a material such as Nicrofer 3220 - alloy 800 (1.4876), which contains aluminium and titanium only in amounts of a few tenths of 1% in each case. Molybdenum further reduces solubility, as does cobalt. Niobium is also absorbed into the $Ni_3(Al,Ti)$ phase, which is then either present as $Ni_3(Al,Ti,Nb)$ or, at relatively high additive levels as an independent Ni_3Nb compound.

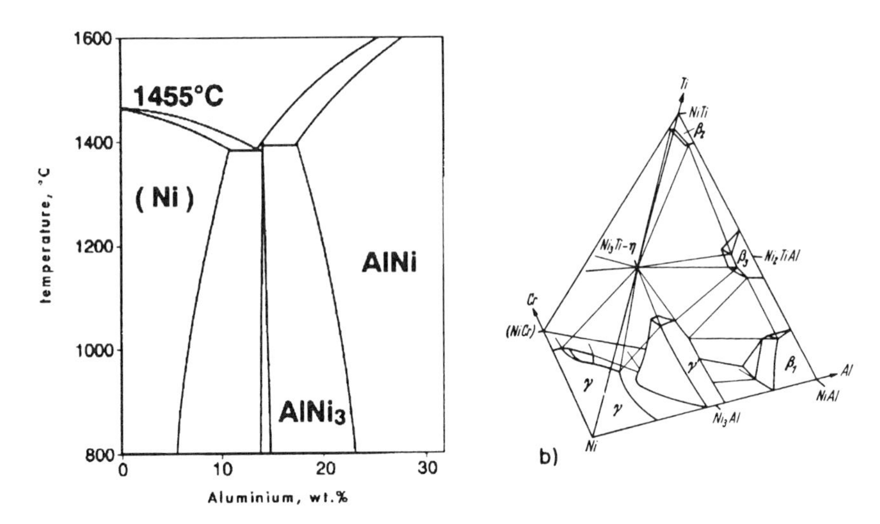

Figure 1.11: a) Ni-Al phase diagram [65]
b) Ni-Cr-Al-Ti quaternary system at 750°C [66]

In the case of a more significant departure from the nickel-iron-chromium ternary system, perhaps by means of the larger-scale addition of elements such as cobalt, molybdenum and tungsten for the purpose of increasing high-temperature strength, the formation of sigma phase and other related hard intermetallic phases such as Mu, Chi, Pi and Laves must be prevented. Such phases induce ductility losses and take up the elements added for the purpose of increasing strength or corrosion resistance of the alloy matrix. Represented in the quaternary systems, of which Figure 1.12 shows an example, these phases separate the face-centered cubic austenite from the body-centered cubic chromium-molybdenum-tungsten volume in a band-type arrangement. *Beyond the ternary*

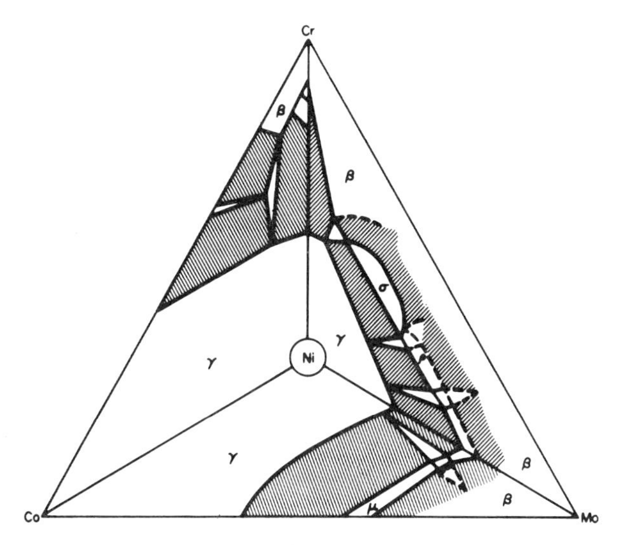

Figure 1.12: Nickel-chromium-molybdenum-cobalt quaternary system at 1200°C [67]

systems, it is possible to use mathematical methods (phase computation = Phacomp) with reasonable success to delimit the gamma volume. [68].

The binary system nickel-copper shows an uninterrupted solid solution series [65]. The iron, however, added to improve the alloy's corrosion resistance, exhibits a solubility which decreases with temperature [69], which has to be borne in mind. With the addition of aluminium and titanium a precipitation-hardening is achieved, equal to that of the nickel-chromium-iron alloys [70].

1.4 Time-temperature precipitation/sensitization diagrams and conclusions for heat treatment

While phase diagrams show equilibrium conditions, time-temperature transformation and precipitation diagrams describe, as a function of temperature and time, the steps required to attain those equilibrium conditions. The initial condition must, of course, be clearly defined. Figure 1.13 shows two *time-temperature precipitation diagrams* for the steel Cronifer 1713 LCN - alloy 317 LN (1.4439), alloyed firstly with 0.04 % and secondly with 0.145 % nitrogen. The multitude of precipitating phases and the strong delay in the precipitation sequence caused by the nitrogen addition can be seen. If greater detail is required, it is possible, using such diagrams, to differentiate further on the basis of the times at which such precipitations start at certain locations in the structure. $M_{23}C_6$ precipitation

thus usually establishes itself at the grain boundaries, which are particularly favourable for nucleation, then subsequently at the twin grain boundaries and, not until some considerable time has elapsed, in the interior of the grain. Commencement of $M_{23}C_6$ precipitation at the grain boundaries signifies that the area around the boundaries is chromium-deficient, since chromium is not capable of diffusing as rapidly as carbon. Not until a longer annealing time has elapsed can sufficient chromium be transferred from the interior of the grain and balance out the initial chromium deficiency. Figure 1.14 shows this, using Nicrofer 7216 - alloy 600 (2.4816) as example. It can be seen that, after 0.5 hours at 700°C, the chromium content in the grain boundary has dropped from an original 15% to some 6%, and falls after 10 hours to less than 4%. With a further increase in annealing time, however, the chromium concentration at the grain boundary begins to rise again, reaching more than 12% after 100 hours.
Corrosion tests, conducted in parallel, revealed that intercrystalline corrosion was observed in the Streicher and Huey test when the chromium content in the

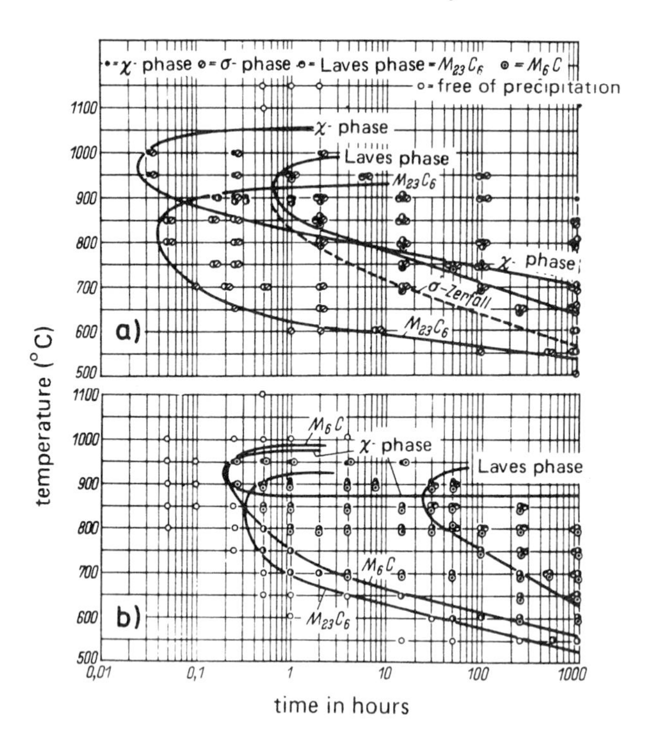

Figure 1.13: Time-temperature precipitation diagram for Cronifer 1713 LCN - alloy 317 LN (1.4439) with:
a) 0.04 and b) 0.145% nitrogen after quenching from 1100°C (a) and 1150°C (b), from [71]

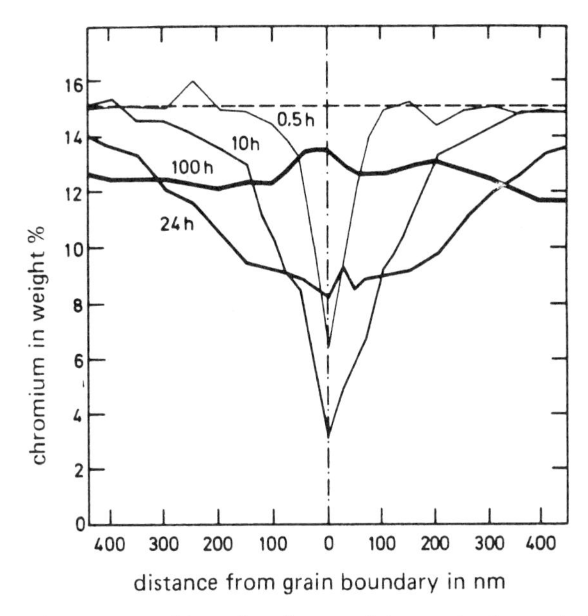

Figure 1.14: Chromium impoverishment in the grain-boundary after annealing of Nicrofer 7216 - alloy 600 (2.4816) at 700°C after solution annealing at 1100°C and water quenching [72]

grain boundary was below approximately 9%. This clearly demonstrates that while *time-temperature sensitization diagrams are indeed related to time-temperature precipitation diagrams, they are not identical with them.* The intercrystalline corrosion accompanying the formation of grain-boundary carbides commences later than carbide precipitation and comes to an end after a certain time and given sufficiently high temperature, while carbide precipitates remain. Figure 1.15 shows these conditions in the form of a generally applicable diagram. Figure 1.16 shows as a practical example an actual *time-temperature sensitization diagram* for the 6% molybdenum containing steel Cronifer 1925 hMo - alloy 926 (1.4529) (dashed line), compared with an earlier variant of this material with the lower nitrogen content of only 0.14% and with a 6% molybdenum steel with only 18% nickel [43]. It is apparent that both lower nickel as well as a lower nitrogen content displaces the sensitization to shorter periods. It should be pointed out that sensitization of this alloy type is not the result of carbide precipitations but of the precipitation of the intermetallic Chi-phase [43]. This precipitation of intermetallic phases, as well as of carbide, can also be the cause of sensitization of other high-alloyed materials [68].

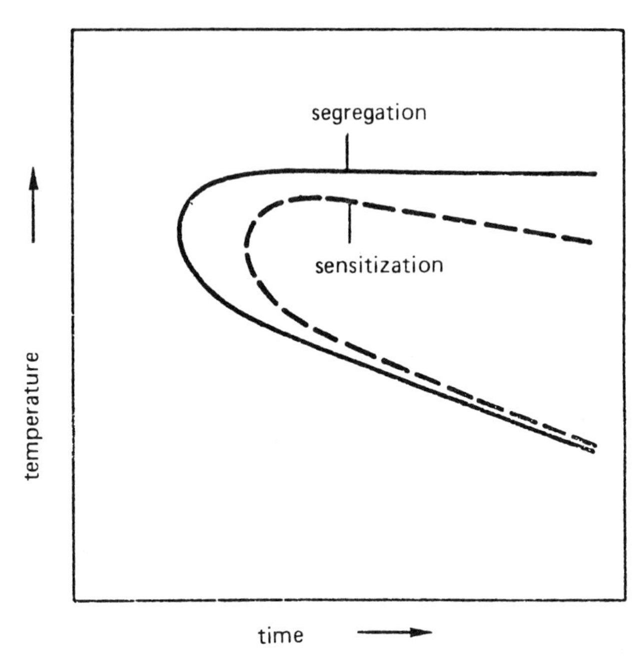

Figure 1.15: Time-temperature precipitation and time-temperature sensitization curves resulting from chromium-carbide precipitation

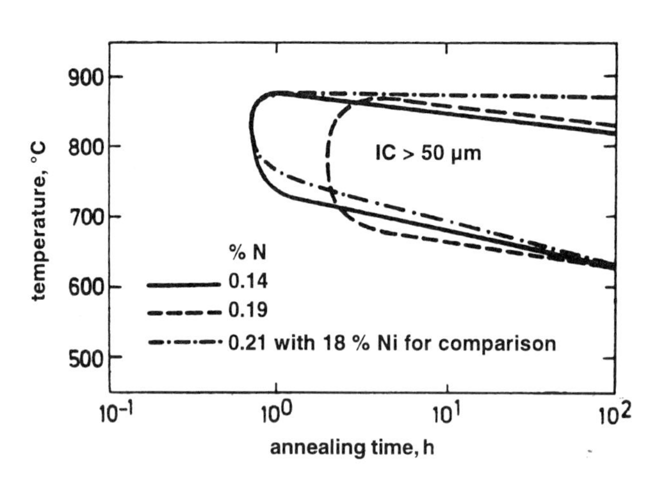

Figure 1.16: Time-temperature sensitization diagram for Cronifer 1925 hMo - alloy 926 (1.4529) as determined by intercrystalline corrosion tests according to SEP 1877/II (dashed line), compared to 6% molybdenum steels with only 0.14% nitrogen or only 18% nickel [43]

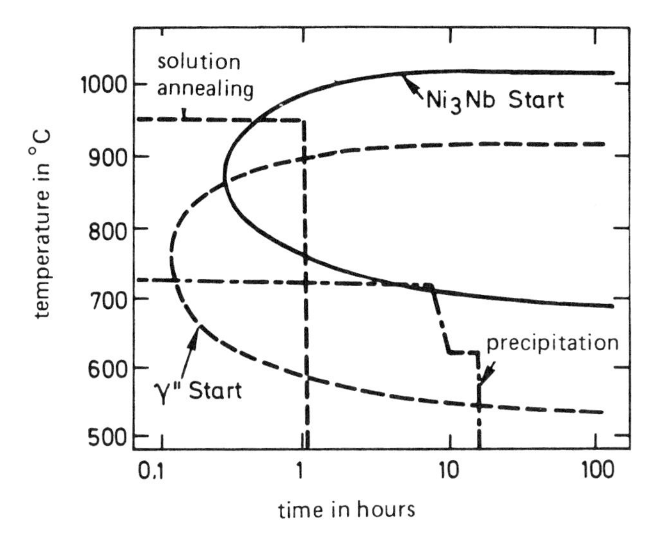

Figure 1.17: Time-temperature precipitation diagram for Nicrofer 5219 Nb - alloy 718 (2.4668), applicable to hot-rolled bars as initial condition, from [73]

Fig. 1.17 shows, as a further practical example, *the time-temperature-precipitation diagram of the precipitation-hardenable nickel superalloy Nicrofer 5219 Nb - alloy 718 (2.4668).* The Ni_3Nb phase is precipitated directly in its orthorhombic equilibrium form in the upper temperature range, commencing at the grain boundaries. In the lower temperature range, precipitation takes place via a metastable body-centred cubic intermediate form, which forms as finely distributed particles within the mixed crystal, and is thus the vehicle for the precipitation-hardening process. The maximum possible utilisation temperature of this high-strength, high-temperature nickel alloy can also be seen from the diagram. It is found to be around 700°C, since above this temperature the precipitation-hardenable metastable phase transforms within finite time into the Ni_3Nb equilibrium phase, the material thus losing its high degree of high-temperature strength.

Also shown in Figure 1.17 are the time-temperature paths for heat treatment. As for some other engineering materials, *solution annealing is carried out as a partial solution annealing.* It brings the material into the range of globular and flaky Ni_3Nb grain-boundary precipitation, which restricts grain growth and is thus decisive for the ductility of this alloy even under creep loading in the 650/700°C range. Subsequent *precipitation annealing* is carried out somewhat above these temperatures and is concluded as a graduated treatment. In other precipitation-hardenable nickel superalloys, *stabilization of the grain boundaries* is achieved by means of appropriate carbide precipitation. For example, in the alloy Nicrofer 7520Ti - alloy 80A (2.4952), the carbide Cr_7C_3 is precipitated at the grain boundaries in globular form during solution annealing at 1080°C [74] followed by further $Cr_{23}C_6$ precipitation in the course of subsequent precipitation-hardening at 700°C. To obtain the requisite carbide reaction at the grain boundaries, intermediate heat-treatment at a medium temperature, such as 845°C in the case of Nicrofer 7016 TiNb - alloy X-750 (2.4669) can also be applied [3, 4].

Non-precipitation hardenable high-temperature materials are also solution annealed before use [74]. Carbon is thus dissolved and grain size adjusted to the required value. As examples, 1130 and 1160°C are the solution-annealing temperatures for the materials Nicrofer 7216 H - alloy 600 H (2.4816) and 3220 H - alloy 800 H (1.4958) respectively [74]. During subsequent service in the high-temperature range, carbon rapidly precipitates out again, for example, as grain-boundary carbide. However, where these materials are used in a temperature range below 600°C, or for wet-corrosion applications, soft annealing to achieve stabilization must be carried out as a final treatment at a significantly lower temperature of e.g. 960°C. Therefore, while carbide precipitation at the grain boundaries must be accepted for high-temperature materials, or even turned to good use by means of appropriate procedures, *in the case of wet-corrosion application, the guiding principle of any type of heat treatment must be the avoidance of sensitization* caused by carbide precipitation at the grain boundaries. This applies both to the temperature at which final heat treatment is carried out and to the subsequent rate of cooling. The latter must be greater, the more the material tends to precipitation of carbide or other phases at the grain boundaries, and where the tendency to nucleation is particularly high; that is, the more the respective time-temperature precipitation diagram indicates that precipitation will commence within a very short time.

1.5 Impurities

1.5.1 External Impurities

The highest possible degree of resistance to corrosion and oxidation can only be attained by a given material if it is free of external impurities when used in its designated application. Even when welding semi-finished products for the manufacture of equipment components, particular attention should be paid not only to ensuring a metallically bright surface in the technical sense, but also to the cleanness of the materials being joined. Where hot working or heat treatment is carried out, even slight contamination of the surface as, for example, with sulphur-bearing oils or greases, can lead to extensive intercrystalline destruction in these high-nickel containing materials [74]. Nickel has a high affinity for sulphur and the nickel/nickel sulphide (Ni_3S_2) eutectic composition melts at 645°C. But even at temperatures as low as 550°C, sulphur (and sulphur from sulphur-bearing annealing gases) penetrates into the interior of the nickel, especially along the grain boundaries [70]. Furthermore, other surface impurities originating, for example, from lead-bearing paint markings, may produce similar damage.

1.5.2 Internal impurities

Apart from the carbides already dealt with in detail, austenitic chromium-nickel steels and nickel-based materials contain, according to type and manufacturing process, greater or lesser amounts of dispersed non-metallic inclusions, such as sulphides, oxides, nitrides and carbonitrides. Their amount, nature and distribution determine the so-called *'degree of cleanness'* of the material. This degree of cleanness is classified and evaluated by comparing the inclusion dispersion identified under a microscope on an unetched, polished section with a standardized series of micrographs specially created for this purpose, for instance according with Stahl-Eisen Prüfblatt 1570 (Steel/Iron Testing Sheet 1570) or ASTM E 45-63. Limiting the sulphide inclusions to a value specified by a maximum permissible sulphur content of 0.005 wt.% has proved absolutely necessary if a negative influence on the pitting- and crevice-corrosion resistance of the material is to be avoided [75].

In addition to undissolved impurities, the materials under discussion here also contain small amounts of dissolved alloy components such as phosphorus, tin, lead, bismuth, antimony and arsenic. These *companion elements* are unavoidably introduced from raw materials, but should only be regarded as impurities to the extent that they impair the alloy's processing and service properties. The permissible limit values should therefore be set at differing levels according to the material and its application, where experience indicates that their specification is necessary at all, since any and every unjustified or exaggerated requirement increases the manufacturing costs unnecessarily. In fact, very low limits, as for phosphorus, for example, are only required to meet special requirements of weldability or corrosion resistance. With regard to corrosion resistance, this might be necessary where attack by particularly severe oxidizing nitric acid solutions are to be anticipated [8]. With this in mind the phosphorus content of the nitric-acid-quality Cronifer 2521 LC (1.4335), for example, is restricted to a maximum of 0.015 wt.%.

Unavoidable companion elements may also have an adverse influence on high-temperature properties, in particular on creep-rupture strength [76, 77]. Figure 1.18 shows by way of example the manner in which traces of bismuth and lead shorten the service life of a nickel superalloy Nicrofer 5219 Nb - alloy 718 (2.4668), at approximately 650°C under a load of 690 N/mm². Due to their low solubility in nickel, these impurities tend to concentrate preferentially at the grain boundaries, as sites of increased solubility, resulting in premature decohesion under mechanical loading at elevated temperatures. There is an obvious need to severely limit the presence of both elements, in the case of bismuth down to the limit of analytical detectability, a maximum of 1 g/t. Fortunately, bismuth and lead can be largely removed by vacuum treatment of the melt. However, the degree of material impairment due to undesirable impurities is also dependent on the type of alloy. No generally applicable indications have yet been published on negative influences exercised by tin, antimony or arsenic in amounts of up to, for example, 0.01%. Before limitations can sensibly be specified, for instance with regard to Nicrofer 3220 H - alloy 800 H (1.4958), conclusive experimental data on this material must be documented.

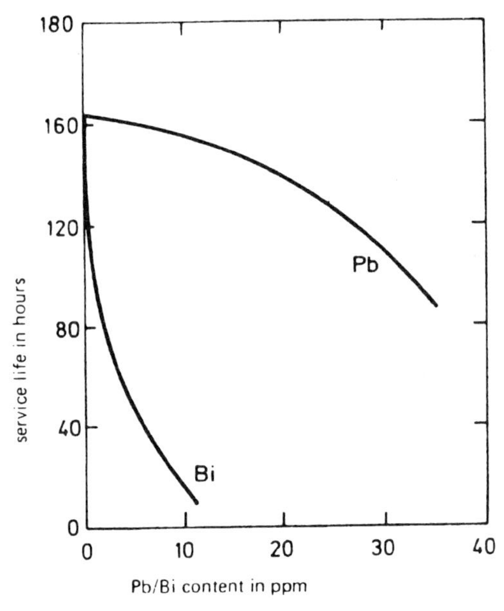

Figure 1.18: Influence of bismuth and lead on the service life of Nicrofer 5219 Nb - alloy 718 (2.4668) at 649°C and 690 N/mm², according to Kent [76]

1.6 Production, processing, heat-treatments and standards

The materials under consideration are normally produced in electric arc or induction furnaces. From the electric-arc furnace, and in some cases also from the induction furnace, the molten and predecarburized heat is transferred in its liquid state to an argon-oxygen decarburizing (AOD) converter or to a vacuum-oxygen decarburizing (VOD) unit. Using the AOD and VOD technologies, developed in the sixties, it is possible to produce nickel-chromium-molybdenum-iron-based alloys and stainless steels as LC (low-carbon) grades, with very low carbon contents of less than 0.03%, on a large scale.

After decarburization and deoxidation, extensive desulphurization is also possible in each case, but not dephosphorization. In the VOD unit, volatile impurities such as bismuth and lead are simultaneously evaporated. Where high requirements are made with regard to low-segregation characteristics of the ingot, subsequent electro-slag remelting (ESR) or remelting in a vacuum-arc furnace may be necessary. After hot and, where necessary, cold forming, including any homogenization and intermediate annealing necessary, heat treatment is usually the final stage in the manufacture of semi-finished products.

Very strict requirements regarding corrosion resistance or long-term high-temperature strength require the material state to be predefined, for which the nature and method of final heat treatment is of decisive importance. Such heat treatment includes in particular soft annealing, solution annealing and, to a lesser extent, precipitation hardening. The preconditions for these heat-treatment measures, and their basic principles, are detailed in Sections 1.3 and 1.4. They are summarized for certain materials by way of example in Table 1.10. It can be seen that the low-carbon materials Cronifer 1815 LCSi (1.4361), 1925 LC - alloy 904 L (1.4539), 1925 hMo - alloy 926 (1.4529) and Nicrofer 3127 hMo - alloy 31 (1.4563) are solution-annealed at temperatures of 1100°C on average and above. In accordance with the explanations given in Section 1.4, the titanium-bearing alloy Nicrofer 3220 - alloy 800 (1.4876) in its low-carbon (LC) wet corrosion-resistant variant is stabilization-annealed at 960°C. Also in its heat-resistant variant, Nicrofer 3220 - alloy 800 (1.4876) is soft-annealed at 960°C, whereas the high-temperature variant Nicrofer 3220 H - alloy 800 H (1.4958) requires solution-annealing at 1160°C. This also applies to the variants of Nicrofer 7216 - alloy 600 (2.4816), and to the niobium-alloyed Nicrofer 6020 hMo - alloy 625 (2.4856), which is solution-annealed as a high-temperature material and stabilization-annealed as a wet-corrosion material (62). The materials LC Nickel - alloy 201 (2.4068), Nicorros - alloy 400 (2.4360) and Cunifer, which contain no carbide-forming elements, are all soft-annealed [74].

In this context, precipitation-hardening is described in sufficient detail in Section 1.4. Stress-relief annealing is sometimes of significance for processing equipment and power-plant engineering, particularly when tubes are used in a slightly redrawn condition or, exceptionally, in their hard condition, or where unallowable internal stresses have arisen. It is generally true in such cases that a slight reduction in internal stresses can be achieved, even at relatively low annealing temperatures, as for example in the case of Nicrofer 3220 - alloy 800(1.4876) by means of annealing at 540°C, but complete removal of internal stresses can only be achieved by means of recrystallization and/or soft-annealing [74]. The temperatures stated in Table 1.10 should therefore be regarded only as approximate guide. It should, moreover, always be born in mind that no possible but unacceptable sensitization of the material may be allowed to occur during stress-relief-annealing.

The wide range of national and international standards and technical regulations for the materials discussed here is listed in publications of the material manufacturers (78), so that there is no need to discuss them here individually.

Table 1.10: Typical heat-treatment temperatures for various materials. In each case, only one temperature (°C) is given out of a wide ranging temperature scale. In case of doubt, the relevant VdTÜV-Werkstoffblatt or the manufacturer's data sheet should be consulted

Krupp VDM trade name	**German Material No. DIN**	**Solution-heat treatment °C**	**Annealing or stabilizing-heat treatment °C**	**Stress relieving °C**
Cronifer				
1815 LCSi	1.4361	1140	---	
1925 LC - alloy 904 L	1.4539	1120	---	
1925 hMo - alloy 926	1.4529	1160		
Nicrofer				
3127 hMo - alloy 31	1.4562	1180	---	
3220 LC - alloy 800 L	1.4558	---	960	700
3220 - alloy 800	1.4876	---	960	
3220 H - alloy 800 H	1.4958	1160	---	
4221 - alloy 825	2.4858	---	940	
5923 hMo - alloy 59	2.4605	1120	----	
6020 hMo - alloy 625	2.4856	1120	980	
6616 hMo - alloy C-4	2.4610	1070	---	
7216 LC - alloy 600 L	2.4817	---	960	
7216 H - alloy 600 H	2.4816	1130	---	
Nickel				
VDM LC-Nickel 99.2 - alloy 201	2.4068	---	780	600
Nicorros - alloy 400	2.4360	---	800	550
Cunifer 10 - alloy CuNi90/10	2.0872	---	780	400
Cunifer 30 - alloy CuNi70/30	2.0882	---	850	400

1.7 References Chapter 1

[1] D. Peckner, I. M. Bernstein: Handbook of Stainless Steels, McGraw-Hill Book Comp., 1987

[2] Nichtrostende Stähle, Eigenschaften, Verarbeitung, Anwendung, Normen, 2. Aufl., Verlag Stahleisen mbH, Düsseldorf, 1989

[3] W. Betteridge, J. Heslop (Editors): The Nimonic Alloys, 2nd Edition, Edward Arnold, London, 1974

[4] C. T. Sims, W. C. Hagel (Editors): The Superalloys, John Wiley & Sons, New York, 1972; C. T. Sims, N. S. Stoloff, W. C. Hagel: The Superalloys II, John Wiley & Sons, New York, 1987

[5] E. F. Bradley: Superalloys, A Technical Guide, ASM International, Metals Park, Ohio, 1988

[6] M. Verneau, J.-P. Audouard, J. Charles: Special Stainless Steels and Alloys for the Storage and Transport of Chemical Products, Proc. Int. Conf. Stainless Steels '96, Verlag Stahleisen, Düsseldorf, 1996, 163 - 170

[7] E. Altpeter, R. Kirchheiner, F.-E. White: Sulphuric Acid Corrosion of Some Special Stainless Steels and Nickel Alloys, - Laboratory Tests and Plant Experience, NACE Int. Italia Section, Monza, 1995, 391 - 399

[8] G. Herbsleb: Der Einfluß von Legierungselementen auf das Passivierungsverhalten nichtrostender Stähle, VDI-Z 123 (1981), No. 12, 505 - 512

[9] M. B. Rockel, M. Renner: Critical Review of Laboratory Methods Testing the Resistance of High Alloyed Stainless Steels and Ni-Alloys against Localized Corrosion in Chloride Solutions, Paper presented at European Congress of Corrosion, Nizza, Nov. 1985

[10] M. B. Rockel, M. Renner: Pitting, crevice and stress corrosion resistance of high chromium and molybdenum alloy stainless steels, Werkstoffe und Korrosion 35 (1984), 537 - 542

[11] U. Heubner, M. B. Rockel, E. Wallis: The Corrosion Behaviour of Copper-Alloyed Stainless Steels in Reducing Acids, ATB Metallurgie XXV, No. 3 (1985) 235 - 241

[12] A. J. Sedriks: Metallurgical aspects of passivation, Stainless Steels '84, The Institute of Metals, London (1985) 125 - 133

[13] K. Lorenz, G. Medawar: Über das Korrosionsverhalten austenitischer Chrom-Nickel-(Molybdän)-Stähle mit und ohne Stickstoffzusatz unter besonderer Berücksichtigung ihrer Beanspruchbarkeit in chloridhaltigen Lösungen, Thyssenforschung 1 (1969) 94 - 108

[14] M. Renner, U. Heubner, M. B. Rockel, E. Wallis: Temperature as a pitting and crevice corrosion criterion in the $FeCl_3$ test, Werkstoffe und Korrosion 37 (1986) 183 - 190

[15] U. L. Heubner, E. Altpeter, M. B. Rockel, E. Wallis: Electrochemical Behavior and Its Relation to Composition and Sensitization of NiCrMo Alloys in ASTM G-28 Solution, CORROSION 45 (1989) 249 - 259

[16] U. Heubner, R. Kirchheiner, M. Rockel: Alloy 31, A New High-Alloyed Nickel-Chromium-Molybdenum Steel for the Refinery Industry and Related Applications, CORROSION 91, Paper No. 321, National Association of Corrosion Engineers, Houston, Texas, 1991

[17] R. Kirchheiner, M. Köhler, U. Heubner: A New Highly Corrosion Resistant Material for the Chemical Process Industry, Flue Gas Desulfurization and Related Applications, CORROSION 90, Paper No. 90, National Association of Corrosion Engineers, Houston, Texas, 1990

[18] D. C. Agarwal, U. Heubner, R. Kirchheiner, M. Köhler: Cost Effective Solutions to CPI Corrosion Problems with a New Ni-Cr-Mo Alloy, CORROSION 91, Paper No. 179, National Association of Corrosion Engineers, Houston, Texas, 1991
[19] R. Kirchheiner, M. Köhler, U. Heubner: Nicrofer 5923 hMo, ein neuer hochkorrosionsbeständiger Werkstoff für die Chemische Industrie, die Umwelttechnik und verwandte Anwendungen, Werkstoffe und Korrosion 43 (1992) 388 - 395
[20] E. Altpeter, E. Wallis, U. Heubner: Erprobung verschiedener metallischer Legierungen für die Abwasseraufbereitung hinter Rauchgasentschwefelungsanlagen, Werkstoffe und Korrosion 45 (1994) 539-549
[21] G. Riedel, U. Heubner: Werkstoffe für die Eindampfung von Abwasser aus der Rauchgasentschwefelung, Beitrag zur Werkstoffwoche '96, Stuttgart, 28.- 31.05.1996
[22] R. Kirchheiner, W. Schalk, H. Müller, S. M. Palomino: Qualification of Nicrofer 3127 LC for the preconcentration of phosphoric acid, Werkstoffe und Korrosion 40 (1989) 545 - 551
[23] M. Köhler, U. Heubner, K.-W. Eichenhofer, M. Renner: Alloy 33, a New Corrosion Resistant Austenitic Material for the Refinery Industry and Related Applications, CORROSION '95, Paper No 338, NACE International, Houston, Texas, 1995; Progress with Alloy 33 - A New Corrosion-Resistant Chromium Based Austenitic Material, CORROSION '96, Paper No 428, NACE International, Houston, Texas 1996; Alloy 33, A New Nitrogen-Alloyed Chromium-Based Material for Many Corrosive Environments, Proc. Stainless Steels '96, Düsseldorf, 1996, pp. 178 - 181; Alloy 33: a New Mayterial for the Handling of HNO3 / HF Media in Reprocessing of Nuclear Fuel, CORROSION 97, Paper No 115, NACE International, Houston, Texas, 1997
[24] D.C. Agarwal, U. Heubner, M. Köhler, W. Herda: UNS N 10 629: A New Ni-28% Mo Alloy, Materials Performance 33 (1994) No 10, 64-68
[25] U. Heubner, R. Kirchheiner: High-alloy special stainless steels and nickel-based materials for nuclear fuel reprocessing, in U. Heubner et al., Nickel alloys and high-alloy special stainess steels, expert-verlag, Sindelfingen, 1987, S. 223 - 240
[26] R. Kirchheiner, U. Heubner, F. Hofmann: Increasing the lifetime of nitric acid equipment using improved stainless steels and nickel alloys, CORROSION 88, Paper No. 318, National Association of Corrosion Engineers, Houston, Texas, 1988
[27] E. M. Horn, K Schoeller: Temperatur- und Konzentrationsabhängigkeit der Korrosion nichtrostender austenitischer und ferritischer Werkstoffe in 20- bis 75%iger Salpetersäure, Werkstoffe und Korrosion 41 (1990) 97 - 112
[28] U. Heubner, M. B. Rockel, E. Wallis: Ein neuer hochlegierter Nickel-Chrom-Molybdän-Stahl für den Chemie-Apparatebau, Werkstoffe und Korrosion 40 (1989) 418 - 426
[29] E. Horn, A. Kügler: Entwicklung, Eigenschaften, Verarbeitung und Einsatz des hochsiliziumhaltigen austentischen Stahls X2CrNiSi1815, Z. Werkstofftech. 8 (1977) 362 - 370, 410 - 417
[30] R. Kirchheiner, F. Hofmann, Th. Hoffmann: A silicon alloyed stainless steel for highly oxidizing conditions, CORROSION 86, Paper No. 120, National Association of Corrosion Engineers, Houston, Texas, 1986
[31] C. Miola, H. Richter: Utilization of stainless steels and special metals in nitric acid and urea production plants, Werkstoffe und Korrosion 43 (1992) 396 - 401
[32] M. Köhler, R. Kirchheiner, U. Heubner: Nicrofer 2509 Si 7, ein neuer korrosionsbeständiger Werkstoff für die Handhabung heißer, hochkonzentrierter Mineralsäuren, Werkstoffe und Korrosion 46 (1995) 18-26
[33] Das Verhalten von austenitischen nichtrostenden Stählen, Nickellegierungen und Nickel in Natronlauge, Druckschrift der Mannesmann Edelstahlrohr GmbH, 1988
[34] W. Z. Friend: Corrosion of Nickel and Nickel-Base Alloys, J. Wiley & Sons, New York

- Chichester - Brisbane - Toronto, 1980
[35] A. J. Sedriks: Corrosion of Stainless Steels, , J. Wiley & Sons, New York - Chichester - Brisbane - Toronto, 1980
[36] H. Donat: Möglichkeiten, Grenzen und Besonderheiten bei der Verarbeitung von hochlegierten Stählen, Ni-Legierungen und Sonderwerkstoffen, Vortrag zum 15. DECHEMA-Konstruktionssymposium, Frankfurt/Main, 15./16.01.1986
[37] A. H. Tuthill: Guidelines for the use of copper alloys in seawater, Materials Performance 26 (1987), No. 9, 12 - 22
[38] W. R. Herda, M. R. Jasner: Werkstoffe für Seewasser-Rohrleitungssysteme, 3R international 27 (1988) 258 - 264
[39] A. H. Tuthill: Usage and Performance of Nickel-Containing Stainless Steels in Both Saline and Natural Waters and Brines, Materials Performance 26 (1988), No. 7, 47 - 50
[40] M. Köhler, U. Heubner: Alloy 24, A New Seawater Resistant High Manganese Stainless Steel for Offshore Application, CORROSION 93, Paper No 500, National Association of Corrosion Engineers, Houston, Texas, 1993
[41] P. Lövland: Use of advanced steels and alloys mounts in seawater process systems, Nickel 2 (1986), No. 1, 10 - 11
[42] A. J. Sedriks: New Stainless Steels for Seawater Service, NACE-CORROSION 45 (1989) 510 - 518
[43] U. Heubner, M. B. Rockel, E. Wallis: Das Ausscheidungsverhalten von 6 % Mo-Stählen und sein Einfluß auf die Korrosionsbeständigkeit, Werkstoffe und Korrosion 40 (1989) 459 - 466
[44] M. Jasner, U. Heubner: Alloy 31, A New 6 Moly Stainless Steel with Improved Corrosion Resistance to Seawater, CORROSION '95, Paper No 279, NACE International, Houston, Texas 1995
[45] S. W. Borenstein: Microbiologically influenced corrosion handbook, Industrial Press Inc., New York, 1994
[46] M. Renner, G. Wagner: Microbiologically Induced Corrosion (MIC) of Stainless Steels, Proc. Int. Conf. Stainless Steels '96, Verlag Stahleisen, Düsseldorf, 1996, 411-417
[47] C. Voigt, H. Werner, R. Kirchheiner, L. Schambach: Die Bedeutung der pH-abhängigen Durchbruchsreaktionen bei NiCrMo-Werkstoffen in Medien der Chemie- und Umwelttechnik mit hohem Redoxpotential, Werkstoffe und Korrosion 46 (1995) 27 - 32
[48] U. Brill, U. Heubner, M. Rockel: Hochtemperaturkorrosion handelsüblicher hochlegierter austenitischer Werkstoffe im geschweißten und ungeschweißten Zustand, METALL 44 (1990) 936 - 946
[49] U. Brill: Neue warmfeste und korrosionsbeständige Nickel-Basis-Legierung für Temperaturen bis 1200° C, METALL 46 (1992) 778 - 782
[50] U. Brill, G. Giersbach, H.-W. Kettler: Effizienzsteigerung kontinuierlicher Wärmebehandlungsanlagen durch den Einsatz ungekühlter Ofenrollen aus dem neuen Werkstoff Nicrofer 6025 HT (2.4633), VDI-Berichte No 1151 (1995) 65-88
[51] U. Brill, J. Klöwer: Cronifer III-TM und Nicrofer 45 TM - Zwei neue Werkstoffe für den Einsatz in Müllverbrennungsanlagen, METALL 46 (1992) 921-926
[52] D.C. Agarwal, U. Brill, J. Klöwer: Nicrofer 45 TM - Results of Various Practical Exposure Tests, Proc. 2nd Int. Conf. Heat Resistant Materials, ASM Int., Materials Parks, Ohio, 1995, pp. 327 - 331
[53] U. Heubner, F. Hofmann: Special Aspects of Localized High-Temperature Corrosion, Werkstoffe und Korrosion 40 (1989) 363-369
[54] J. Klöwer: High Temperature Corrosion Behaviour of Commercial High Temperature

Alloys under Deposits of Alkali Salts, Proc. 2nd Int. Conf. Heat Resistant Materials, ASM Int., Materials Park, Ohio, 1995, pp. 245 - 252
[55] J. Klöwer, G. Sauthoff, D. Letzig: Alloy 10 Al - A New Sulphidation and Carburization Resistant Alloy For Fuel Combustion and Conversion, CORROSION / 96, Paper No 144, NACE International, Houston, Texas 1996
[56] High Performance Materials, Publication No N 524 92-11 of Krupp VDM GmbH
[57] D.C. Agarwal, U. Brill, H.W. Kettler, G. Giersbach: Innovations in Alloy Metallurgy for Furnace Rolls and Other High Temperature Applications, Proc. 2nd Int. Conf. Heat Resistant Materials, ASM Int., Materials Park, Ohio, 1995, pp. 63 - 66
[58) U. Brill, D.C. Agarwal: Alloy 602 CA - A New Alloy for the Furnace Industry, Proc. 2nd Int. Conf. Heat Resistant Materials, ASM Int., Materials Park, Ohio, 1995, pp. 153 - 162
[59] U. Brill, U. Heubner, K. Drefahl, H.-J. Henrich: Zeitstandwerte von Hochtemperaturwerkstoffen, Ingenieur-Werkstoffe 3 (1991) No. 4, 59 - 62
[60] U. Heubner: in Materials Science and Technology, Vol. 8: Structure and Properties of Nonferrous Alloys, Chapter 7: Nickel-Based Alloys, VCH, Weinheim, 1996, 349 - 397
[61] P. Dahlmann, U. Brill, T. Haubold: Entwicklung massiv aufgestickter Ni-Basis-Superlegierungen für leistungsgesteigerte Gasturbinen, paper presented at Werkstoffwoche '96, Stuttgart, May 28-31, 1996
[62] M. Köhler: Effect of the Elevated-Temperature-Precipitation in Alloy 625 on Properties and Microstructure, Superalloys 718, 625 and Various Derivatives, E. A. Loria Edt., TMS, Warrendale, Pennsylvania, (1991), 363 - 374
[63] V. G. Rivlin, G. V. Raynor: Critical evaluation of chromium-iron-nickel-system, Internat. Metals Reviews 1980, No. 1, 21 - 38
[64] J. Orr: A Review of the Structural Characteristics of Alloy 800, in Alloy 800, Proc. Petten Internat. Conf., North Holland Publishing Comp., Amsterdam - New York - Oxford, 1978, Paper 2 - 1
[65] T. B. Massalski: Binary Alloy Phase Diagrams, 2nd Edition, ASM International, Materials Park, Ohio (1991)
[66] A. Taylor: Constitution of Nickel-Rich Quarternary Alloys of the Ni-Cr-Ti-Al System, Trans. AIME 206 (1956) 1356 - 1362
[67] L. R. Woodyatt, C. T. Sims, H. J. Beattie: Prediction of Sigma Phase Occurrence from Composition in Austenitic Superalloys, Trans. AIME 236 (1966) 519 - 527
[68] U. Heubner, M. Köhler: Das Zeit-Temperatur-Ausscheidungs- und das Zeit-Temperatur-Sensibilisierungs-Verhalten von hochkorrosionsbeständigen Nickel-Chrom-Molybdän-Legierungen, Werkstoffe und Korrosion 43 (1992) 181 - 190
[69] Kupfer-Nickel-Legierungen, Informationsdruck, herausgegeben vom Deutschen Kupfer-Institut (DKI) e.V., Berlin
[70] K. E. Volk: Nickel und Nickellegierungen, Springer-Verlag, Berlin - Heidelberg - New York, 1970
[71] H. Thier, A. Bäumel, E. Schmidtmann: Einfluß von Stickstoff auf das Ausscheidungsverhalten des Stahles X5CrNiMo1713, Arch. Eisenhüttenwes. 40 (1969) 333 - 339
[72] G. S. Was, H. H. Tischner, R. M. Latanision: Influence of Thermal Treatment on the Chemistry and Structure of Grain-Boundaries in Inconel 600, Met. Trans. 12 A (1981) 1397 - 1408
[73] D. R. Muzyka: The Metallurgy of Nickel-Iron Alloys, in (4), (1972), S. 113 - 143
[74] U. Heubner: Wärmebehandlung von Nichteisenmetallen, in: Handbuch der Fertigungstechnik, Edt. G. Spur u. Th. Stöferle, Band 4/2 Wärmebehandeln, Carl Hanser Verlag, München - Wien, 1987, S. 972 - 1011
[75] U. Heubner, M. Rockel: Entwicklung neuer korrosionsbeständiger Werkstoffe für den

Chemieapparatebau, Werkstoffe und Korrosion 37 (1986) 7 - 12
[76] R. T. Holt, W. Wallace: Impurities and trace elements in nickel-base superalloys, Internat. Metals Reviews, 1976, No. 3, S. 1 - 24
[77] M. McLean, A. Strang: Effects of trace elements on mechanical properties of superalloys, Metals Technology 11 (1984) 454 - 464
[78] see (56)

2 Corrosion behaviour of nickel alloys and high-alloy special stainless steels

M. Rockel

2.1 Introduction - types of corrosion and of corrosion resistance

According to DIN 50 900 Part 1 corrosion is defined as the reaction of a metallic material with its environment causing a measurable alteration in the material and being capable of producing corrosion damage, which impairs the function of the corresponding component. [1]

Although high-alloy special stainless steels and nickel materials have a significantly better corrosion resistance than standard stainless steels (for example AISI 316) they still undergo various types of corrosion, which are summarized in Figure 2.1.

The simplest form of corrosion is uniform or general corrosion where corrosion is practically uniform across the entire surface of a component. Uniform corrosion causes less problems than other types, since it can be quantified by determining the weight loss over a given period of time (g/m^2d) or the rate of removal (mm/a).It allows the long-term corrosion behaviour of a component to be estimated. The extent of uniform corrosion depends on the nature and degree of aggressiveness of the medium, on its concentration (C) and on the temperature (T).

Weight-loss time curves and ISO-corrosion curves in temperature/concentration diagrams are used to elaborate test results – see DIN 50 905 [2].

Pitting corrosion may occur with alloys whose resistance to uniform corrosion is caused by passivation particularly in chloride-containing aqueous media. In practise chloride is present not only in most process media but also in cooling water from a wide range of sources (river, brackish and sea-water) therefore this type of corrosion cannot usually be excluded. The susceptibility to pitting corrosion rises with increasing chloride concentrations (C_{Cl^-}), higher temperatures (T) as well as reducing pH values and lower/approaching zero flow rates (V) [3, 4]. If copper-nickel alloys or stainless steels with poor pitting corrosion resistance are brought into contact with for example sea-water, stationary conditions must be avoided e.g. in sea-water desalination plants.

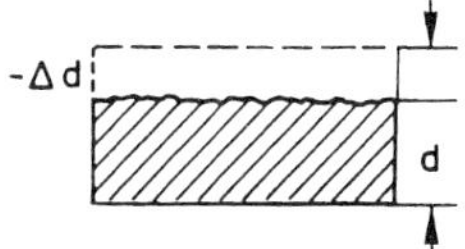

GENERAL CORROSION

Avaluable corrosion rate
$g/m^2 \cdot d$ (mm/a)
f(medium, conc., T)

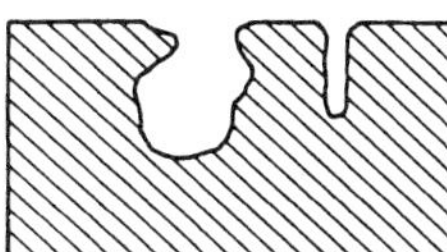

PITTING CORROSION

Local corrosion at passivable alloys, preferentially by chlorides (Br^-, J^-) also at inclusions as MnS
f(chloride conc., sol. velocity, T)

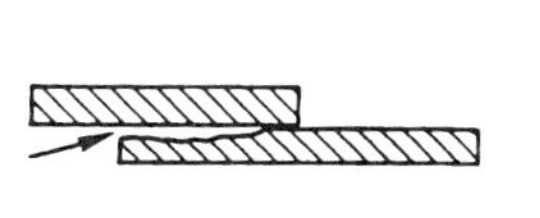

CREVICE CORROSION

In chloride media, in narrow crevices, under deposits, in sealings; at passivable alloys
f(chloride conc., sol. veloc., T, O_2)

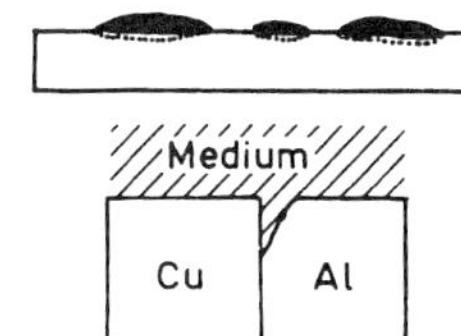

GALVANIC CORROSION

Galvanic series of metals, the less noble metal dissolves
f(potentialdiff., conduct., T)

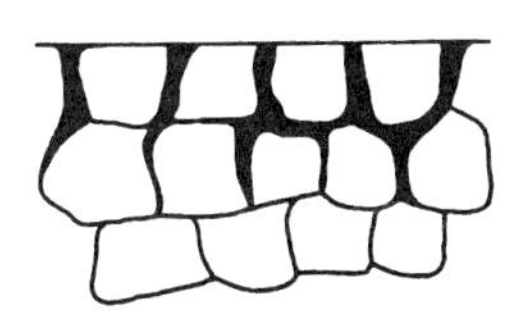

INTERGRANULAR (INTERCRYSTALLINE) CORROSION

Preferential attack of precipitations containing grain boundaries; especially stainless steels (unstabilized; too high carbon)
f(medium, annealing, welding, T)

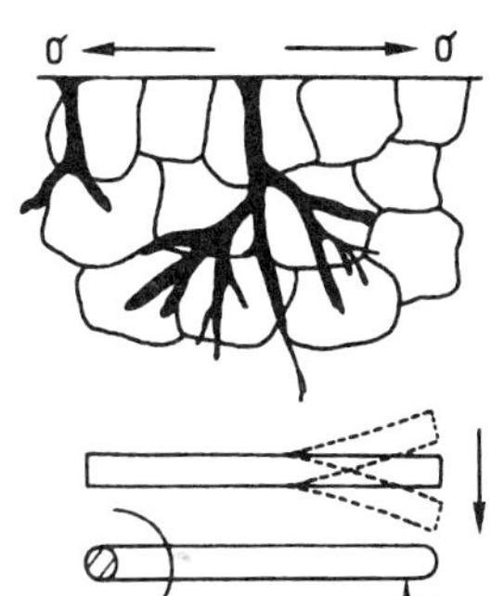

STRESS CORROSION CRACKING

Specific corrosion media and alloy system; high mechanical (also internal) stress
f(medium, stress, T)

CORROSION FATIGUE

Fatigue under corrosive conditions
f(frequency, stress, conc., T)

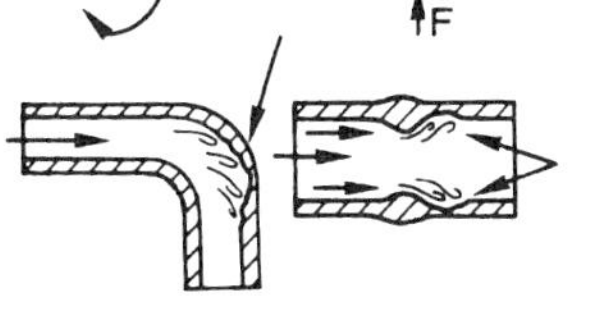

EROSION

Attack in solutions of high velocity; i.e. with turbulences and sharp bents

Figure 2.1: Summary of different types of corrosion

Where crevices are present in a structure, around flanges, at expanded tube-ends and around gaskets, for example, local corrosion can occur. Such local corrosion is caused by corrosion elements formed in the medium as a result of varying concentration (due to differing aeration for example). We also speak of crevice corrosion where corrosion occurs under deposits.

Since crevice corrosion is a similar phenomenon to pitting corrosion it usually takes place before the latter, i.e., it occurs under less corrosive conditions. Pitting and crevice corrosion are further described in Section 2.3.

Galvanic corrosion, also known as contact corrosion, occurs when two metals positioned relatively widely apart in the galvanic electrochemical series are in direct (metallic) contact and are surrounded by a corrosive medium of sufficient ionic conductivity. Corrosion then occurs preferentially on the baser metal (lower corrosion potential) [5]. As well as metal/metal combinations, galvanic corrosion may also be caused by the combination metal/electron-conductive solid (e.g. graphite). As high-alloy special stainless steels and nickel alloys have a relatively high free corrosion potential they rarely undergo contact corrosion.

Intercrystalline corrosion (IC) or intergranular attack (IGA), respectiveley, is a selective type of corrosion in which mainly the areas adjacent to the grain boundaries are attacked. Nickel alloys and stainless steels may be susceptible to IC as a result of excessive carbon contents and sensitizing heat treatments. This is due to the fact that chromium-rich carbides form in the grain-boundary areas causing chromium concentration to drop to values of less than 13 %, which results in loss of passivity in microareas of the alloy.

Whereas intercrystalline corrosion used to be a severe problem for the weldability of standard stainless steels, today it is largely controlled. Improvements in melting metallurgy have contributed to this extensively. AOD and VOD processes permit carbon contents to be reduced to 0.005 %. In addition time-temperature sensitization diagrams have been systematically produced in recent years for all special stainless steels and nickel alloys. These diagrams show the IC sensitive areas which enable the behaviour after welding to be deduced. Measures to avoid intercrystalline corrosion are further described in Section 2.5.

Where a metal or an alloy exposed to a specific corrosive medium is subjected simultaneously to a high level of tensile stress, cracking and thus premature failure of the component may occur without any sign of uniform corrosive attack [6]. The cracks follow an intercrystalline or transcrystalline course, perpendicular to the tensile stress but with a low degree of deformation, similar to brittle fracture. Even natural stresses within components, for instance as a result of a high degree of cold-forming or after welding, can be sufficient to cause stress-corrosion cracking (SCC).

Whereas standard austenitic stainless steels may suffer from SCC particularly in chloride-containing aqueous solutions at elevated temperatures, high-alloy special stainless steels with higher Nickel and Molybdenum contents show substantially longer resistance times to SCC in all laboratory test media and practical media. According to Copson [7] nickel alloys have the highest SCC resistance, see Section 2.3.

Loss of strength in a material due to cyclically changing or oscillating stresses is known as material fatigue. One refers to oscillating stress corrosion cracking (OSCC) or corrosion fatigue when the material is simultaneously subjected to a corrosive medium [8]. In contrast to SCC, for the occurance of which the material must be susceptible to SCC and a specific medium must be present, all metallic materials can principally suffer from OSCC in all corrosive media. Transcrystalline cracking then occurs with minimal deformation.

One speaks of erosion corrosion when simultaneously with erosion caused by mechanical abrasion by solid materials (sea-water flowing in brass tubes), chemical or electrochemical corrosion occurs. Erosion corrosion can be observed on relatively soft materials in particular, for instance on brass and Copper-Nickel 90/10 tubing, carrying sea-water at very high velocities. In such cases the flow velocity has to be restricted to a maximum of approximately 3m/s for brass, 3.5 m/s for Copper-nickel 90/10 (CuNi10Fe) and 4.5 m/s for Copper-nicke! 70/30 (CuNi30Fe). On the other hand neither high-alloy stainless steels nor nickel alloys suffer from erosion corrosion in practical media so that in some cases flow velocities up to 20m/s may be permitted [9].

Corrosion Resistance: It must be emphasized that corrosion resistance – this applies to all types of corrosion – is not a material property, i.e., a given property characteristic of the material, like physical or mechanical properties. Instead, corrosion resistance is understood to be the behaviour of a material in a surrounding aqueous or gaseous medium. The results of all standard tests should be interpreted and evaluated within this context. That is that corrosion data obtained in this way apply in the first instance only to the interaction of the material with this test medium. The extent to which the results of such standard tests can be directly applied to corrosion behaviour in practise is frequently open to questions. Therefore the results of such investigations in laboratory media should not be overvalued. The results of pitting and crevice corrosion tests in 10% $FeCl_3x6H_2O$ [10] test should be considered with this in mind (Chapter 2.3). The transfer for example of the limiting temperatures measured to the practical application of the material in sea-water or high chloride-containing cooling water is inadmissible. The $FeCl_3$ test selected is a more aggressive medium, rarely encountered in practise, which represents the most extreme corrosive conditions with its high chloride fraction, its low pH value as well the oxidizing action of the iron III ions. Only testing a number of materials enables the corrosion behaviour between these materials to be differentiated so that it is possible to rank the materials into poor, moderate, good and very good behaviour. The transfer of such a differentiation or ranking sequence to practical applications however is not necessarily possible for each application.

A similar example is found in the results of IC tests on nickel alloys. The nickel-chromium-molybdenum alloy Nicrofer 5716 hMoW - alloy C-276 (2.4819) gave very unfavourable results in the ASTM G-28 method A IC test (Chapter 2.4.2). Therefore, it can be concluded that there is some structural instability and that it is necessary to pay particular attention to the welding – the basically outstanding resistance of this material in numerous less oxidizing high-corrosive media in practical application is not indicated. This shows that with such an IC standard test, which was originally introduced for the IC susceptible stainless steels and later – from todays view point too carelessly – simply transferred to the high nickel containing materials. A corrosion behaviour is thus obtained that actually only applies to the media specific oxidizing conditions. To draw a general statement on the corrosion behaviour under completely different conditions in practise is not possible.

2.2 Corrosion resistance of nickel alloys [11]

2.2.1 Nickel

Nickel's high degree of corrosion resistance is due to the fact that it is a relatively noble metal within the galvanic electrochemical series of metals. Compared, for example, to the substantially less resistant iron (steel), nickel possesses a free corrosion potential greater by 250mV in the positive ("nobler") direction. In addition, nickel – like nearly all of its alloys – is a so-called "passiviable" metal, that is, with increasing polarization, the anodic dissolution of the metal decreases due to the formation of a protective oxide or hydrogenated-oxide layer of NiO and Ni_2O_3 (Ni_3O_4, NiO_2) or $Ni(OH)_2$, resulting in a decrease in the corresponding rates of corrosion to negligibly low figures. These properties make nickel highly resistant to:

- Water, contaminated and aerated.
  Corrosion rates are generally ≤ 0.1mm/a. This is one of the reasons for the use of nickel for the storage of high-purity water, since water contamination by the metal is extremely low.

- Sea-water under flowing and uniformly aerated conditions.
  However, since nickel is not toxic (poisonous) to marine organisms, there is, in contrast to copper-nickel alloys, a tendency to severe encrustration under static conditions. This leads to local destruction of passivity, with the result that crevice and/or pitting corrosion may occur under deposits.

– Salt solutions.
  Especially neutral and alkaline solutions such as carbonates, phosphates, sulphates, chlorides and nitrides, even at high concentrations and temperatures. For this reason, nickel is often used as material for crystallizers, where corrosion rates are generally ≤ 0.1 mm/a.

Nickel is also resistant to certain acidically reacting salts, such as ammonium and zinc chloride. Oxidizing salts, such as iron-III-chloride, copper-II-chloride and hypochlorites, attack nickel severely, as do chromates and nitrates in acidic solutions.

Nickel is only moderately resistant (i.e. its use should be limited to very diluted solutions and moderate temperatures, below 50 °C), in

– inorganic acids, such as sulphuric, phosphoric, hydrochloric, hydrofluoric and nitric acids. In all these corrosion increases relatively severely with increasing aeration

– certain organic acids, of which acetic, formic, and tartaric acids can be extremely aggressive at very high temperatures and concentrations, and under aeration.

Compared to nickel the nickel-chromium-iron-molybdenum alloys offer considerably better resistance to these acids.

Nickel, however, possesses excellent resistance to lyes at all concentrations, up to very high temperatures. This property is the reason for the popularity enjoyed by nickel in chemical engineering equipment, especially the manufacture and processing of sodium and potassium hydroxide solutions. Figure 2.2 shows a resistance diagram for nickel in sodium hydroxide solution.

At temperatures above 300 °C and concentrations above 70 % low carbon nickel LC-Nickel 99.2 – alloy 201 (2.4068) is used, since, due to its very low carbon content (0.03 % maximum), it stops the formation of graphite at the grain boundaries and thus prevents SCC, which can occur locally in nickel.

Nickel's high resistance to lyes is due to the formation of an initially very thin layer of $Ni(OH)_2$, which is transformed via β-NiO Σ OH into a non-stochiometric layer under further oxidation and can be assumed to contain Ni^{3+}, Ni^{4+}, O^{2-} as well as OH^-.

In conclusion, it should be noted that for many applications pure nickel is selected not merely because of its excellent corrosion resistance (in many cases, there are nickel alloys of equivalent or even better resistance which could be used), but also because

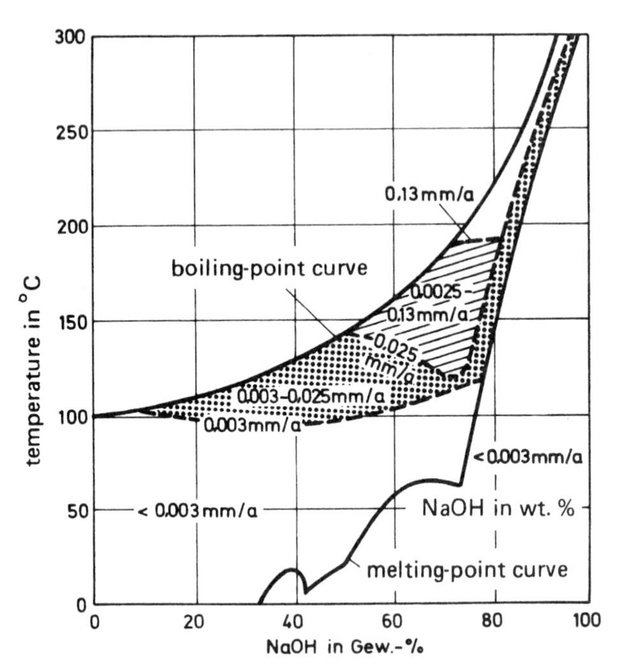

Figure 2.2: Corrosion resistance of pure nickel in sodium hydroxide, related to concentration and temperature, according to [11]

- nickel does not affect food products, thus, e.g., vitamins in fruit and vegetable juices are not damaged or destroyed;

- nickel, in contrast to metals such as iron and copper, does not affect sensitive products by decomposition, discoloration or alteration of physical characteristics, a factor of great importance in the manufacture of artificial silk (rayon), phenol and synthetic resins.

2.2.2 Nickel-copper alloys

Nicorros - alloy 400 (2.4360), consisting of approximately 65 % nickel, 30 % copper and some iron (Table 1.4), is of primary importance. A variant, Nicorros Al – alloy K-500 (2.4375), contains some 2.8 % aluminium, producing increased strength and resistance to wear, for products such as stirring shafts. In the broader sense, the alloys Copper-nickel 70/30 (CuNi30Fe) and Copper-nickel 90/10 (CuNi10Fe) (Table 1.4), which are on the copper side in alloy terms, should also be included. In all nickel-copper alloys, iron added at relatively low concentrations produces an increase in the quasi-passive protective layer.

Due to the high nickel content of Nicorros - alloy 400 (2.4360), it can be assumed that the remarks made with regard to the corrosion resistance of pure nickel also apply in large measure to this alloy. For this reason, only certain corrosion properties responsible for the popularity of Nicorros - alloy 400 (2.4360) are dealt with in detail below.

- Sea-water
  Nicorros – alloy 400 (2.4360) is used extensively under conditions of high flow velocity and erosion, for instance for propeller shafts, propellers, pump-impeller blades, casings, condenser tubes and tubes for heat exchangers. Static conditions, under which marine organisms could be deposited or form encrustations, should be avoided since, under conditions of locally differing aeration, pitting and crevice corrosion may occur.

- Hydrochloric acid
  Nicorros - alloy 400 (2.4360) has excellent resistance to dilute hydrochloric acid solutions, provided such solutions are kept free of air – the same applies to sulphuric-acid solutions. Under these air-free conditions, the alloy can be used up to concentrations of approximately 15 % at room temperature, and up to 2 % at somewhat higher temperatures at or above 50 °C. A significant factor is that, due to its relatively good resistance to highly diluted hydrochloric acid solutions, Nicorros - alloy 400 (2.4360) can be used in processes in which chlorinated solvents may form hydrochloric acid by hydrolysis, which would lead to failure in other alloys.

- Hydrofluoric acid
  Nicorros - alloy 400 (2.4360) is one of the few metallic materials which show good corrosion resistance at ambient temperatures to all HF concentrations and which achieve figures of 0.05 mm/a corrosion or less in the absence of air. It can be seen from ISO-corrosion diagrams [11] that maximum erosion rates of approximately 0.25 mm/a, and thus good corrosion resistance, can be achieved in the absence of air, provided temperatures of 100 °C for concentrations up to 30 %, and temperatures of 60 °C for concentrations above approximately 50 %, are not exceeded. At elevated temperatures in aerated solutions, stress-corrosion cracking has been observed, but this can be combated by means of deaeration of the medium or by means of stress-relief annealing of plant component. Nicorros - alloy 400 (2.4360) is equally corrosion-resistant to silicofluoric acid.

- Salt solutions
  Neutral and alkaline salt solutions, such as chlorides, carbonates, sulphates, nitrates and acetates have only a slight corrosive effect on Nicorros - alloy 400 (2.4360), even at high concentrations and temperatures up to their boiling points. As a result, rates of corrosion are generally 0.1mm/a or less. For this reason, this alloy is used in plants for the crystallization of salts from

saturated brines. Nicorros - alloy 400 (2.4360) is also used in a very large number of evaporator plants and crystallizers. Oxidizing acidic salts and some alkaline salts, such as hypochlorates, attack Nicorros - alloy 400 (2.4360) severely.

- Sodium hydroxide solution
  Nicorros - alloy 400 (2.4360) has a good resistance to sodium hydroxide and potassium hydroxide solutions similar to that possessed by pure nickel. Erosion rates of 0.05 mm/a or less are observed up to a concentration of 75 % NaOH and a temperature of 130 °C. Under evaporative concentration of 75 % NaOH to water-free NaOH, corrosion rates are greater than for pure nickel, and stress-corrosion cracking is observed in severely mechanically stressed (highly cold-formed) components in molten NaOH. For such conditions the components should be stress-relief annealed, or pure nickel should be used.

The use of Cunifer 10 (2.0872) and Cunifer 30 (2.0882) (Table 1.4) is restricted mainly to marine engineering and, in particular, to sea-water desalination, in which they are the classic tube materials for heat exchangers. The good corrosion resistance of both these materials is improved by small additions of iron.

2.2.3 Nickel-molybdenum alloys [11]

A typical representative of this group of alloys is Nimofer 6928 - alloy B-2 (2.4617) with approximately 28 % molybdenum and about 1.7 % iron (Table 1.1). The very high molybdenum content gives excellent resistance to reducing acids, the iron addition ensures good workability and processing of the material by improving its ductility [12].

Nimofer 6928 - alloy B-2 (2.4617) offers excellent resistance to hydrochloric acid at all concentrations and at temperatures up to the boiling point if reducing conditions prevail, that is in the absence of oxidizing additions such as oxygen, air and heavy metal ions. Fe^{3+}, Cu^{2+} can significantly increase corrosion rates, even at concentrations < 20 ppm.

The alloy also shows good resistance to sulphuric acid, particularly at concentrations up to about 90 % and temperatures up to approximately 120 °C. Here, too, reducing conditions must definitely exist; oxygen, aeration, heavy metal ions and chlorides (> 100 pm Cl^-) accelerate corrosion.

The alloy's good behaviour to organic acids, particularly to acetic acid is dependent on similar conditions.

As well as good resistance to uniform corrosion in the above-mentioned applications, Nimofer 6928 - alloy B-2 (2.4617) exhibits outstanding resistance to pitting and stress corrosion cracking. The resistance to intercrystalline and knife-edge corrosion is provided by its low carbon content (C = max. 0.01 %), which prevents the precipitation of carbides at the grain boundaries. This also ensures the material's good weldability. To approve the structural stability, several corrosion tests are customarily performed:

- Steel-Iron Test Sheet 1877/III: test pieces of semi-finished products are tested in boiling 10 % HCl for 24 hours. The IC depth must not exceed 50 µm (0.05 mm). Average values found on sheets: 10 µm (0.01 mm).
- Monsanto specification No. M3, Rev O 8/82 or Lummus specification No. 9575: testing in 20 % HCl at 149 °C (300 °F) for 100 hours, maximum permissable IC depth 175 µm (0.175 mm), mean values found: 60-80 µm (0.06-0.08 mm).
- Trent tube specification 122, Rev H: testing in boiling 20 % HCl for 24 h, maximum requested corrosion rate is 0.5 mm/a.

If semi-finished products meet these requirements, the excellent resistance of the alloy to the above-mentioned applications is guaranteed.

2.2.4 Nickel-chromium-iron alloys

Typical representatives of this class are Nicrofer 7216 - alloy 600 (2.4816) and Nicrofer 3220 - alloy 800 (1.4876) (Table 1.6). Both alloys are primarily intended for use in corrosive gases at extremely high temperatures under oxidizing conditions. Their use in conditions involving wet corrosion is restricted to a few special applications. Due to their chromium contents of 16 and 20 % respectively, these two alloys are considerably more resistant to oxidizing conditions than are pure nickel-iron alloys, since they possess a more marked tendency to passivation. Their high nickel content ensures resistance to stress-corrosion cracking in aqueous chloride-bearing media at high temperatures. For this reason, both alloys are used for steam-generation tubes in nuclear power stations. Here the LC grades, possessing very low carbon contents, are selected, in order to avoid the possibility of intercristalline corrosion (IC). Further applications for Nicrofer 7216 - alloy 600 (2.4816) include preheaters and turbine condensers, with maximum service temperatures being around 450°C. Nicrofer 7216 - alloy 600 (2.4816) has extremely good resistance to sodium hydroxide solution, up to 80 % boiling NaOH. This material is used instead of nickel where greater mechanical strength at high temperatures is required by the design, or where sulphides are present in the lye.

2.2.5 Nickel-chromium-molybdenum alloys and nickel-chromium-iron-molybdenum-copper alloys

2.2.5.1 General

Nickel-chromium-molybdenum alloys combine the good resistance properties of nickel-molybdenum alloys under reducing conditions with the good resistance of nickel-chromium alloys under oxidizing conditions. Nickel-chromium-iron-molybdenum-copper alloys represent the connecting link between the nickel-chromium-molybdenum alloys and the austenitic stainless steels. All these alloys are characterized by good resistance under wet-corrosion conditions, even in the presence of oxidizing salts such as Fe^{3+} and Cu^{2+}, which may also be present as chlorides, the alloys thus also having good resistance to pitting and crevice corrosion.

2.2.5.2 Nickel-chromium-molybdenum alloys

Nicrofer 5716 hMoW - alloy C-276 (2.4819) and Nicrofer 6616 hMo - alloy C-4 (2.4610) constitute the higher alloyed grades with 16 % chromium and 16 % molybdenum, the remainder being mainly nickel (Table 1.1). Good resistance to sulphuric acid, hydrochloric acid, phosphoric acid and other reducing acids is achieved due to these alloys' high molybdenum contents. With 9 % molybdenum, Nicrofer 6020 hMo - alloy 625 (2.4856) has a slightly lower resistance to these media but, thanks to its higher chromium content (Table 1.1), has greater resistance to more severely oxidizing media. In principle this also applies to Nicrofer 5621 hMoW - alloy 22 (2.4602), an alloy, compared to Nicrofer 5716 hMoW - alloy C-276 with an approximately 5 % higher chromium content - the molybdenum content, however, reduced to 13 % - exhibiting good resistance in numerous media of oxidizing characteristics. This alloy, however, has been surpassed in many applications by the more recently developed Nicrofer 5923 hMo - alloy 59 (2.4605) with an even higher chromium and molybdenum content, not only for the reason of increased corrosion resistance but also for an improved structural stability and thereby improved fabrication characteritics, welding in particular. This is discussed in detail in Section 2.4.2.

Resistance of this alloy group to certain selected media is described in the following paragraphs:

- Absolute resistance to various atmospheres - in some cases, even to tarnishing. This also applies to fresh water and severely gas (O_2, H_2)- and chloride-contaminated cooling, heating, process and pit-waters.

- Sea-water causes an extremely low degree of corrosion attack on these alloys, even at very low flow velocities (static conditions), when encrustations may be formed. Pitting and crevice corrosion resistance are therefore at their relatively greatest. Operating temperature can be well above 100 °C, provided the design does not give rise to crevices.

- The corrosion rate for these materials in sulphuric acid at all concentrations is below 0.05mm/a at room temperature. The behaviour in sulphuric acid at higher temperatures is shown in the ISO-corrosion diagrams of Figures 8.7 and 8.8.

- Hydrochloric acid at all concentrations attacks these alloys only slightly (maximum 0.25mm/a) at room temperature and without aeration. At higher temperatures, the higher molybdenum-alloyed materials have better resistance. In hydrochloric acid solutions, the influence of aeration on the corrosion rate is similar to that experienced with sulphuric acid.

- These alloys have a similar corrosion behaviour with respect to hydrofluoric acid. There is good resistance (maximum 0.25mm/a) up to 40 % HF and approximately 60 °C.

- Phosphoric acid, particularly when technically contaminated, i.e. produced using the wet-extraction process (sulphuric acid reacts with phosphate-bearing minerals), is aggressive if it contains free hydrofluoric acid, sulphuric acid or acidic iron-III-salts (oxidizing). Nicrofer 6020 hMo - alloy 625 (2.4856), 5716 hMoW - alloy C-276 (2.4819) and 6616 hMo - alloy C-4 (2.4610) possess a high resistance against phosphoric-acid corrosion under all concentrations at temperatures up to 110 °C.

- Nitric acid, a strongly oxidizing acid, attacks the higher chromium-alloyed Nicrofer 6020 hMo - alloy 625 (2.4856) less severely than it does Nicrofer 5716 hMoW - alloy C-276 (2.4819), which has a greater molybdenum content. Both materials have good resistance to HNO_3 at all concentrations (maximum 70 %) up to a temperature of approximately 65 °C. However, boiling HNO_3 solutions are aggressive and austenitic stainless steels (Table 1.2) with substantially better resistance are available for such cases.

- Of the organic acids, even the relatively aggressive acetic acid cause scarcely any corrosion attack. Nickel-chromium-molybdenum alloys of the Nicrofer 5716 hMoW - alloy C-276 (2.4819) type exhibit a maximum corrosion rate of 0.15 mm/a in boiling solutions of all concentrations. As compared to stainless steels, which possess relatively good resistance to pure acetic acid, Nicrofer 6020 hMo - alloy 625 (2.4856) and Nicrofer 5716 hMoW - alloy C-276 (2.4819) continue to have excellent corrosion resistance even when inorganic impurities or formic acid are present in the medium.

- Salt solutions, such as $CaCl_2$, $MgCl_2$ and $AlCl_3$, have only a slight aggressive effect even at very high concentrations and temperatures. These alloys can in particular be used for oxidizing salts such as $FeCl_3$, $CoCl_2$, chlorites and hypochlorites.

- The resistance to sodium hydroxide and potassium hydroxide solutions is very good up to concentrations of 50 % and temperatures of 150 °C. Pure nickel has better resistance at even higher concentrations and/or temperatures. Nicrofer 6020 hMo - alloy 625 (2.4856) or Nicrofer 5716 hMoW - alloy C-276 (2.4819) are preferred when, for example, hypochlorites are present as impurities. Under these conditions they possess better pitting-corrosion resistance than does pure nickel.

2.2.5.3 Nickel-chromium-iron-molybdenum-copper alloys

These materials occupy an intermediate position amongst the high-alloy steels, that is to say they are more limited in their application than the nickel-chromium-molybdenum alloys already discussed, but are substantially more resistant than stainless steels in the majority of practical media. Due to their relatively high nickel content - a factor of approximately three, as compared to 18Cr10NiMo-based stainless steels - they possess excellent resistance to stress-corrosion cracking in chloride-ion-bearing aqueous and acidic media. In particular, all these alloys can be used in the reducing sulphuric and phosphoric acids (Table 1.1). Their resistance to these media is essentially attributable to the addition of copper. Development commenced with Nicrofer 4221 - alloy 825 (2.4858) (40Ni,23Cr,3Mo,2Cu) which required improvement due to its sometimes inadequate resistance to pitting and stress-corrosion cracking in highly chloride-bearing solutions. This was achieved by raising the molybdenum content to approximately 7%, the result being Nicrofer 4823 hMo - alloy G-3 (2.4619).

Since chromium increases resistance to oxidizing media, the alloy Nicrofer 3127 LC - alloy 28 (1.4563), with a particularly high chromium content (Table 1.1), has proved very successful for use in technical phosphoric acid, which may contain large amounts of, for instance, iron-III ions. A further development is the alloy Nicrofer 3127 hMo - alloy 31 (1.4562). Due to its relatively high chromium and molybdenum contents, this material has a relatively high resistance to pitting and crevice corrosion in chloride-ion-bearing aqueous media, as is evidenced by high pitting-corrosion potentials in hot sea-water. This material thus has the advantage of being extremely resistant, not only on the process side, but also on the cooling-water side in heat exchangers (Table 1.5). This explains the fact that it is often preferred to high-alloy stainless steels for such applications. This alloy is discussed in detail in Section 2.4.2.

2.3 Corrosion behaviour of high-alloy stainless steels

In the following, the reader's familiarity with basic austenitic stainless steels is assumed for the purpose of the following discussion. Attention is paid instead to the even more highly alloyed materials developed in recent years, which essentially feature increased chromium and molybdenum contents. Here one should differentiate between two groups: those materials for use in acidic reducing and/or chloride-bearing media, and those used mainly for oxidizing media (most often nitric acid). Tables 1.1 and 1.2 list certain steels important for these applications.

However, a few results obtained in potentiodynamic pitting-corrosion tests on such materials should first be presented [13]. Figure 2.3 shows polarization curves for Cronifer 1925 hMo - alloy 926 (1.4529) in aerated ASTM sea-water at various temperatures. With increasing voltage (from left to right in the diagram), the current density increases slowly above a certain potential, until, with further increasing voltage, a very sudden increase in current occurs above a critical limit potential, the so-called breakthrough potential, this being a result of the commencement of pitting corrosion. For this reason, this breakthrough potential is known as the pitting-corrosion potential. It can also be seen from Figure 2.3 that the pitting-corrosion potential decreases with increasing temperature. This clearly indicates a decrease in resistance to pitting corrosion as the temperature increases. Also of importance in Figure 2.3 is the course of the current-density/potential curves during the subsequent reduction of voltage. This is shown in the diagram with the

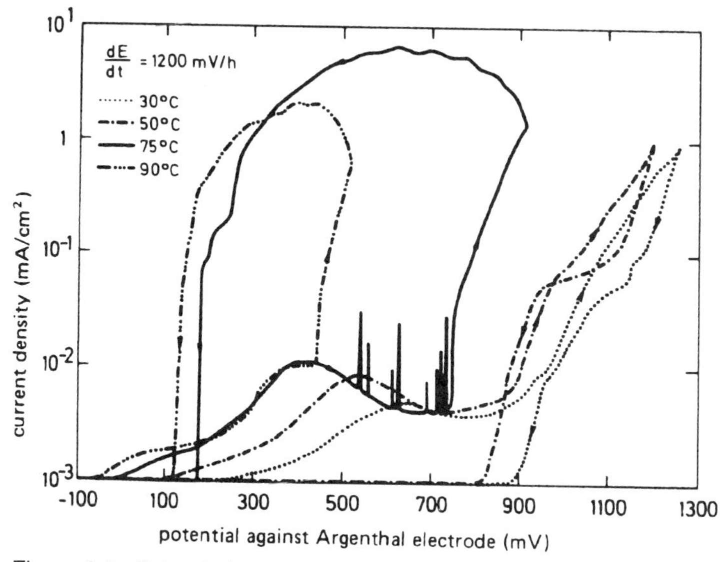

Figure 2.3: Potentiodynamic current-density potential curve for Cronifer 1925 hMo - alloy 926 (1.4529) in ASTM sea-water, aerated.

curves running back from right to left as indicated by the arrows. Here, the current density drops back to very low figures at a low potential. This so-called repassivation potential is in the vicinity of the pitting-corrosion potential at around 30 and 50 °C, but considerably lower at 75 and 90 °C. More highly nickel-alloyed materials such as Nicrofer 4823 hMo - alloy G-3 (2.4619) and Nicrofer 6020 hMo - alloy 625 (2.4856) do not exhibit such a large difference between pitting-corrosion potential and repassivation potential. For this reason, pitting-corrosion may more readily terminate under alternating operating conditions with differing levels of corrosive attack. The same applies to other media, for instance chloride-contaminated sulphuric acid such as that found in flue-gas scrubbers.

While such potentiodynamic measurements are capable of providing preliminary results very quickly, the potentiostatically-determined pitting-corrosion potential provides more information on the behaviour of the materials under alternating operating conditions. It generally lies in the vicinity of the pitting-corrosion potential determined by the potentiodynamic method, and is that potential at which passive current density begins to rise within 24 hours. Figure 2.4 illustrates the manner in which high-alloy stainless steels and some nickel-chromium-iron-molybdenum alloys behave in ASTM sea-water. With the exception of Cronifer 1810 Ti (1.4571), all the materials examined have a high pitting-corrosion potential at 30 °C. However, with a slight increase in temperature, there is, as shown in Figure 2.4, a steep drop in the case of nearly all the materials. Cronifer 2205 LCN - alloy 318 LN (1.4462), in particular, is impermissibly low at 50 °C, followed by Cronifer 1713 LCN - alloy 317 LN (1.4439). In the case of

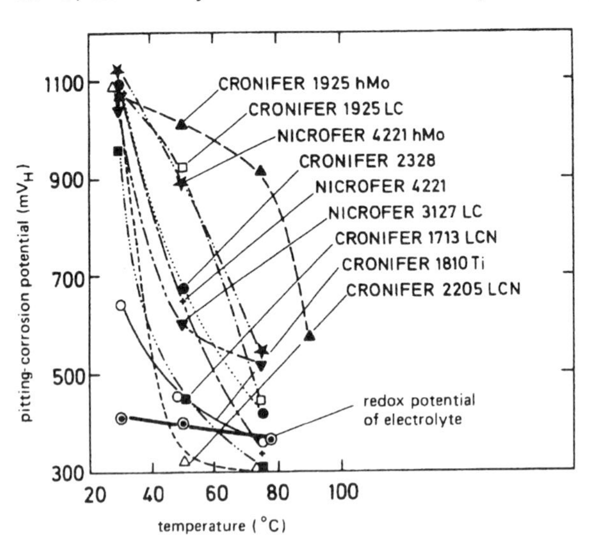

Figure 2.4: Pitting-corrosion potential of high-alloy, corrosion-resistant stainless steels and nickel alloys in ASTM sea-water, agitated and air-saturated

Nicrofer 3127 LC - alloy 28 (1.5463), this drop is slighter as the test temperature of 75 °C is approached, while Cronifer 1925 LC - alloy 904 L (1.4539), and above-all Cronifer 1925 hMo - alloy 926 (1.4529) show a slighter drop in pitting-corrosion potential as soon as the temperature exceeds 30 °C.

The chromium and molybdenum contents of the high-alloy stainless steels and nickel-chromium-iron-molybdenum alloys are of decisive importance across a broad concentration range in the attainment of a high degree of resistance to pitting corrosion and the related crevice corrosion [13]. The critical pitting- and crevice corrosion temperatures determined in a 24-hour immersion test in 10 % $FeCl_3$ solution, the magnitude of which is an indication of resistance to both forms of corrosion, can thus be plotted as function of the so-called effective sum, respectively the Pitting Resistance Equivalent (PRE). The latter is nothing more than the sum of the proportions by mass or weight of chromium and molybdenum contained in the respective alloy, expressed as percentages, the molybdenum content being more heavily weighted than the chromium, and therefore multiplied by a factor of 3.3.

A corrected PRE takes into account the positive influence of nitrogen with a factor of 30, so that PRE_N = %Cr + 3.3 %Mo + 30 x %N. As shown in Figure 2.5 the pitting-corrosion resistance determined in the 10 % $FeCl_3$-test is in fact linearly dependent on this PRE_N within a certain scatter band. Austenitic stainless

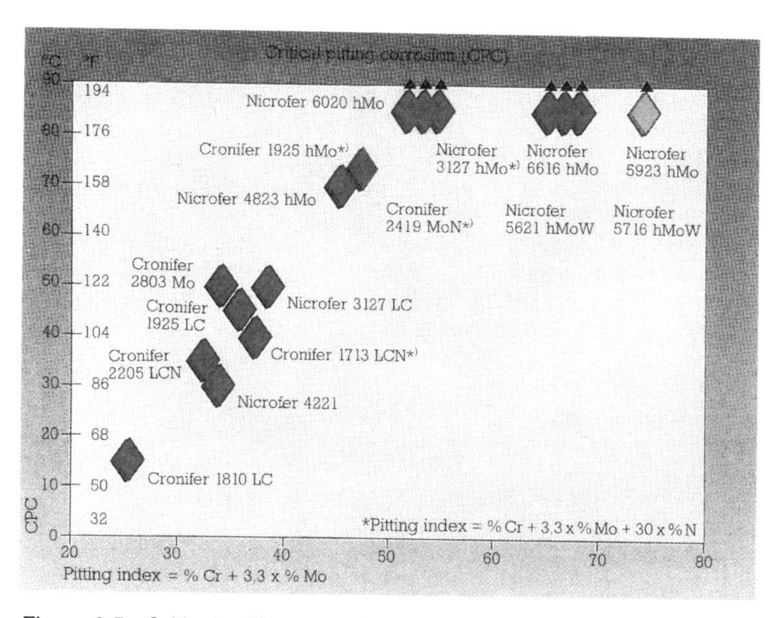

Figure 2.5: Critical pitting-corrosion temperature in 10 % $FeCl_3$ solution as a function of the pitting resistance equivalent (PRE)

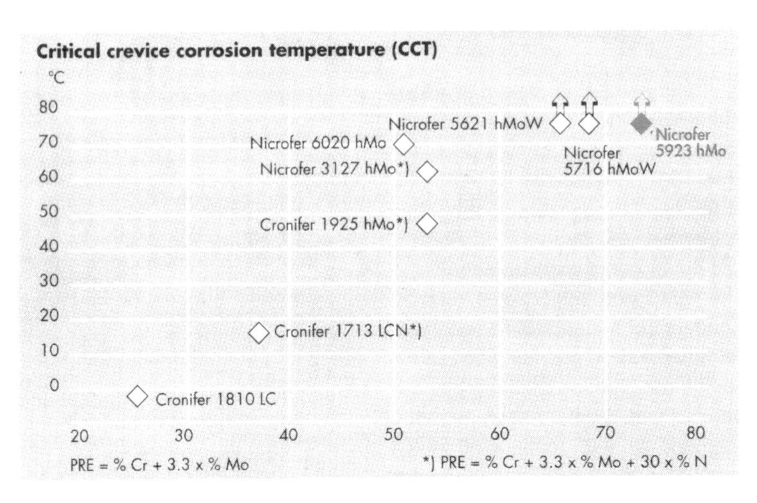

Figure 2.6: Critical crevice-corrosion temperature in 10 % $FeCl_3$ solution as a function of the pitting resistance equivalent (PRE)

steels, the superferritic Cronifer 2803 Mo (1.4575) and the austenitic/ferritic Cronifer 2205 LCN - alloy 318 LN (1.4462) all comply with this correlation. Figure 2.6 clearly shows that this applies in a similar way to crevice-corrosion resistance. Stainless steels in chloride-bearing media may suffer stress-corrosion cracking (SCC) as well as pitting and crevice corrosion [6, 8]. There has been no lack of experiments to demonstrate the susceptibility of stainless steel to SCC and to use the results of these more or less suitable laboratory tests to differentiate between individual steel qualities. Boiling solutions, such as 44 % $MgCl_2$ (very aggressive, and not encountered too often in practice), 40 – 60 % $CaCl_2$, 20 % NaCl, with other variables including concentrations, pH-values and oxidizing additives, are used in tests. The service period up to the occurrence of the first cracks (beat-beam specimen) or up to fracture is determined and entered as a function of mechanical load, as shown in Figure 2.7. This is then used to ascertain the critical stress, below which SCC does not occur. Further important values are the critical SCC temperature as a function of chloride concentration, the relationship of service life to electrochemical potential and the relationship of service life to alloy composition. SCC resistance increases with the rising nickel content, as shown in Figure 2.8.

The diagram indicates, that in the close to practice $CaCl_2$ solution, "surviving specimens" occur already with a nickel content of above 40 %. A test method only recently developed, in which a solution very close to practise (0.5m NaCl), drops onto a mechanically-stressed, heated material, allows a clear differentiation between various stainless steels as shown in Figure 2.9. Whereas standard stainless steels crack after a short period of time, higher nickel- and molybdenum-containing steels exhibit much longer service lives. The special stainless

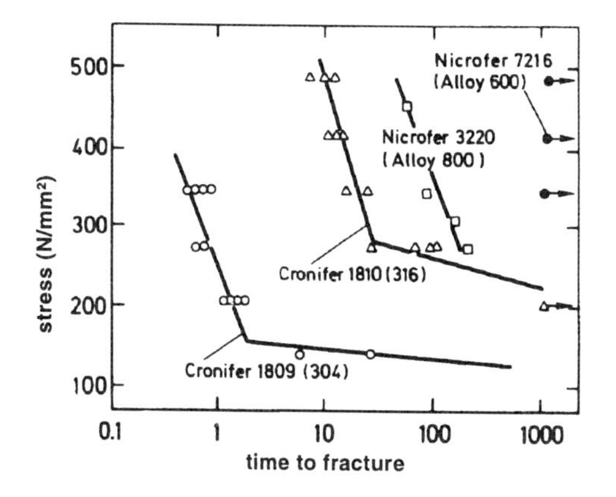

Figure 2.7:
Effect of stress on service life up to fracture for various materials in a $MgCl_2$ solution boiling at 154 °C, according to Denhard and Gough [7]

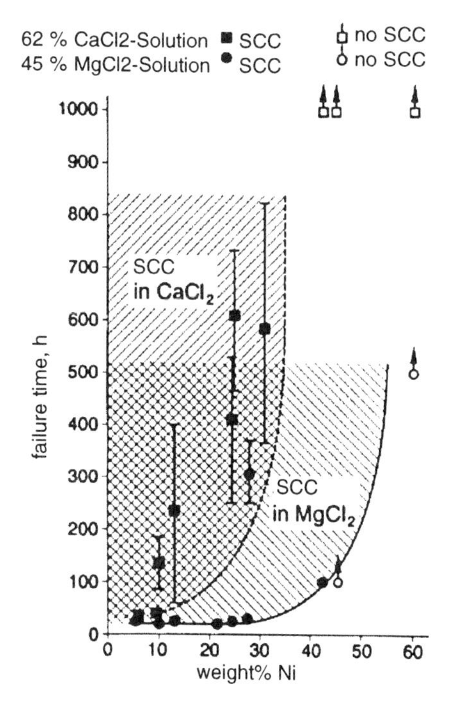

Figure 2.8:
Influence of nickel content in stainless steels and nickel alloys on service life up to fracture under testing for stress corrosion cracking in a 62 % $CaCl_2$- and 45 % $MgCl_2$ solution [13]

steel Cronifer 1925 hMo - alloy 926 (1.4529) shows the longest service life, which, according to these results can be explained not only with the alloy's high nickel content (25 %) but also with its very high molybdenum content (approx. 6.5 %). Therefore, if in practical media one has to consider besides pitting- and crevice-corrosion conditions also SCC, then Cronifer 1925 hMo - alloy 926 (1.4529) should be the first priority if a stainless steel is to be used.

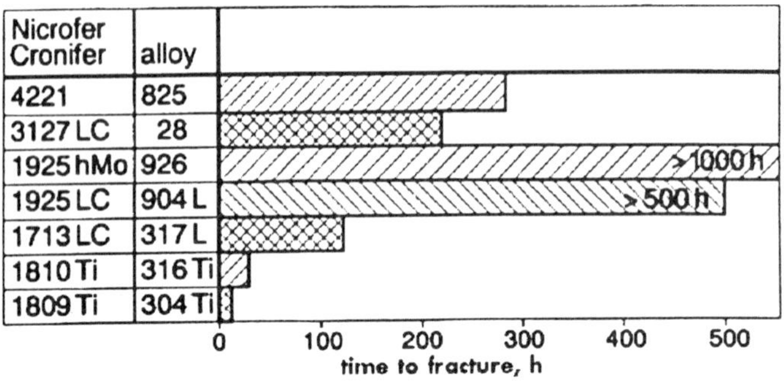

Nicrofer Cronifer	alloy	
4221	825	
3127 LC	28	
1925 hMo	926	>1000 h
1925 LC	904 L	>500 h
1713 LC	317 L	
1810 Ti	316 Ti	
1809 Ti	304 Ti	

0 100 200 300 400 500
time to fracture, h

Figure 2.9: Service life up to fracture under stress-corrosion cracking conditions by dropping a 0.5m NaCl solution onto a mechanically stressed material specimen, heated up to 300 °C [13]

2.4 New corrosion-resistant materials

R & D activities at Krupp VDM GmbH have concentrated in recent years on the development of a special stainless steel with an even better corrosion behaviour than the service-established Cronifer 1925 hMo - alloy 926 (1.4529) and to overcome the weakness of high-alloyed nickel-chromium-molybdenum materials by a version, significantly improved in its corrosion behaviour and fabrication characteristics. Result of these activities are the materials Nicrofer 3127 hMo - alloy 31 (1.4562), Nicrofer 5923 hMo - alloy 59 (2.4605), Nicrofer 3033 - alloy 33 (1.4591), Nicrofer 2509 Si - alloy 700 Si (1.4390), Nimofer 6629 - alloy B-4 (2.4600) and Nimofer 6224 - alloy B-10 (2.4604). The following test results clearly show that the new developments can be considered very successful.

2.4.1 Nicrofer 3127 hMo - alloy 31 (1.4562)

The criteria for an optimum analysis of this new material, its mechanical properties and fabricating characteristics, welding in particular, are described in detail elsewhere [14].

In the following, the altogether very positive corrosion results of laboratory-based investigations are summarized.

To test the alloy's resistance to pitting corrosion, the pitting-corrosion potential was determined in artificial ASTM sea-water. In Figure 2.10 the pitting-corrosion potential is plotted as a function of the temperature in comparison to two other high-alloyed austenitic stainless steels, Nicrofer 3127 LC - alloy 28 (1.4563), with 31 % nickel, 27 % chromium and 3.5 % molybdenum, and Cronifer 1925 hMo - alloy 926 (1.4529), with 21 % chromium, 25 % nickel and 6.5 % molybdenum. The potentials measured at room temperature were all well above the redox potential of sea-water, therefore the differences are negligible for the application. At increased temperatures, however, they drop significantly for the two reference steels, whereas for Nicrofer 3127 hMo - alloy 31 (1.4563) the pitting corrosion potential remains almost unchanged even at temperatures up to 90 °C. This indicates the alloy's outstanding resistance to pitting corrosion in media such as sea-water, brackish water and chloride-contaminated cooling water.

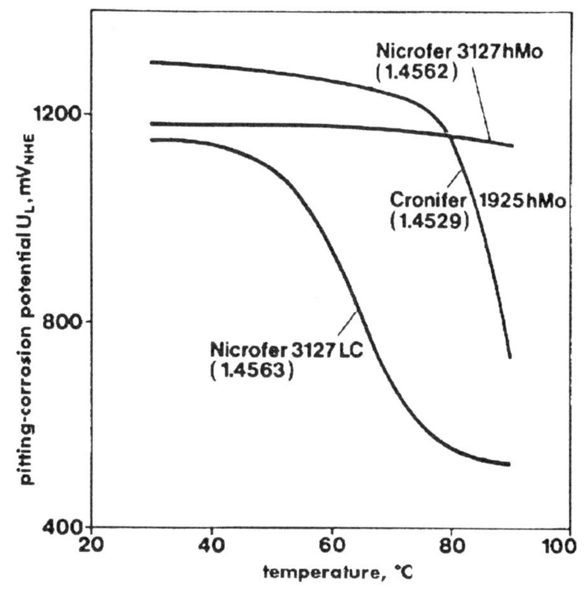

Figure 2.10: Pitting-corrosion potential of 3 high-alloy, corrosion-resistant steels in ASTM sea-water, agitated and air-saturated. The curve for Cronifer 1925 hMo - alloy 926 (1.4529) differs from that in Figure 2.4 because chromium, molybdenum and nitrogen contents of this material have been increased in the meantime.

To determine the alloy's resistance to crevice corrosion the critical crevice-corrosion temperature (CCT) was determined in a 10 % $FeCl_3$ solution. Figure 2.11 shows these values together with the critical pitting-corrosion temperatures (CPT) in comparision with those of reference materials. Nicrofer 3127 hMo - alloy 31 (1.4562) thereby possesses the best resistance to chloride media of all high-alloyed stainless steels and exhibits a resistance similar to that of the high-molybdenum containing nickel material Nicrofer 6020 hMo - alloy 625 (2.4856).

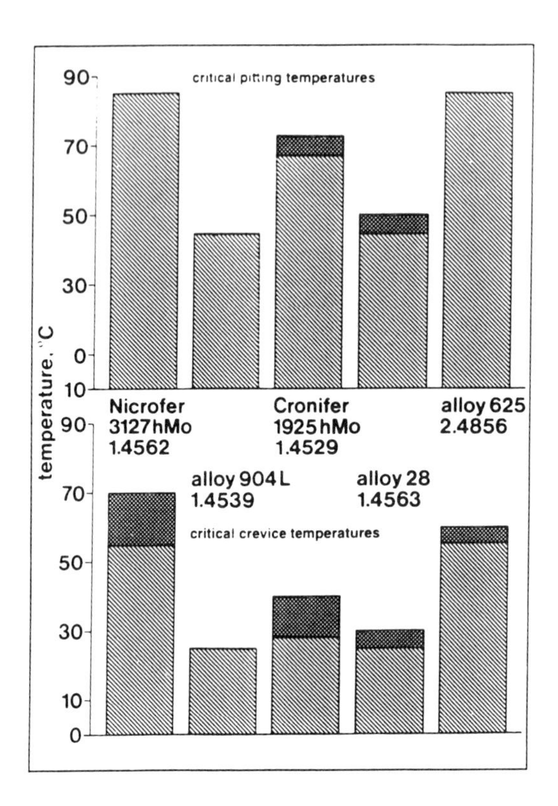

Figure 2.11:
Critical pitting- and crevice-corrosion temperatures of several materials in the 10 % $FeCl_3$ test [14]

Nicrofer 3127 hMo - alloy 31 was also tested in artificial sea-water [15]. The determination of crevice corrosion resistance and repassivation behaviour of alloy 31 in artificial sea-water was carried out on sheet specimens connected to PTFE crevice formers. Alloy 926 and alloy 24 were also included in these tests as reference materials. Chlorinated sea-water is known to be more corrosive than unchlorinated natural sea-water. This is usually associated with a potential increase of approximately 600 mV with reference to stainless steels and nickel alloys. As laboratory tests with chlorinated sea-water are very difficult to perform, these conditions were simulated with aerated artificial sea-water, the specimens being potentiostatically polarized at 600 mV SCE. This potential was kept constant throughout the test period. The temperature was then increased in 5°C steps every two hours until a distinct increase in current was recorded. This point was defined as the critical crevice corrosion temperature (CCT). The critical crevice corrosion temperatures obtained are shown in Figure 2.12. Both alloy 926 and alloy 24 failed at 50 °C whereas alloy 31 exhibited a considerably higher CCT, namely 70 °C, which is 40 % above that of the two lower-alloyed stainless steels.

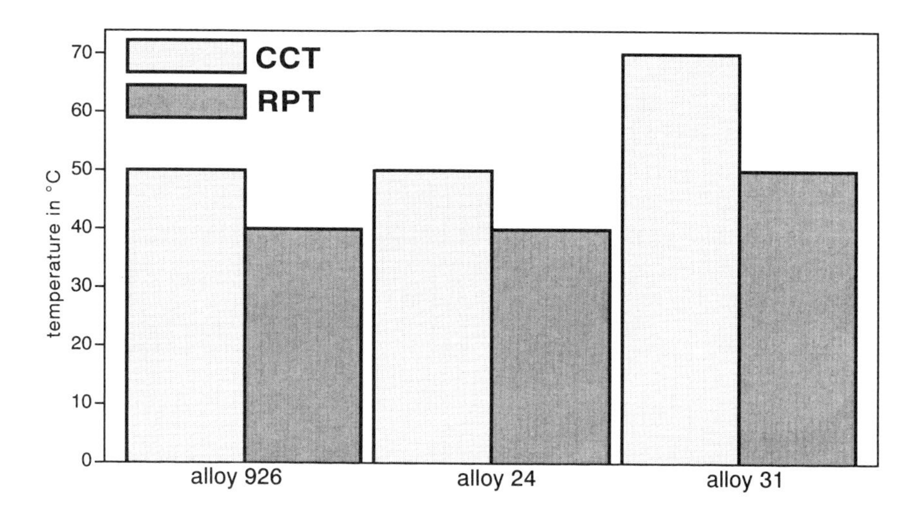

Figure 2.12: Critical crevice corrosion temperature (CCT) and repassivation temperature (RPT) in artificial sea-water - 600 mV SCE

To determine the repassivation behaviour, the temperature of the electrolyte was maintained for 20 hours at the CCT. The temperature was then decreased in 5°C steps until a distinct drop in current was observed. The temperature at which this drop in current occurred was defined as the repassivation temperature (RPT). Figure 2.12 shows the RPTs obtained. Alloys 926 and 24 repassivate at 40 °C, whereas alloy 31 already repassivates at 50 °C, which is again 25 % superior to the temperature for the two lower-alloyed stainless steels.
Various field tests in natural sea-water have been performed at different locations worldwide [15]. The three which seem to provide the most reliable results are summarized.
In 1991 a one-year test was carried out in German North Sea-water on Helgoland at 50 °C and 70 °C. Six test racks, each containing 14 sheet specimens from one material grade, were exposed to filtered sea-water in a 500 l overflow tank. The material grades investigated were alloys 317, 904, 926, 28, G-3, 625, C-276, 59 and 31.
The welded specimens were connected to PTFE crevice formers. Another specimen had two PTFE crevice formers in the upper part, while the lower part was connected to a sheet of the same material grade but was sealed with a PTFE and a metal gasket. With these crevice arrangements the flange connection of a real piping system should be well simulated. Whereas in the welded condition the lower-alloyed reference materials and even UNS 06985/alloy G-3 are affected by crevice corrosion, alloy 31 is not, nor are the higher-alloyed nickel-base materials.

Comprehensive corrosion tests in sodium chloride solutions and natural Arabian Gulf water to determine the suitability of various alloys for sea-water desalination plants have been carried out. The crevice corrosion behaviour was investigated by means of Teflon crevice formers in natural sea-water at 50 °C and at room temperature. After 150 days testing time at 50 °C, alloy 31 showed the highest corrosion resistance of all the austenitic materials tested.
These investigations were complemented by accelerated tests at 50 °C using critical crevice corrosion solutions (CCSpH), defined as the point where active corrosion occurs, i.e. when an anodic current peak of 10 μA/cm² is produced with potential runs.

The plot of the active peak height (APH) and the pH of the critical solution recorded in these accelerated tests indicates the excellent resistance of alloy 31 to more aggressive crevice-forming solutions and the susceptibility of standard stainless steels to crevice corrosion attack. Figure 2.13 shows a plot of PRE_N versus CCSpH in natural sea-water and NaCl solution. Alloy 31 presents the highest crevice corrosion resistance of all the alloys tested.

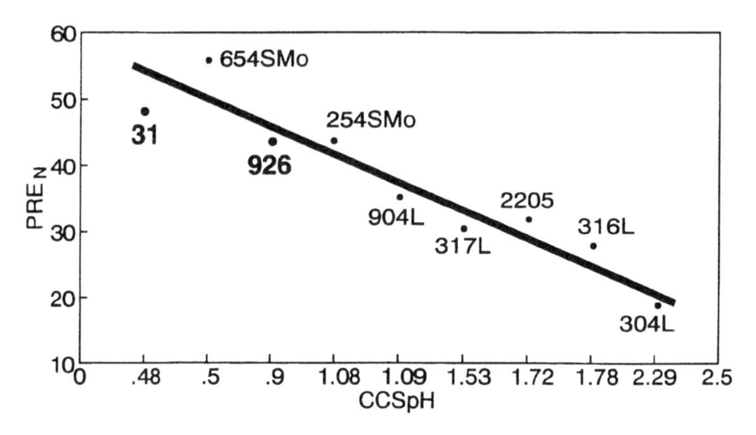

Figure 2.13: Plot of PRE_N Vs. CCSpH in natural sea-water and NaCl solution for alloy 31 and various reference materials (after Malik et al.). CCSpH = critical crevice solution

Of the various test conditions and test arrangements selected, the ones where chlorinated sea-water is pumped through pipe loops are regarded as being closest to real-life conditions. Such tests were performed in Baltic Sea-water and North Sea-water. Results obtained in Baltic Sea-water show that alloy 31 is resistant up to 40 °C and 2 ppm chlorine. At these parameters 6Mo steels are expected to show crevice corrosion attack. Only at 50 °C at 2 ppm chlorine did alloy 31 exhibit slight crevice corrosion on 2 out of 10 flanges.

North Sea-water has about five times the strength of Baltic Sea-water. Thus North Sea-water can be expected to present more severe conditions than Baltic Sea-water. This was confirmed by the results. Alloys 926 and 24 showed no crevice corrosion in loop tests in North Sea-water up to 30 °C and 1 ppm chlorine. At higher temperatures or at higher chlorine levels, initiation of crevice corrosion occurred on these lower-alloyed materials, whereas alloy 31 was resistant up to 45 °C and 1.5 ppm chlorine. Again, alloy 31 exhibited the best resistance to crevice corrosion. The first slight crevice corrosion attack was found at 50 °C and 0.5 ppm chlorine on 3 out of 10 flanges. All the results from the North Sea-water investigations are condensed in Figure 2.14, which presents the threshold conditions of alloys 24, 926 and 31.

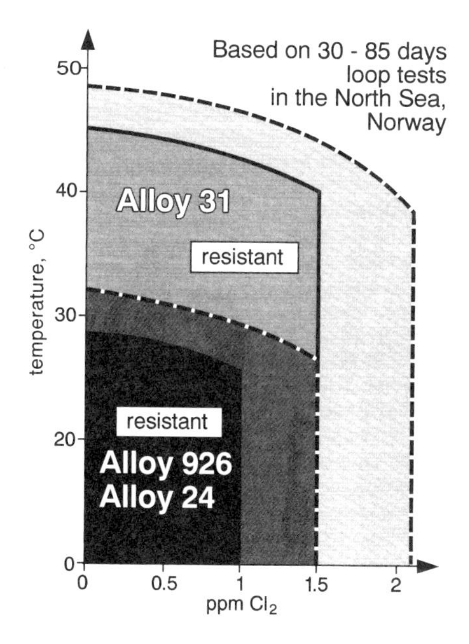

Figure 2.14:
Guidelines for selection of flange material made from alloys 31, 926 and 24 in chlorinated sea-water

At the same time the material possesses an outstanding resistance to oxidizing media such as nitric acid. This is shown in Huey test investigations, where the material is subjected to a 67 % boiling HNO_3 solution. Even after 10 boiling periods of 48 hours each, Nicrofer 3127 hMo - alloy 31 (1.4562) demonstrates outstandingly low corrosion rates of about 0.1 mm/a. Therefore, despite its high molybdenum content, this stainless steel is also outstandingly resistant to strongly oxidizing media.

Remarkably good is the alloy's corrosion resistance to reducing acids. Sulphuric acid with no impurities and in concentrations of 20 to 80 % represents such reducing conditions, under which standard stainless steels fail with severe corrosion. Complementary to the ISO-corrosion diagram in Figure 8.6, Figure 2.15, with the extremely low corrosion rate of less than 0.025 mm/a, shows the exceptionally good resistance of Nicrofer 3127 hMo - alloy 31 (1.4562) in H_2SO_4 up to 60 % and 100 °C and in 80 % H_2SO_4 up to 80°C.

The irregular behaviour of alloy 31 under conditions of 80 %/80 °C was studied in more detail and impurities were found to have a pronounced effect [16]. Table 2.1 summarizes the results of a series of tests on one material, alloy 31, at a single temperature of 80 °C and using highly pure acids of one concentration, 80 %.

Table 2.1: Effect of minor variations in sulphuric acid composition on the corrosion rate of alloy 31

Acid concentration: 80 %	Temperature: 80 °C
Acid grade and test conditions	Corrosion rate, mm/y
„Analysis grade", no agitation, no aeration (<2 ppm Fe)	4.34
„Analysis grade", nitrogen bubbled	2.67
„Analysis grade" +100 mg/l Fe^{3+}, nitrogen bubbled	0.03
„Chemical grade", no agitation, no aeration (<30 ppm Fe, <10 ppm N_2O_3)	0.02
„Chemical grade", nitrogen bubbled	2.33

The only differences lie in the presence or absence of minor traces of various oxidizing species. It is immediately apparent that it is these trace impurities which determine the material's behaviour in 80 % H_2SO_4 at 80 °C. We previously saw that alloy 31 has a corrosion rate of only 0.019 mm/a at 80 °C in H_2SO_4 of technical grade purity. At the same concentration and temperature, however, the corrosion rate in analytical grade acid, which is essentially free from oxidizing impurities, rises to over 4 mm/a, a 200-fold increase. The addition of only 100 ppm of Fe^{3+} to this analytical grade acid reduces the corrosion rate to a level comparable to that obtained in the industrial grade acid used for other lab tests, as does the presence of slight traces of nitrogen oxides. The practical significance of these observations is important in the discussion of industrial experience.
Finally, the behaviour of alloy 31 in high-temperature, highly concentrated sulphuric acid has also been studied and plotted in a conservative isocorrosion curve; cf. section 2.4.4 and especially Figure 2.20.

In hydrochloric acid of 3 % HCl, the corrosion rate is 0.13 mm/a up to 80 °C according to Figure 1.3, for 8 % HCl its use is restricted to room temperature. Nicrofer 3127 hMo - alloy 31 (1.4562) therefore exhibits good resistance to chemical processes in which traces of HCl may develop. Organic acids important to the chemical industry contain chlorine atoms which can separate from their organically bond-state and appear either as traces of hydrochloric acid or as salts therefrom, raising general corrosion or - at localized concentrations – result in pitting- or crevice corrosion. A monochloracetic acid with small amounts of water (less than 2 % H_2O) was chosen as representative test solution. The test was carried out at a temperature of 80 °C over 12 days. Together with Nicrofer 3127 hMo - alloy 31 (1.4562), which was tested also in its welded condition, two other high-alloyed special stainless steels were tested as reference materials. Whereas the corrosion rate of the two reference materials Cronifer 1925 hMo - alloy 926 (1.4529) and Nicrofer 3127 LC - alloy 28 (1.4563) was about 0.2 mm/a, it was significantly less than 0.1 mm/a for the new material, even in its welded condition.

Depending on the origin of the phosphate rock, impurities with differing concentrations may arise in the production of phosphoric acid, such as chlorides, fluorides, oxidizing metal ions and, in certain cases, other corrosion-accelerating impurities. Therefore only very high chromium- and molybdenum-alloyed special stainless steels should be considered for the concentration section (50 % H_3PO_4). It had to be established, whether Nicrofer 3127 hMo - alloy 31 (1.4562) possesses good resistance also under these conditions. Besides laboratory tests in simulating media field tests were performed with specimens fixed on rake arms of acid digestors. As can be seen from the results in Table 2.2, Nicrofer 3127 hMo - alloy 31 (1.4562) shows the best results under all conditions, compared

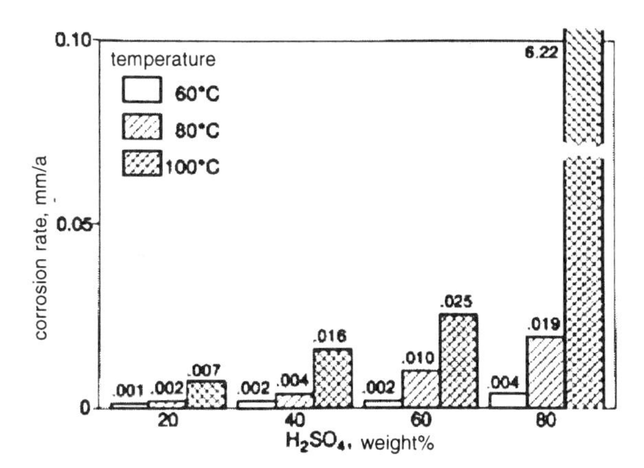

Figure 2.15: Corrosion rate of Nicrofer 3127 hMo - alloy 31 (1.4562) in hot sulphuric acid of different concentrations and temperatures [16]

Table 2.2: Corrosion rates in mm/a of Nicrofer 3127 hMo - alloy 31 (1.4562) and of the reference material Nicrofer 3127 LC - alloy 28 (1.4563) in technically-relevant phosphoric-acid solutions

Test medium	Temperature °C	Nicrofer 3127 hMo - alloy 31 (1.4562)	Nicrofer 3127 LC - alloy 28 (1.4563)
52% P_2O_5 + 4.5% H_2SO_4 + 0.9% H_2SiF_6 + 1.5% Fe_2O_3 + 400 ppm Cl^-	80	0.02	0.0075
	120	0.78	-
52% P_2O_5	116	0.08	1.2
30% P_2O_5 + 2.4% H_2SO_4 + 2.3% H_2SiF_6 + 1% Fe_2O_3 + 1000 ppm Cl^-	80	0.015	-
54% P_2O_5	120	0.05	1.4
54% P_2O_5 + 2000 ppm Cl^-	120	2.04	2.3
	100	1.30	-

with Nicrofer 3127 LC - alloy 28 (1.4563) and other stainless steels - not shown here in detail - which are usually used for the production and processing of phosphoric acid.
The material was also tested under sour-gas conditions (10 bar H_2S at 232 °C). Uniform corrosion was 0.01mm/a after 35 days with no signs of pitting- or stress-corrosion cracking at a load of 95 % Rp 0.2.

With regard to the material's resistance to intercrystalline corrosion (IC), the TTS diagram first shows IC after 2h at a temperature range of 650 – 700 °C. Complementary to this, individual tests were carried out also on welded specimens (sheet thickness 6mm) according to ASTM G-28 test method A (120h) and SEP 1877/method II (24h) IC respectively. In Table 2.3 the results are compared with those of the two high-alloyed special steels Cronifer 1925 hMo - alloy 926 (1.4529) and Nicrofer 3127 LC - alloy 28 (1.4563). In both tests the new alloy exhibits the lowest corrosion rate. Weld, heat-affected zone and parent metal are free from IC. The IC resistance after welding, required in practise, is thus guaranteed.

Table 2.3: Intercristalline-corrosion (IC) resistance of Nicrofer 3127 hMo - alloy 31 (1.4562) specimens and of reference materials in welded condition
a) Test according to ASTM G-28, method A. 120h
b) Test according to SEP 1877, method II. 24h
TIG welding
Surface ratio weld/base metal of the corrosion specimens 1:6.

Table 2.3a:

	Base material mm/a	Weld specimen mm/a
Cronifer 1925 hMo - alloy 926 (1.4529)	0.43; 0.37	0.45; 0.58
Nicrofer 3127 LC - alloy 28 (1.4563)	0.18; 0.14	0.20; 0.23
Nicrofer 3127 hMo - alloy 31 (1.4562)	0.18; 0.13	0.17; 0.20

Table 2.3b:

	Base material mm/a	Weld specimen mm/a
Cronifer 1925 hMo - alloy 926 (1.4529)	0.13	0.24
Nicrofer 3127 LC - alloy 28 (1.4563)	0.03	0.08
Nicrofer 3127 hMo - alloy 31 (1.4562)	0.04	0.05

The results of the comprehensive corrosion tests described above indicate the following typical application areas for Nicrofer 3127 hMo - alloy 31 (1.4562):

- equipment for the production and handling of phosphoric acid
- piping systems and heat exchangers for sulphuric acid
- condensers, coolers, piping systems for sea-water-related industries, such as shipbuilding and offshore oil and gas production
- oil and gas production under sour-gas conditions
- components for pulp and paper industries [17]
- flue-gas desulphurization of fossil-fired power generation

2.4.2 Nicrofer 5923 hMo - alloy 59 (2.4605)

Considerations that gave rise to the development of this new nickel-chromium-molybdenum alloy as well as comments with regard to its metallurgy, fabrication and processing characteristics, welding in particular, are presented in detail elsewhere [18-20]. Besides outstanding resistance to oxidizing and reducing acidic media, the improvement of the thermal stability was a further strong objective. Criterion for this is the time-temperature-sensitization diagram. Figure 2.16 summarizes the results of investigations to determine the range of resistance to intercrystalline corrosion for important nickel-chromium-molybdenum alloys [20]. It can be seen that under such criteria the well known nickel alloy C-276 (2.4819) is very susceptible to intercrystalline corrosion. Welding of at least thicker dimensions in this material is difficult. The new nickel alloy Nicrofer 5923 hMo - alloy 59 (2.4605), however, possesses significantly more favourable properties, even when compared to the alloys C-4 (2.4610) and 22 (2.4602). Sensitization in the sense of the 50 µm IC-criterion commences only after 2 hours at the earliest. This time is adequate to avoid intercrystalline corrosion after hot-forming, heat treatment and welding even of thicker dimensions.

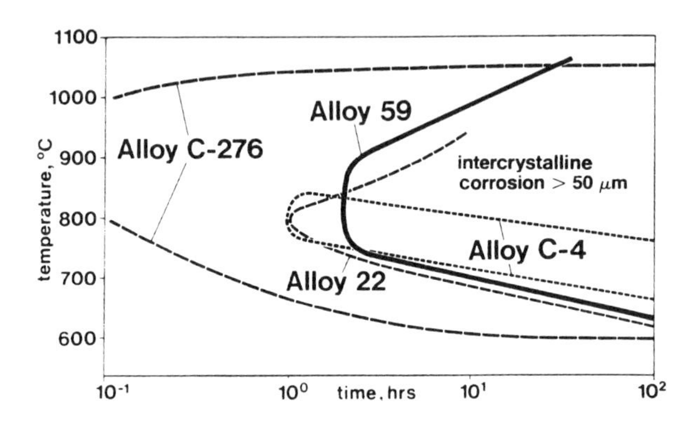

Figure 2.16: Time-temperature-sensitization diagram for different nickel-chromium-molybdenum alloys [20]

Figure 2.17 compares the results of corrosion tests according to ASTM G-28 methods A and B on all C-alloys in solution-annealed condition. Also here the new alloy Nicrofer 5923 hMo - alloy 59 (2.4605) shows the lowest corrosion rate.

To determine the resistance to pitting and crevice corrosion, other test media have been used because nickel-chromium-molybdenum alloys without exception are resistant to the $FeCl_3$ test and a differentiation is therefore not possible. A more practical relationship (chemical processing industry, pulp and paper in-

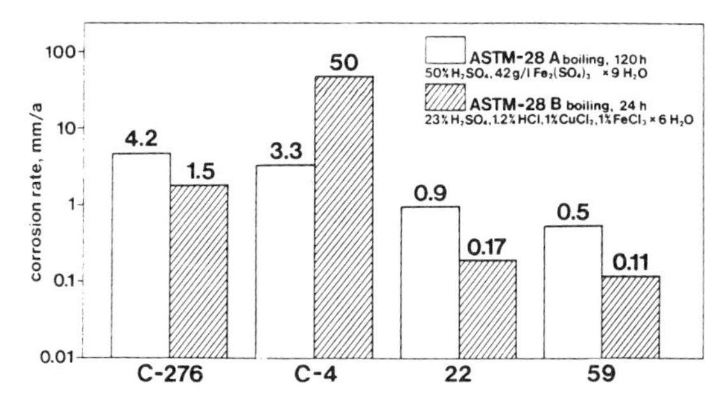

Figure 2.17: Corrosion rate of different nickel-chromium-molybdenum alloys in sulphuric-acid standard tests [18]

dustries, exhaust gas desulphurization) is reached with strongly acidic, high-chloride-containing media. The result of tests in a non-standard test solution of this kind, simulating extreme, below dew point acidic – and high-chloride-containing - as well as oxidizing conditions of a flue-gas desulphurization plant (FGD) is shown as columnar diagram in Figure 2.18. The alloy shows resistance to pitting corrosion up to a temperature of 120 °C and to crevice corrosion up to 115 °C.

Complementary to these analyses the nickel-chromium-molybdenum alloys were additionally tested in welded condition (inert gas and manual metal arc) in a high-chloride-containing and simultaneously acidic media at boiling tempera-

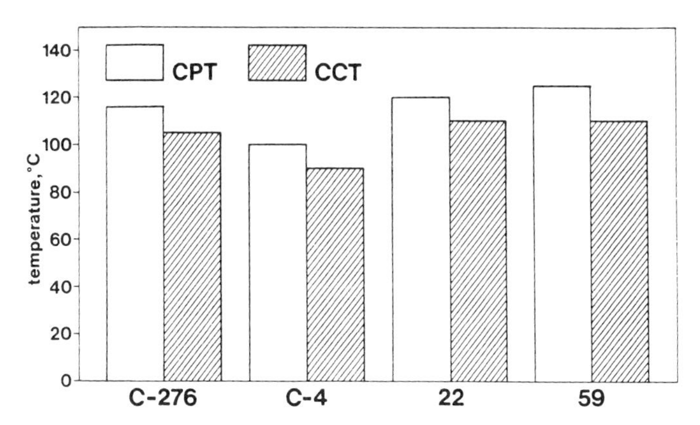

Figure 2.18: Critical pitting and crevice-corrosion teperatures of different nickel-chromium-molybdenum alloys in the green-death solution (7 % H_2SO_4, 3 vol% HCl, 1% $CuCl_2$, 1% $FeCl_3x6H_2O$) over 72 hours [19]

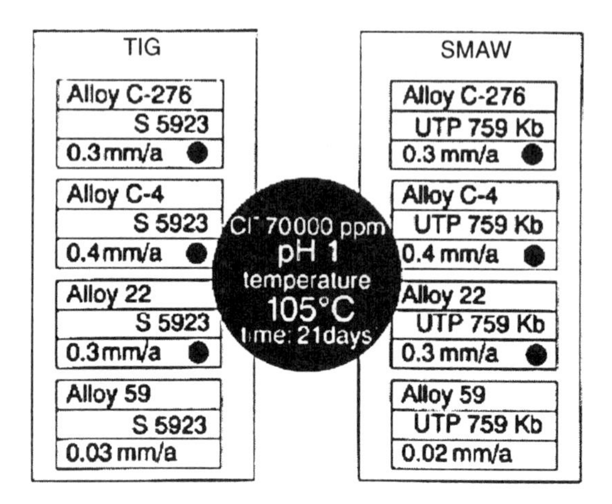

Figure 2.19: Corrosion resistance of different nickel-chromium-molybdenum alloys in welded condition in the chloride-pH-temperature screen

ture. As seen in Figure 2.19 the new material Nicrofer 5923 hMo - alloy 59 (2.4605) with a corrosion rate of 0.003 mm/a shows not only outstanding resistance to uniform corrosion, but contrary to the other three materials it is free from every local corrosion. IC is observed neither in the weld nor in the heat-affected zone.

Testing the resistance of Nicrofer 5923 hMo - alloy 59 (2.4605) to reducing hydrochloric acid in a simple immersion test resulted in the ISO-corrosion curves in the temperature-concentration screen of Figure 1.4. Accordingly, the material has good resistance at all concentrations up to at least 40 °C. This, however, implies good corrosion resistance to those organic-process media where hydrochloric acid is released in traces which give rise to pitting or stress-corrosion cracking of high-alloyed stainless steels. Nicrofer 5923 hMo - alloy 59 (2.4605) in this case is superior to all other nickel-chromium-molybdenum alloys [20]. The ISO-corrosion diagram of Figure 8.8 shows the corrosion performance of Nicrofer 5923 hMo - alloy 59 (2.4605) to sulphuric acid of various concentration rates and at different temperatures.

All corrosion-test results, based on both laboratory tests and on field tests in process media showed for the new nickel alloy Nicrofer 5923 hMo - alloy 59 (2.4605) excellent resistance to uniform corrosion as well as to local corrosion in acidic, high-chloride-containing media with reducing and oxidizing characteristics. This, in particular, also applies to welded material, which underlines the unusual thermal stability of this new nickel alloy. Therefore, this material can be considered for a wide range of applications, including:

General applications:

- Organic processes where chlorides or HCl is released.
- Pulp and paper industry, digesters and bleach-washer drums.
- Flue-gas desulphurization, waste incineration, scrubbers, gas preheaters, ventilators.
- Oil and gas production under sour gas conditions.
- Equipment and components.
- Acidic chloride solutions, heat exchangers.

Specific applications:

- Flue-gas desulphurization: absorber inlet, scrubbers, waste incineration, overlay-welding of superheater tubes.
- Combustion of chlorinated hydrocarbons, flare booms.
- Citric acid production: reactors.
- Gold refining, HCl-releasing drying processes.
- Electrolytic galvanizing of steel strip, electricity-carrying rolls.

2.4.3 Nicrofer 3033 – alloy 33 (1.4591)

This alloy with a nominal composition of (wt. %) 33 Cr, 32 Fe, 31 Ni, 1.6 Mo, 0.6 Cu and 0.4 N has been developed especially for service in highly oxidizing media [21]. This is the reason why alloy 33 additionally has an unusually high Cr content, making the material very resistant to localized corrosion in chloride media. To stabilize the austenitic microstructure, the alloy contains 31 % Ni and also 0.4 % nitrogen. Its high mechanical strength is almost 40 % higher than that of the 6Mo stainless steels. Because of its structural stability, this new alloy clearly has advantages where fabrication problems are concerned, e.g. problems with welding of superferritic stainless steels which are also in use for highly oxidizing media. Thermal stability of alloy 33 is verified by its high resistance to the nitric-acid-based Huey test, even after up to eight hours sensitization in the temperature range 600 – 1000 °C. Low corrosion rates and no intergranular attack were observed; the same applies to weldments.
Good weldability is given by commercial welding processes such as GTAW, PAW and SMAW. Weldments with a matching filler do not show degradation of ductility in the heat-affected zone (HAZ) according to ISO V-notch impact strength data for 15 mm plates. When welded with a non-matching filler, e.g. with an alloy C-4 based filler which has a lower solubility for nitrogen, porosity or indications of cracks were not observed. There is obviously no loss of nitrogen in the weld seam with respect to the base material since ductility data for the weld seam and the HAZ compare with those of the parent metal.
Concerning the corrosion resistance of alloy 33, attention was focused on oxidizing mineral acids, priority being given to sulphuric acid. Preceding laboratory tests [21] had already demonstrated its outstanding corrosion resistance in the

temperature range 100 – 200 °C. Moreover, in field tests under flowing conditions in 99.1 % H_2SO_4 at 150 °C alloy 33 performed best, as shown in Table 2.4. The same applies in a 96 % H_2SO_4 solution saturated with SO_2 at 80 °C (see Table 2.5); this was confirmed by another test in a sulphuric acid plant with a nominal regular process concentration of 98.5 %, but with detrimental fluctuations down to 96 % at a temperature of 135 – 140 °C. The reference material alloy C-276, which has many applications in sulphuric acid plant heat exchangers, confirms the application potential of alloy 33. Finally, good resistance is obtained in 96 % sulphuric acid with additions of nitrosylsulphuric acid at 240 °C as well as in oleum up to 33 % (107.5% H_2SO_4) at 150 °C.

Table 2.4: Corrosion resistance of alloy 33 and various other materials tested in a sulphuric acid plant
99.1 % H_2SO_4, velocity of flow $\geq$ 1.2 m/sec., 150 °C, test duration: 134 days

Alloy	Corrosion Rate mm/a
825	1.46
316 Ti	0.81
690	0.09
33	<0.01

Table 2.5: 96% H_2SO_4 saturated with SO_2, velocity of flow $\geq$ 1 m/sec., 80 °C, test duration: 65 days

Alloy	Corrosion Rate mm/a
654 SMO	3.1
28	0.10
C-276	0.05
33	0.06

As already mentioned in connection with structural stability and sensitization behaviour, alloy 33 also exhibits excellent resistance in a strongly oxidizing mineral acid such as boiling azeotropic HNO_3 (67 %). In addition, data from tests in 75, 80 and 85 % HNO_3 at temperatures of 25, 50 and 75 °C indicate that the alloy can be used in nitric acid up to 85 % HNO_3 at temperatures up to 75 °C or even somewhat more [21].

Table 2.6: Corrosion resistance of alloy 33 and various other materials in 1M HCl (1.3 % HCl) at 40 °C, corrosion rate in mm/a, test duration: 4 x 7 days

Alloy	FM	Base Material	Welded Sample
904 L	3040 E	0.47	0.54
926	625	<0.01	0.42
SAF 2507	25.10.4.L	0.02	2.58
625	625	<0.01	0.09
33	33	<<0.01	<0.01

Table 2.7: Alloy 33, critical pitting temperature (CPT) and critical crevice corrosion temperature (CCT) when tested in 10 % $FeCl_3 \times 6H_2O$ solution, over 24 hours, compared to other materials

Alloy	CPT		CCT	
	°F	°C	°F	°C
316 Ti	50	10	28	-2.5
310 L	77	25	≤68	≤20
904 L	113	45	77	25
28	140	60	95	35
926	158	70	104	40
G-30	167	75	122	50
33	185	85	104	40

Table 2.8: Results of stress-corrosion cracking test in saturated $CaCl_2$ at 135 °C, test equipment after J.A. Jones, average time to failure of at least 3 specimens

Alloy	Time to failure, hours
304	73.1
316	491.3
904L	>1000
926	>1000
33	>5000

Testing of alloy 33 in mixed nitric/hydrofluoric acid was done at 90 °C together with three other high chromium alloyed materials for comparison. Here, at 12 % HNO_3 and with increasing additions of HF up to 3.5 % alloy 33 exhibits the lowest corrosion rate. This effect is still more pronounced if larger amounts of HNO_3 are added to a 0.4 % HF solution.
Laboratory corrosion tests have been performed in hydrochloric acid of molarity 1 at 40 °C. Table 2.6 indicates the corrosion resistance of alloy 33 and of the nickel-base alloy 625 as well as of three special steels in unwelded and in welded condition. Stainless steel 904 L is incompatible with hydrochloric acid. The superduplex grade 2507 and the superaustenitic 6 % molybdenum grade 926 show severe attack in the weldment, whereas alloy 33 is free of any attack.

In phosphoric acid alloy 33 compares favourably with some established nickel alloys and will therefore be a cost-effective solution for the process of digestion of rock phosphates and for the concentration stage of the same process.
Concerning the use of alloy 33 in alkaline solutions, corrosion resistance is present up to the boiling points of 20 % and 50 % NaOH solutions [22].
With regard to chloride-bearing cooling water, natural sea-water and chloride-containing chemical and other process media, good resistance to localized corrosion, i.e. pitting and crevice corrosion, is another outstanding feature of this alloy. This is based on a high pitting resistance equivalent (PRE); because of the very high Cr content and the high nitrogen content, it is in the order of PRE $\approx$ 50, which compares favourably with the PRE of the 6Mo austenitic stainless steels as well as the nickel-base alloy 625. Correspondingly, test results in 10% $FeCl_3 \times 6H_2O$ reveal comparable resistance to crevice corrosion and much better pitting resistance, as shown in Table 2.7. Pitting potentials from polarization curves in artificial (ASTM) sea-water substantiate the excellent pitting behaviour since even at 95 °C the repassivation potential is still above the redox potential of the sea-water test solution. Also, crevice corrosion tests in artificial sea-water do not reveal any crevice corrosion attack for alloy 33 after a testing time of more than three months at temperatures up to 100 °C.
Finally, the stress-corrosion cracking behaviour was tested by the usual method in a saturated $CaCl_2$ solution at 125 °C at various stress levels and under constant load. As shown in Table 2.8, alloy 33 shows the longest time to failure when compared with two commercial and two high alloyed stainless steels. Thus it can be concluded that in real-life cooling and process media, there is a high probability of resistance to stress-corrosion cracking, even in solutions with a high chloride content and at higher temperatures.
In conclusion, this new alloy 33 has a high corrosion resistance in oxidizing mineral acids such as highly concentrated sulphuric acid, nitric acid, mixed acids and diluted hydrochloric acid. Its pitting, crevice corrosion and stress-corrosion cracking resistance in chloride-bearing media is excellent. Thus, with its high mechanical properties, good thermal stability, resistance to sensitization and therefore good weldability alloy 33 will respond to many challenges arising in the chemical process industry, the oil and gas industry and the refining industry.

2.4.4 Nicrofer 2509 Si 7 – alloy 700 Si (1.4390)

For handling highly concentrated nitric acid, an austenitic stainless steel containing 4 % silicon on the basis of 18 % Cr and 15 % Ni (cf. Chapter 1.2.2/Table 1.2 above) has been widely used in the past. Since it was known that with increasing Si content the corrosion resistance is improved for higher acid concentrations and/or temperatures, efforts were made to develop a stainless steel grade with a higher Si content.

Furthermore, Si grade stainless steels show excellent corrosion resistance in hot highly concentrated sulphuric acid, having the same positive effect with increasing Si content.

Normally, for the industrial production of sulphuric acid 98 % to 98.5 % is a typical acid concentration. For its handling, the aforementioned stainless steels with a silicon content of 4.5 – 5.0 % would be sufficient. However, the problem is the anomalous nature of sulphuric acid, i.e. the increase in its corrosiveness with decreasing concentration, e.g. from 98 % to 96 %, which may occur in a production process due to water dilution, at least part of the time or in cycles.

Therefore, for applications under strongly oxidizing mineral acid conditions, such as highly concentrated nitric acid and especially highly concentrated hot sulphuric acid a new austenitic stainless steel containing around 7 % Si, Nicrofer 2509 Si 7 - alloy 700 Si, has been developed [23]. As the nominal analysis shows (cf. also Table 1.2 above), the steel contains 25 % Ni, only 9 % Cr, some Mn (1.5 %) and extra-low carbon (0.01 % C). The reason for the greatly reduced Cr content (in comparison to commercial stainless steels with 18 % Cr) is hot formability, which presents problems in the case of high Cr and Si contents. In order to stabilize the austenitic structure, a sufficiently high Ni content is necessary. On the other hand, too high a Ni content may lead to precipitation of iron-nickel silicides of the type $Fe_5Ni_3Si_2$.

The mechanical properties of alloy 700 Si are higher than for a 4 % Si steel ($R_{p0.2}$: 265 N/mm², R_m: 624 N/mm²). The elongation is unusually high, being in all cases above 70 %, which indicates good cold formability, e.g. for plate heat exchangers. As time-temperature-precipitation/sensitization behaviour is a criterion for the workability of a stainless steel, corresponding TTT and TTS diagrams were plotted for alloy 700 Si. According to these the steel is sensitive to precipitation in the temperature range 650 – 900 °C, verified in a modified Huey test. While intergranular corrosive attack is observed in sensitized condition, there is no detrimental effect on ductility. A final solution heat treatment at 1080 °C obviously dissolves the silicide precipitates.

Hot forming of alloy 700 Si should be carried out between 1100 and 950 °C, followed in all cases by solution annealing at 1020 – 1100 °C and subsequent water quenching. The steel is weldable without degradation of its corrosion resistance, as is shown by extended laboratory tests. However, to obtain sufficient ductility of the weldments special measures might be required, e.g. welding with very low heat input, solution heat treatment of the welded components, or use of rollbond-clad material where the new 7 % silicon steel is present as a corrosion-

resistant cladding on a load bearing structure of carbon steel, the latter having the advantage of substantial cost savings. Welding with very low heat input (e.g. laser beam welding) is most effective on thin sections, e.g. when manufacturing welded tubes with a wall thickness of a few mm, whereas the use of rollbond-clad material has to be considered for larger installations.
In addition to the first laboratory heats, the corrosion resistance of alloy 700 Si was finally tested on material from a 30 t industrial heat. Immersion tests were performed in the concentration range 90 to 98.5 % sulphuric acid and at temperatures between 75 and 200 °C. Corrosion rates of 0.10 – 0.15 mm/a indicate that alloy 700 Si can be used in industrial 98.5% sulphuric acid up to a temperature of 200 °C. At the lower concentration of 96 % an average corrosion rate of 0.15 mm/a is found at a temperature of 150 °C. Thus alloy 700 Si can be considered resistant under these conditions without the danger of catastrophic corrosion if in the production process the regular 98 % concentration deviates down to 96 % in cycles or temporarily.

Figure 2.20 summarizes the corrosion data obtained in an ISO-corrosion diagram. It can be seen that the 0.1 mm/a isocorrosion line starts at 85 °C for a 90 % sulphuric acid, goes up to 126 °C at 94 %, to 137 °C at 96 % and reaches

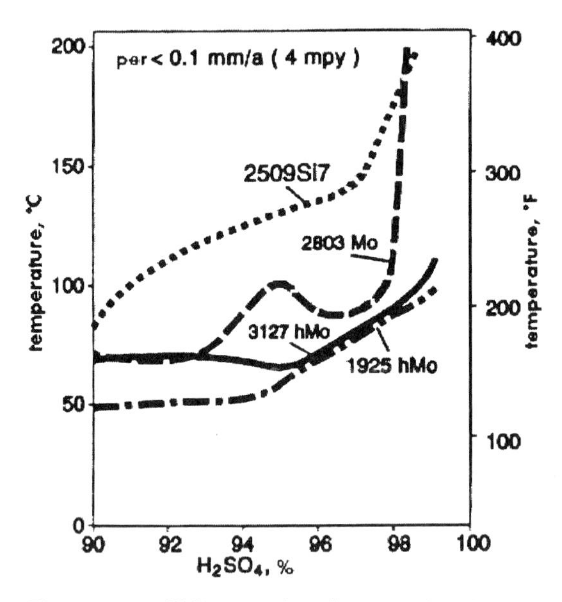

Figure 2.20: ISO-corrosion diagram of various materials in highly concentrated sulphuric acid. Indicated are the 0.1 mm/a isocorrosion lines of: Nicrofer 2509 Si 7 - alloy 700 Si (1.4390); Cronifer 2803 Mo/ superferrite (1.4575); Nicrofer 3127 hMo - alloy 31 (1.4562); Cronifer 1925 hMo - alloy 926 (1.4529)

195 °C at 98.5 %. In comparison to other stainless steels, alloy 700 Si has a much higher application temperature – definitely higher than the often-used superferrite Cronifer 2803 Mo (1.4575).
Concerning the behaviour of the new steel in highly concentrated boiling nitric acid, the data in Table 2.9 confirm the excellent corrosion resistance for base metal and weld material. The extremely low corrosion rate of < 0.02 mm/a makes alloy 700 Si at least 10 times more corrosion resistant than common 4 % Si stainless steels (1.4361).

Table 2.9: Corrosion resistance of alloy 700 Si in highly concentrated (~100 %) boiling nitric acid, for the solution-annealed homogeneous condition, for samples which have been sensitized at 600 and 700 °C and for welded samples

Condition 5mm plate	Average corrosion rate over 15 x 48 h testing time mm/a
Solution annealed at 1075 °C/ water quench	0.005
Sensitized 20 min / 600 °C 10 min / 700 °C	 0.003 0.003
GPAW square butt weld without filler	0.020
GTAW, one pass fusion test	0.009

2.4.5 Nimofer 6629 – alloy B-4 (2.4600)

Nickel-molybdenum alloys of the type Ni28Mo/alloy B-2, basically containing 28 % Mo, 1.5 – 2.0 % Fe and 0.4 – 1.0 % Cr, exhibit excellent corrosion resistance in reducing media, especially in hydrochloric and sulphuric acid over a wide range of concentrations and temperatures. This is based on the positive effect of Mo which, for example, reduces the corrosion rate of NiMo alloys in 10 % hydrochloric acid to a negligible value if the Mo content is in the range of 26 – 30 % Mo.
In spite of its almost binary constitution and therefore simple phase equilibria, difficulties in workability have been encountered with the alloy in the past, creating considerable problems when the alloy had to be manufactured into components for chemical plants. Also, stress-corrosion cracking has occasionally been

observed, sometimes after long periods of service and in positions where a high heat input may have occurred during manufacturing of components, e.g. due to repeated welding operations.
Metallurgical, metallographic and corrosion studies have shown that the formation of ordered intermetallic compounds and here especially the Ni_4Mo beta phase can cause embrittlement. On the other hand, small quantities of iron and chromium influence the kinetics of phase precipitation. Therefore, by increasing the alloy's iron and chromium contents to 2 – 5 wt. % and 0.5 – 1.5 wt. % respectively a new alloy is created with increased resistance to embrittlement, sensitization, intercrystalline corrosion and stress-corrosion cracking as well as „fabricability" (e.g. welding, hot forming) without impairing the overall corrosion behaviour [24].
The new alloy B-4 was developed by systematic studies of laboratory heats with different Fe and Cr contents (cf. also Table 2.10).
Time-temperature 100-J impact curves were plotted from the data generated by V-notch impact tests. As Figure 2.21 reveals, increasing the Fe content suppresses the kinetics of ordered beta phase Ni_4Mo formation in Ni28Mo alloys, shifting the nose of the curve significantly to the right.

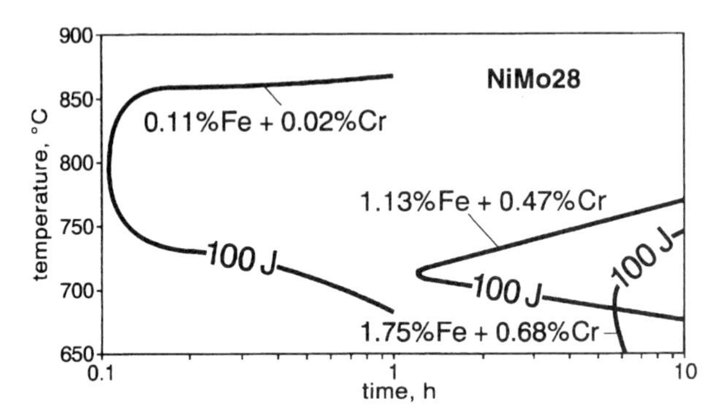

Figure 2.21: Impact 100-J contours for heats A, B and C

It was also found that the iron and chromium contents affected resistance to stress-corrosion cracking (SCC), as is indicated in Table 2.10. No fracture occurred in 100 hours for any heat in solution-annealed condition when tested in boiling 10 % H_2SO_4 solution, However, ageing for 1 and 6 hours at 700 °C increased the susceptibility to SCC for the lower-Fe heats A, B and C, whereas heats containing more than 3 % Fe, e.g. heats D, E and F, did not show any evidence of SCC. Heat C, which contains 1.75 % Fe and 0.68 % Cr, is safe regarding embrittlement during fabrication but is not resistant to SCC in 10 % boiling H_2SO_4 and ageing conditions of 700 °C.
Testing the various heats in boiling 10 % and 20 % hydrochloric acid over 24 hours according to SEP 1877 III and DuPont SW 800 M did not reveal any sig-

Table 2.10: Results of stress-corrosion cracking tests on Ni28Mo alloys according to ASTM G30 in 10% boiling sulphuric acid

Heat: Name/Fe+Cr/wt%			Time to fracture (hours) for 3 conditions		
Name	Fe	Cr	solution annealed	aged 1h 700°C	aged 6h 700°C
A	0.08	≤.01	>100	3	2
B	1.13	.47	>100	>100	>100
C	1.75	.68	>100	>100	>100
D	3.17	1.42	>100	>100	>100
E	3.23	.72	>100	>100	>100
F	5.86	.78	>100	>100	>100

Table 2.11: Corrosion resistance of Ni28Mo alloys in standard immersion tests: SEP 1877 III and DuPont SW 800 M

Heat: Name/Fe+Cr/wt%			Corrosion rate, mm/a	
Name	Fe	Cr	SEP 1877 III, 10% HCl, boiling	DuPont SW 800 M, 20% HCl, boiling
A	0.08	≤.01	.29	.58
B	1.13	.47	.31	n.d.
C	1.75	.68	.37	.60
D	3.17	1.42	.25	.56
E	3.23	.72	.24	.46
F	5.86	.78	.38	.58

nificant difference with regard to the rate of uniform corrosion which, according to Table 2.11, is below 0.4 mm/a (16 mpy) in boiling 10 % HCl and 0.6 mm/a (24 mpy) max. in boiling 20 % HCl. Intergranular penetration remained below 50 μm (2 mils) without exception. Additionally, the Lummus test, which was introduced to qualify industrial heats of Ni28Mo alloys, was applied to compare the corrosion behaviour of the weldments of the compositions investigated. 12 mm plate material had been TIG-welded using alloy B-2 filler metal. The surface ratio of base metal to weldment was about 5:1. Tests were performed for 100 hours in 20 % HCl at 149 °C. The limiting criterion of this test is an intercrystalline penetration of less than 175 μm in the heat-affected zone. All the samples passed this test. The overall corrosion rate of the samples varies between 1.8 mm/a and 3.5

mm/a, which is within the statistical scatter of this test, without any major difference between weldment and base metal.
Finally, systematic and extended investigations of the time-temperature-sensitization behaviour underline the fact that only heats with increased iron and chromium contents show no significant changes in the overall corrosion rate even after ageing for 8 hours at 700 °C and no intergranular penetration of more than 20 μm under all conditions of heat treatment.
In conclusion, a new Ni28Mo alloy grade has been created with the name „alloy B-4“, showing no risk of stress-corrosion cracking (at least in laboratory testing) together with increased high-temperature ductility, as demonstrated in Table 2.12. Hence the new B-4 alloy, being different from the popular B-2 grade with respect to its iron and chromium content, combines excellent cold and hot formability and weldability with significantly increased resistance to stress-corrosion cracking, but without impairing the resistance to overall uniform corrosion and intergranular corrosion, as shown in the preceding tables. Resistance to sensitization is also improved.

Table 2.12: High-temperature ductility and risk of stress-corrosion cracking as indicators of ease of manufacturing into components and of safety during service of those components for three different Ni28Mo alloy grades

NiMo 28	High temperature ductility at 700°C	Risk of stress corrosion cracking
Grade	Tensile elongation after 1h at 700°C, %	Evaluation on U-bend specimens (ASTM G-30) after aging 1h at 700°C in boiling 10% H_2SO_4 up to 100 hrs.
B-2 UNS N 10665 Fe max 2.0 Cr max. 1.0	5	full risk
B-2 UNS N 10665 restricted analysis Fe 1.6 - 2.0 Cr 0.5 - 1.0	42	reduced risk
B-4 new Krupp VDM alloy Fe 2 - 5 Cr. 0.5 - 1.5	48	no risk

2.4.6 Nimofer 6224 – alloy B-10 (2.4604)

This recently developed nickel alloy has a nominal composition of (wt. %) 24 % Mo, 9 % Cr, 6 % Fe, balance Ni [25]. It is intended for service in very corrosive yet almost oxygen-free conditions, such as occur in, for example, FGD plants. Recent field experience with FGD installations has shown that even the established high-alloyed nickel-base alloys like alloy 59 and alloy C-276, both of which contain around 16 % Mo, may fail under very special corrosion conditions. These exist in FGD plant operating ranges where the acidity and chloride concentration are high and where, at the same time, oxygen or aeration is almost absent, as is the case under deposits and incrustations. The same applies to heat recovery ranges in FGD. Another field of application will be recovery of waste sulphuric acid, which is normally carried out under almost oxygen-free conditions. The aforementioned established nickel-base materials move out of their passive state and become active, resulting in unacceptably high corrosion rates of several millimetres per year.
The new alloy B-10 exhibits better corrosion behaviour under the special corrosion conditions described above when tested in comparison with alloy 59 and alloy C-276.
Table 2.13 presents results of testing in a solution simulating almost oxygen-free FGD conditions. As can be seen, general corrosion is lower for the established NiCrMo alloys 59 and C-276; with regard to practical use, however, it is still sufficiently low for the newly developed nickel alloy grade (laboratory heats 3 and 4). In contrast to the general corrosion data, the established alloys suffer from severe crevice corrosion attack with, for instance, 0.17 mm deep local attack after only 30 days of testing. This may lead in practice to breakthrough and therefore to rapid perforation of thin-walled components. The new alloy B-10 shows no signs of crevices and will therefore have a much longer life cycle and will be largely maintenance-free in service.

Table 2.13: Corrosion resistance in a solution simulating flue-gas desulphurisation conditions with pH 1 (H_2SO_4), 7 % chloride ions, 0.01 % fluoride ions, 15 % FGD gypsum, agitated, under a nitrogen blanket at 100 °C after 30 days

Alloy no.	Ni	Mo	Cr	Fe	General corrosion mm/a	Crevice corrosion depth/mm
1	56.25	16.15	15.90	6.19	0.029	0.17
2	60.25	15.50	22.55	0.99	0.006	0.16
3	62.23	24.80	5.95	6.05	0.103	0
4	58.14	24.86	9.93	6.07	0.061	0

Further test media selected were 70 % sulphuric acid at 120 °C and 80 % sulphuric acid at 125 °C. These concentrations and temperatures correspond to field conditions in FGD heat recovery ranges and, to some extent, to waste acid recovery conditions (30 – 90 % sulphuric acid). It can be seen from Table 2.14 that the established alloys 59 and C-276 definitely cannot be used because of extremely high general corrosion rates. The new alloy B-10 does, however, exhibit a relatively low corrosion rate – sufficiently low for industrial applications.

Table 2.14: Corrosion resistance in hot, medium-concentration sulphuric acid

Alloy No.	General corrosion, mm/a	
	70% H_2SO_4 120°C	80% H_2SO_4 125°C
1	5.85	1.36
2	7.61	1.91
3	0.09	0.07
4	0.25	0.12

Finally, the alloy has been tested in a basically reducing acid such as hydrochloric acid with additions of oxidizing heavy metal ions. These conditions occur in pickling bath installations of the metal producing industry. At higher temperatures the normally HCl-based pickling bath is extremely corrosive and components made of high NiCrMo alloyed materials are used, although they have corrosion rates above 5 mm/a and therefore have to be replaced at frequent intervals. As may be concluded from Table 2.15, the new alloy B-10 offers much better corrosion resistance and therefore warrants a much longer life cycle.

Table 2.15: 10 % HCl with and without a N_2 blanket at boiling temperature

Alloy No.	10 % HCl with a N_2 blanket	10 % HCl + 200 ppm Fe^{3+} without a N_2 blanket
1	6.56	7.55
2	8.49	6.31
3	0.34	0.42
4	1.31	1.55

2.5 Corrosion resistance of welded nickel alloys and special stainless steels

2.5.1 The specific structural state of the weld and the heat-affected zone

The corrosion resistance determined for the base material (sheet, strip, tube, forging) must principally be also provided for the material in its welded condition, i.e., for the weld itself and for the weld surrounding area. If this would not be the case, corrosion-resistant equipment- and plant construction would not be possible.

The weld area of a construction can well be the location of preferred corrosion attack, because - in contrast to the homogenous solution- or solid-annealed base material - here a heterogeneous structure may exist and this not only in the weld metal itself (cast structure), but also in the heat-affected zone (HAZ) of the base material immediately adjacent to the weld. The extent of structural changes caused by the welding process in the HAZ depends on the susceptibility of the material to the precipitation of carbides and intermetallic phases on the grain boundaries, as well as on the welding parameters (welding process, heat input, speed of cooling).

To determine the susceptibility of high-alloy stainless steels and nickel alloys to grain boundary precipitation, time-temperature-precipitation- and time-temperature-sensitization diagrams (TTP and TTS) were prepared (see section 1.4). The basic susceptibility of the material can be judged from the course of the sensitization area and the position of the sensitization peak as the temperature with the shortest sensitization time. A material can only be welded without difficulties when by welding processes and -parameters the heat input can be held so low that the cooling curve even for thicker dimensions (above 20mm) does not effect the temperature of the shortest sensitization time in the TTS diagram.

Figure 2.22 shows negative influences on the corrosion resistance in the weld area caused by surface defects. They are further discussed below.

Slag residues on the weld surface are preferred locations for corrosion attack, offering ideal conditions for crevice and pitting corrosion, particularly in chloride-bearing media. The same applies to pores, which, as solidification porosity, cannot be excluded for nickel alloys but are under control by nitrogen-bonding titanium additions in the welding material as well as by welding-technological measurements (short metal-arc). A too rough material surface (ribbed structure) provides space for sediments, deposits and adhesions, having a sponge effect, resulting in local concentrations of corrosion-bearing media that may result in localized corrosion. A rough-ribbed structure of the weld is also location for increased mechanical stresses in micro dimensions, which could eventually in-

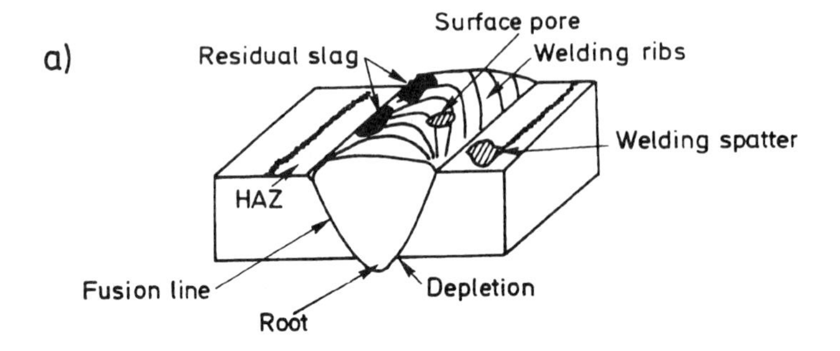

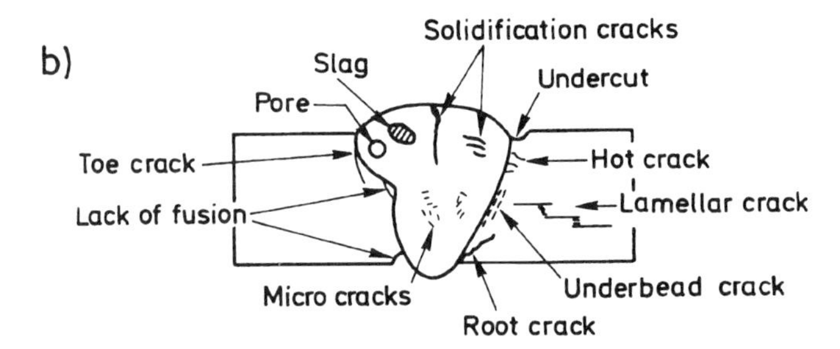

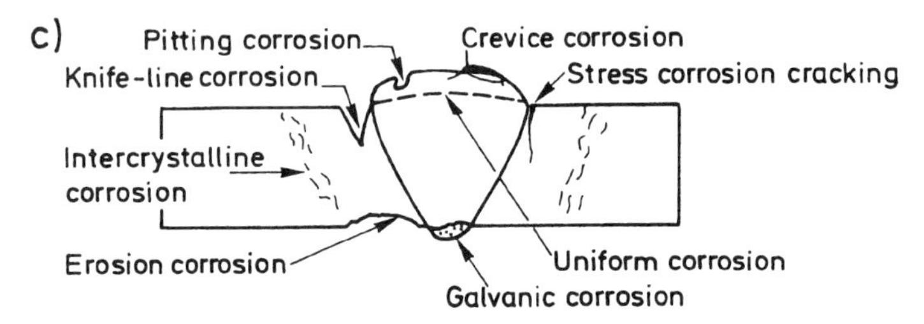

Figure 2.22: Possible defects in the weld area and their effect on the corrosion behaviour [26]
a) in the surface, b) in the welding deposit,
c) resulting types of corrosion

crease the susceptibility to stress-corrosion cracking. Hot cracks in the form of solidification cracks on the metal surface, as intercrystalline cracks in the weld material, transcrystalline microcracks, poor weld joints (weld passage- and root cracks), also if they are just under the surface of the weld material, are ideal crevices to promote pitting corrosion. The molten zones - primarily the root, but also the cover layer - are locations of preferred corrosion, by, for some materials (high-molybdenum-containing special stainless steels) as a consequence of segregations very pronounced proliferation of alloying elements, that have a decisive influence on the corrosion resistance. This category of effects influencing the corrosion resistance within the weld area also includes those resulting from the so-called burning-loss. In electrode welding, their extent can be controlled by optimizing the analysis of the coating.

The introduction of high mechanical stresses into the entire weld area, in the transition area, in particular, significantly increases the susceptibility to stress-corrosion cracking (SCC) mainly observed on stainless steels.
This may involve both classic, i.e. transcrystalline SCC by chloride media, as well as intercrystalline SCC as a result of sensitization in the HAZ. However, it should again be pointed out, that because of their fundamental SCC resistance all nickel alloys can be considered free from problems associated with this type of corrosion.

2.5.2 Intercrystalline corrosion (IC) in the heat-affected zone (HAZ)

Intercrystalline corrosion (IC) belongs to the type of corrosion found in weld areas. It may occur in the parent metal adjacent to the weld, but also directly in the weld metal (at overlapping weld layers). According to DIN 50900 part 1, IC represents a selective corrosion without mechanical stresses in which the area adjacent to the grain boundary is attacked. According to the more usual definition, IC is referred to when the intercrystalline penetration depth of the corrosion exceeds 50 µm. Austenitic stainless steels and nickel alloys can undergo IC when they are kept in the temperature range of approx. 500 - 850 °C and thereby sensitized. This may happen during hot working in the production- or processing state, by an incorrect final heat-treatment or by welding in the heat-affected zone of the parent material (sheet, tube, strip). IC testing therefore has a double purpose:

- testing the effectiveness of final heat-treatment, whereby the material supplied to the end user without sensitization heat treatment is tested for an optimum, precipitation-free, i.e. homogenous structure.

- testing the material-typical resistance to intercrystalline corrosion, whereby either weld specimens or specimens specifically subjected to a sensitization heat-treatment are tested for IC. The graph prepared as a function of sensitization temperature and time, marking the curves enclosing the sensitization area is known as time-temperature-sensitization (TTS) diagram.

Depending on the material, testing can be performed according to various test procedures. The DIN 50914 (Strauss test) is usually taken for commercial stainless steels. High alloy special stainless steels and nickel alloys are tested in accordance with the Steel-Iron Test Sheet 1877 as well as the American standard ASTM G-28.

Chapter 1.4 describes in detail IC and its metallurgical explanation. It is important to know, that in both high-alloyed stainless steels as well as nickel alloys the reduction of the carbon content to approximately 0.01% results in an IC resistance adequate for commercial applications. For the high-alloyed special stainless steels the increased chromium content has a positive effect as does nitrogen, increasingly used as an alloying element in recent years. Nitrogen does increase the IC resistance because it reduces the precipitation speed of the chromium carbides. Accordingly, with increasing nitrogen content, the sensitization zone in the TTS diagram is markedly moved to the right, i.e. towards longer times [27]. Examplary for this is the TTS diagram of Figure 1.16 for the special stainless steel Cronifer 1925 hMo - alloy 926 (1.4529) in two variations. The diagram proves the excellent IC resistance of the high-nickel-containing special stainless steel Cronifer 1925 hMo - alloy 926 (1.4529) with 25% nickel and approximately 0.2 % nitrogen, therefore good weldability in practise is guaranteed even for thicker dimensions. This is confirmed by investigations on specimens TIG-welded as shown in Table 2.16. TTS diagrams have been prepared for a number of commercial Krupp VDM special stainless steels and nickel alloys [26, 28].

Table 2.16: Results of intercrystalline-corrosion (IC) tests on TIG welds of Cronifer 1925 hMo - alloy 926 (1.4529) according to ASTM G-28 A and SEP 1877 (II) using Nicrofer S 6020 - FM 625 (2.4831) as filler metal

Test		Corrosion rate mm/a	Intercrystalline attack µm	
ASTM G-28 A	base material	0.39		2 - 15
	weld	0.37	HAZ	4 - 15
			WM	2 - 40
SEP 1877/II	base material	0.12		10
	weld	0.13	HAZ	10
			WM	10

2.5.3 Susceptibility of the weld zone to localized corrosion

As shown at the beginning, the weld metal can be a corrosion-susceptible vulnerable spot in a construction because of its cast-structure and associated segregation. Herewith the problem of resistance to local corrosion in cooling waters with low and high chloride content as well as to chloride-containing acidic process media is pointed out.

The interpretation of potential-extension-curves can be misleading for the prediction of a possible reduction in corrosion resistance of weld joints [29]. Suitable, however, is a test in 10 % $FeCl_3$ solution. Comprehensive investigation results on high-alloyed stainless steels and nickel alloys are in hand. If one compares the critical pitting-corrosion temperatures determined for weld metal specimens with those of the relevant parent material of nominally the same chemical composition, one finds for such a welding of matching composition that the weld metal of high-alloyed special stainless steels for instance exhibits a critical pitting-corrosion temperature reduced by up to 20 °C. It is therefore imperative to weld a high-alloyed special stainless steel of this kind with a significantly overalloyed filler material, which guarantees a corrosion resistance comparable to that of the parent material. Figure 2.23 confirms this convincingly. Critical pitting- and crevice-corrosion temperatures for the parent material and the weld metal of the special stainless steel Cronifer 1925 hMo - alloy 926 (1.4529) are marked: weld material of filler metal with equivalent composition reaches critical pitting- and crevice-corrosion temperatures below the values of the parent material. Only the welding with the overalloyed filler metal Nicrofer S 6020 - FM 625 (2.4831) with 22 % chromium and 9 % molybdenum provides a local corrosion resistance comparable to that of the parent metal.

Whereas overalloyed welding is necessary for high-alloyed special stainless steels and nickel-chromium-iron-molybdenum-copper alloys, nickel-chromium-molybdenum alloys should be welded with filler materials of matching composition. If the heat input is kept within the required limits the weld metal usually has a corrosion resistance comparable to the parent material, as confirmed by potential-extension-measurements [29].

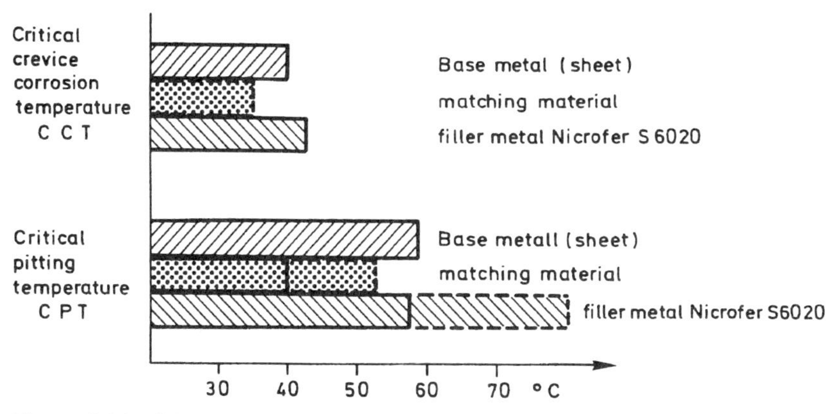

Figure 2.23: Critical pitting- and crevice-corrosion temperatures of welded Cronifer 1925 hMo - alloy 926 (1.4259) as well as in matching and overalloyed welded condition. Dotted lines indicate scatter measurements observed

Similar results are obtained in acidic-chloride media chosen to simulate applications on a practical scale to approve resistance to pitting corrosion. As an example Table 2.17 shows that the special stainless steel Nicrofer 3127 LC - alloy 28 (1.4563) should not be welded under such conditions with a filler metal of matching composition, because of pitting corrosion in the weld zone as well as higher uniform corrosion.

Table 2.17: Corrosion resistance of welded joints of the special stainless steel Nicrofer 3127 LC - alloy 28 (1.4563) in a high-chloride containing acidic solution at 105°C

Cl-concentration: 30.000 mg/l
Temperature: 105°C boiling
pH-value: 1
Test period: 21 days

Welding process	Filler metal	Rate of uniform corrosion mm/a
Base material GTAW manual	 Nicrofer S 3127 - FM 31 (1.4562)	0.01 0.01
GTAW fully mechanised W: GTAW manual	 Nicrofer S 3127 - FM 31 (1.4562)	0.19
GTAW manual	Nicrofer S 6020 - FM 625 (2.4831)	0
GTAW fully mechanised W: GTAW manual	 Nicrofer S 6020 - FM 625 (2.4831)	0.02

W = root

When uniform corrosion is a problem, basic requirement is the filler metal to be at least "equivalent" to the metal being welded, that is to say under no circumstances should it be lower in its alloy composition. In aqueous media the weld would then be less noble and moreover the section of smaller area with the result that the attack on the weld metal would be intensified. However, one should always bear in mind what is ment by "overalloyed". This factor should always be viewed in close relation with the practical service conditions to which a welded component is subjected.

Finally, it should be noted that the extent of segregations is affected by the heat input during welding. Therefore high-alloy weld materials with a tendency to segregation should only be welded with a low heat input. A low specific heat input should therefore be selected, generally not more than 10kJ/cm for materials of the nickel-chromium-molybdenum-iron system.

In conclusion, special care must always be given to the welded joints of high-alloyed materials. Testing just the mechanical-technological properties is not sufficient. In addition, the equivalence of weld and parent material must be guaranteed. This can be varified by suitable weld-filler materials, proven welding technologies as well as suitable welding parameters. In any case, an initial testing, using a test piece for example, is necessary in the sequence - standard laboratory tests for pitting corrosion, tests in a simulated-service medium, by field tests also on welded specimens and finally by a thorough examination of weld joints at regular plant inspections.

2.6 The process and design effects and welding factors [30,31]

2.6.1 Influence of gases dissolved in the process medium

The oxygen in the air has a great influence, both positively and negatively, according to the particular medium and material group. As mentioned in 2.2.1 and 2.2.2, the corrosion resistance of nickel and nickel-copper alloys in reducing acidic media is in some cases severely decreased by aeration. On the other hand, atmospheric oxygen normally has a positive effect on all the classic passiviable nickel-chromium-molybdenum-iron alloys, since it accelerates the formation of the passive layer. However, in aqueous solutions with low chloride contents and high temperatures, it may increase susceptibility to SCC. The effects of oxygen should therefore be differentiated according to circumstances.

Other gases, such as CO_2 and H_2S, cause a slight increase in corrosion by increasing the acidity level of the medium; i.e. they lower the pH value, for instance, from 7 to 4 – 5.

2.6.2 Effects of process medium flow velocity

As is known from experience gained with copper alloys in sea-water, high flow velocities can cause severe erosion corrosion. In the case of nickel alloys, and nickel-chromium-iron-molybdenum alloys in particular, limit velocities are so high that any increase in corrosion is hardly likely at the flow velocities which occur in practice (1 – 3m/sec). It is of significantly greater importance for these alloys to ensure that impermissibly low flow velocities do not occur, thus static conditions must be avoided. Under such conditions deposits may form, with subsequent local spalling. Such deposits prevent adequate passivation of the metal, and the areas exposed after spalling are subjected to intensified corrosion as a result of local element formation (pitting- or crevice corrosion under deposits).

2.6.3 Prevention of static corrosion

Corrosive attack generally increases where remnants of the aggressive medium are not discharged from dead spaces when a plant is shut down. If high temperatures also occur at such points, concentration occurs and a normally only slightly aggressive medium may become locally more aggressive. For this reason, designs should include such provisions as inclined planes in the plant and drainage at the lowest points in a vessel.

2.6.4 Prevention of crevices

Crevice corrosion can to a large extent be prevented by means of design measures. Welded instead of flanged connections should be used for plants in which media which may cause crevice corrosion (such as chloride solutions) are present. Tubes fixed by means of a tube-expander into the tube-sheets of heat exchangers can provide ideal conditions for severe crevice corrosion. The solution is either to modify the method of fixing the tubes to the tube-sheet by means of crevice-free welding, or to ensure that the medium which causes crevice corrosion flows through the tube instead of around the tube shell. For instance, in a heat exchanger using sea-water as the cooling medium for sulphuric acid, the sea-water should flow through the tubes, since there are no crevices there.

2.6.5 Prevention of contact corrosion

Designers should always avoid selection of mixed materials for plant and components exposed to corrosive media. This applies even to plants exposed only to atmospheric conditions. Even in normal industrial and marine atmosphere with relatively high humidity, severe contact corrosion may occur between metals and alloys which are relatively far apart in the galvanic electrochemical series, such as zinc/steel, aluminium/copper and carbon steel/stainless steel.

Nickel alloys, especially nickel-chromium-molybdenum-iron alloys, are generally less susceptible, due to their relatively high potentials. Instead, they tend to cause corrosion of the (less noble) contact metals, because the alloys employed in chemical plant engineering with a substantially higher grade are very few and far between.

Combinations with lower-grade alloys are only permissible where no ion-conducting medium is present, where no atmospheric oxygen is present, and where no aggressive agent is present in the medium. For example: the shell of a heat exchanger, provided the heat exchanger is operated with deaerated, super-heated steam generated from deionized water, or where another heat-transfer medium (such as oil) is used.

2.6.6 Effects of cold-forming during material processing

Cold-forming above even 5 % can impair the corrosion resistance of certain alloys. While the increased uniform corrosion observed in pure metals such as iron and nickel is not present to the same extent in nickel alloys, there is a very noticeable effect on the susceptibility to stress-corrosion cracking. Stresses arising from cold-forming of more than 5 – 15 % should therefore in all cases be relieved by means of subsequent annealing, as prescribed, for example, for certain materials in the relevant VdTÜV Material Sheets. These also call for repetition of soft- and solution-annealing even after hot-forming, and after cold-forming exceeding a certain degree of deformation. Details are given in the VdTÜV Sheets regarding these annealing processes. The limit degrees for cold-forming are 5 % for LC-Nickel 99.2 (VDM LC-Nickel 99.2 - alloy 201), NiCr15Fe (Nicrofer 7216 - alloy 600) and NiCu30Fe (Nicorros - alloy 400), 10 % for X10NiCrAlTi3220/H (Nicrofer 3220 - alloy 800 H), 15 % for NiCr21Mo (Nicrofer 4221 - alloy 825), NiMo16Cr16Ti (Nicrofer 6616 hMo - alloy C-4), NiCr22Mo9Nb (Nicrofer 6020 hMo - alloy 625), NiCr23Mo16Al (Nicrofer 5923 hMo - alloy 59) and - in accordance with the data contained in AD-Merkblatt HP7/3 and W 10 (austenitic steels) - for NiMo16Cr15W (Nicrofer 5716 hMoW - alloy C-276) and NiMo28 (Nimofer 6928 - alloy B-2). In the case of heat treatment of NiMo16Cr15W (Nicrofer 5716 hMoW - alloy C-276) and NiMo28 (Nimofer 6928 - alloy B-2) during processing, an agreement with the material manufacturer is necessary. With the materials LC-Nickel 99.2 (VDM LC-Nickel - alloy 201) and NiCu30Fe (Nicorros - alloy 400) stress-relief annealing may also be carried out after cold-working instead of a repetition of soft-annealing.

2.6.7 Influence of post-weld treatment on the corrosion resistance

Plant manufacturers and operators of welded constructions tend to ask up to what extent a post-weld treatment by grinding, shot-blasting, pickling, etc. does influence the corrosion resistance of the welded joint. In practical plant construction welded joints do occur in untreated as well as mechanically treated, e.g., surface-ground condition – the latter in order to prevent deposits and adhesions. A post treatment of the surface by shot blasting is also common practise. Occasionally even electropolishing is employed. The untreated weld surface can corrode preferentially to the base metal because of surface impurities or weld-technologically related defects (see Figure 2.24) on one side and the treated weld on the other because of the segregated dendrites.

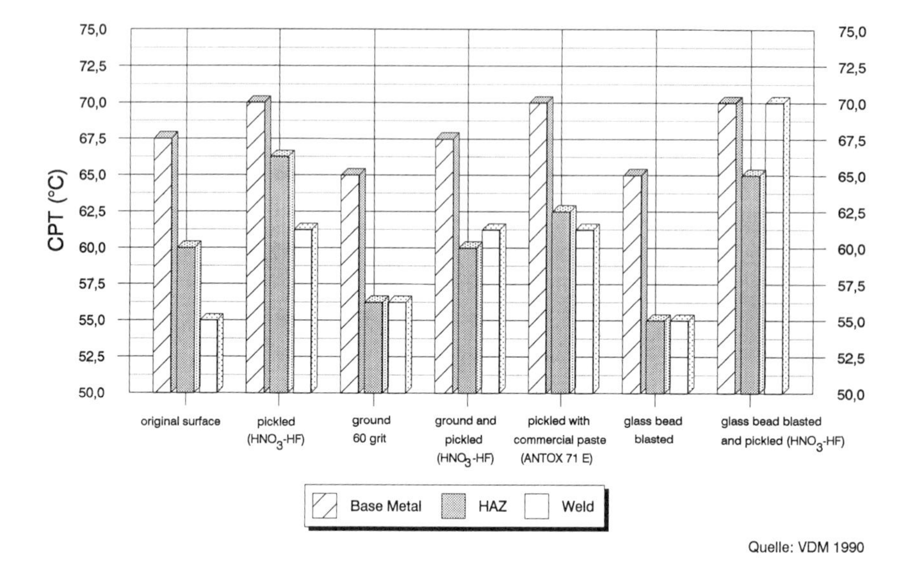

Figure 2.24: Critical pitting-corrosion temperature in the 10 % $FeCl_3$ test of Cronifer 1925 hMo - alloy 926 (1.4529), welded with Nicrofer S 6020 - FM 625 (2.4831) with different post-weld treatments [32]

Investigations on a high-alloyed special stainless steels and a nickel alloy [32] were performed to once and for all systematically investigate this problem.

The critical pitting corrosion temperature in the $FeCl_3$ test was used as criterion for the resistance of these materials to pitting corrosion, and determined for welding specimens with different surface post-weld treatments. Figure 2.24 gives examplarily the results of these investigations on the special stainless steel Cronifer 1925 hMo - alloy 926 (1.4529). The diagram clearly differentiates between base metal (BM), heat-affected zone (HAZ) and weld (W).

The results of these corrosion investigations can be summarized as follows:

- the determination of the critical pitting-corrosion temperature in the 10 % $FeCl_3$ x $6H_2O$ test is a reliable test to determine the corrosion resistance of welded joints of high-alloyed austenitic stainless steels.

- post-weld pickling of welded joints gives rise to increased critical pitting-corrosion temperatures both in the heat-affected zone and the weld in particular.

- grinding of welds with a coarse grit (60 grit, in this case) will bring reduced pitting-corrosion temperatures that will rise again following subsequent pickling.

- shot blasting of the weld areas gives a significant reduction in critical pitting-corrosion temperature that will only be reversed by subsequent pickling.

- electrochemical polishing of the welded joint gives the highest pitting-corrosion resistance.

- pickling itself should be carried out with a pickling liquid based on HNO_3 + HF, or with a good commercial pickling paste.

- these statements apply to austenitic stainless steels based on 1810 CrNiMo (AISI 316 type), to high-alloyed special stainless steels such as Cronifer 1925 hMo - alloy 926 (1.4529) and to nickel alloys such as Nicrofer 6020 hMo - alloy 625 (2.4856).

From the results of these investigations the following practical recommendations can be given.

- post-weld treatment by brushing with stainless steel wire brushes is imperative to largely remove surface impurities, slag residues, tempering colours etc. This mechanically brushing should always be followed by pickling as this significantly improves corrosion resistance.

- ground-welded joints must subsequently be pickled if a coarse grit size is used; otherwise the pitting-corrosion resistance is reduced.

- after shot blasting the weld area, which significantly reduces pitting corrosion resistance, pickling is imperative.

- in the case of critical corrosion conditions such as pitting or crevice corrosion it is necessary to test the equivalence of the corrosion resistance of the weld region to that of the base metal: a mechanical technological testing of the welded joint alone is not sufficient.

2.7 References Chapter 2

[1] DIN 50 900, Teil 1 Korrosion der Metalle, Definitionen
[2] DIN 50 905, Auswertung und Darstellung der Ergebnisse von Korrosionsuntersuchungen
[3] E.M. Horn, D. Kuron and H. Gräfen: Lochkorrosion an passiven Legierungssystemen der Elemente Eisen, Chrom und Nickel, Teil I, Z. Werkstofftechn. 8 (1977), 37 – 55

[4] E.-M. Horn, D. Kuron and H. Gräfen: Lochkorrosion an passiven Legierungssystemen der Elemente Eisen, Chrom und Nickel, Teil II, Z. Werkstofftechn. 8 (1977), 182 – 191
[5] E.-M. Horn, D. Kuron and H. Gräfen: Praktische Kontaktkorrosionstabellen von Konstruktionswerkstoffen des Chemie-Apparatebaues, Metalloberfläche, 26 (1972), 413 – 420
[6] E. Wendler-Kalsch: Stand der Erkenntnisse auf dem Gebiet der klassischen Spannungsrißkorrosion, Werkst. + Korrosion, 29 (19768), 703 – 720
[7] J. A. Sedriks: Corrosion of Stainless Steels, J. Wiley & Sons, New York – Chichester – Brisbane – Toronto (1979)
[8] H. Kaesche: Die Korrosion der Metalle, 2. Aufl., Springer-Verlag, Berlin – Heidelberg – New York, 1979
[9] C. Loss and E. Heitz: Zum Mechanismus der Erosionskorrosion in schnell strömenden Flüssigkeiten, Werkst. + Korrosion, 24 (1973), 38 – 47
[10] M. Renner, U. Heubner, M. B. Rockel, E. Wallis: Temperature as a pitting and crevice corrosion criterion in the $FeCl_3$ test, Werkst. u. Korrosion 37 (1986) 183 – 190
[11] W. Z. Friend: Corrosion of Nickel and Nickel-Base Alloys, J. Wiley & Sons, New York – Chichester – Brisbane – Toronto, 1980
[12] H. J. Büth, M. Köhler: Die Versprödungsneigung der hochkorrosionsbeständigen Legierungen NiMo28 (2.4517) im Chemie-Apparatebau und Maßnahmen zu ihrer Vermeidung, Werkstoff und Korrosion 43 (1992) 421 – 425
[13] M. B. Rockel, M. Renner: Pitting crevice and stress corrosion resistance of high chromium and molybdenum stainless steels, Werkst. u. Korrosion 35 (1984) 537 – 542
[14] U. Heubner, R. Kirchheiner, M. Rockel: Alloy 31, a new high-alloyed nickel-chromium-molybdenum steel for the Refinery Industry and related applications, CORROSION 91, Paper No. 321, National Association of Corrosion Engin., Houston, Texas, 1991
[15] M. Jasner, U. Heubner: Alloy 31, a new 6Mo stainless steel with improved corrosion resistance in sea water, CORROSION 95, Paper No 279, National Association of Corrosion Engineers, Houston, Texas, 1995
[16] E. Altpeter, R. Kirchheiner, F. E. White: Sulphur acid corrosion of some special stainless steels and nickel alloys - laboratory tests and plant experience. International Conference on Corrosion in natural and industrial environments, Grado, Italy, May 23-25, 1995
[17] Angela Wensley et al: Proceedings TAPPI 1991, Engineering Conference, 499 – 506
[18] R. Kirchheiner, M. Köhler, U. Heubner: Alloy 59, a new highly corrosion-resistant material for the Chemical-Process-Industry, Flue Gas Desulphurization and related applications, CORROSION 90, Paper No. 90, National Association of Corrosion Engineers, Houston, Texas, 1990
[19] D.C. Agarwal, U. Heubner, R. Kirchheiner, M. Köhler: Cost Effective Solutions to CPI Corrosion Problems with a New Ni-Cr-Mo Alloy, CORROSION 91, Paper No. 179, National Association of Corrosion Engineers, Houston, Texas, 1991
[20] R. Kirchheiner, M. Köhler, U. Heubner: Nicrofer 5923 hMo, ein neuer hochkorrosionsbeständiger Werkstoff für die chemische Industrie, die Umwelttechnik und verwandte Anwendungen, Werkst. u. Korrosion 43 (1992) 388 – 395
[21] M. Köhler, U. Heubner, K.-W. Eichenhofer, M. Renner: Alloy 33, a new corrosion-resistant material for the refining industry and related applications, CORROSION 95, Paper No. 338, National Association of Corrosion Engineers, Houston, Texas, 1995
[22] M. Köhler, U. Heubner, K.-W. Eichenhofer, M. Renner: Progress with Alloy 33 (UNS 20033), a new corrosion-resistant chromium-based austenitic material, CORROSION

96, Paper No. 428, National Association of Corrosion Engineers, Houston, Texas, 1996
[23] M. Köhler, R. Kirchheiner, U. Heubner: Alloy 700 Si, a new corrosion-resistant material for handling of hot, highly concentrated acids, Werkstoffe und Korrosion 46, 18 – 26 (1995)
[24] M. Köhler, U. Heubner: Stress corrosion cracking behaviour of Ni28Mo Alloys – recent research data, CORROSION 94, Paper No. 230, National Association of Corrosion Engineers, Houston, Texas, 1994 and
D.C. Agarwal, U. Heubner, M. Köhler, W. Herda: UNS N 10629 - a new Ni28%Mo alloy, Material Performance, Vol. 33, 1994, No. 1, 64 – 67
[25] U. Heubner, R. Kirchheiner, M. Köhler: Deutsche Patentanmeldung P 44 46 266.2-24 Nickellegierung Nimofer 6224 – alloy B-10, May 1996
[26] M. Rockel: Korrosionsberständigkeit von Schweißverbindungen hochlegierter Werkstoffe, Metall 42 (1988) 338 – 351
[27] U. Heubner, M. Rockel, E. Wallis: Das Ausscheidungsverhalten von hochlegierten austenitischen Stählen mit 6 % Molybdän und sein Einfluß auf die Korrosionsbeständigkeit, Werkst. u. Korrosion 40 (1989) 459 – 466
[28] U. Heuber, M. Köhler: Das Zeit-Temperatur-Ausscheidungs- und das Zeit-Temperatur-Sensibilisierungs-Verhalten von hochkorrosionsbeständigen Nickel-Chrom-Molybdän-Legierungen, Werkst. u. Korrosion 43 (1992) 181 – 190
[29] E. Altpeter, U. Heubner, M. Rockel: Potential-Sonden-Messungen und Potential-Profile der Schweißverbindungen hochkorrosionsbeständiger Nickel-Basis-Werkstoffe und Sonderstähle, Werkst. u. Korrosion 43 (1992) 96 – 105
[30] Gräfen: Berücksichtigung der Korrosion bei der konstruktiven Gestaltung von Chemieapparaten, Werkst. + Korrosion, 23 (1972), 247 – 254
[31] Spähn and K. Fäßler: Zur konstruktiven Gestaltung korrosionsbeanspruchter Apparate in der chemischen Industrie, Konstruktion 24 (1972), 249 – 258
[32] M. Rockel, W. Herda: Die Oberflächenbehandlung von Schweißverbindungen hochkorrosionsbeständiger 6 % Mo-Stähle und Nickel-Basis-Legierungen, Werkst. u. Korrosion 43 (1992) 354 – 357

3 Welding of nickel alloys and high-alloy special stainless steels

T. Hoffmann

3.1 Welded joints

3.1.1 General

The question of weldability has a direct bearing on the use of nickel alloys and high alloy special stainless steels in the manufacture of chemical plant and process equipment, offshore technology, as well as environmental and energy technologies. The safety and service life of equipment and plant is largely determined by the quality of the welded joints. The requirements specified by the operators are usually demanding. The selection of the materials to be used must take into account the subsequent operating conditions such as the process medium, temperature and pressure. However, optimum results can only be obtained if design and fabrication are harmonized with the welding characteristics of the materials to be used. The selection of suitable welding filler materials and welding processes plays a decisive role and must be considered from the very beginning of component design. An ideal welding design provides for trouble-free assembly in the workshop and also on site. The weld arrangement and geometry must be selected so that the welder has sufficient freedom of movement during his work to make welds with flat profiles, no slag inclusions and without excessive root penetration whilst maintaining the correct heat input for the material. If the number of welds in a component is high, it is advantageous to use mechanized welding processes. Experience is available, for example, from critical applications on the use of the GTAW process with cold or hot filler wire. This is discussed in more detail below. In general, however, preliminary tests have to be carried out in regions subjected to severe corrosion attack to determine the welding process and suitable filler. With manual processes the qualifications of the personnel must be considered in addition to the welding process. It is essential that well trained and reliable personnel are used, preferably with qualifications in accordance with DIN 8561 for non-ferrous metal welders.

3.1.2 Preparatory work

Design drawings should be checked at an early stage, to ensure that the designer has taken sufficient account of the technical necessities of welding. In case of doubt, qualification tests can be used to verify satisfactory weldability and procedures, given the material combinations, weld profiles and welding positions demanded by the design drawing. Reference should be made to DIN 8563, Parts 1 and 2: Quality Assurance of Welding Procedures.

The semi-finished product to be fabricated should normally be in a soft, metallurgically clean and grease-free condition (see also Section 1.5.1). If cold-forming procedures such as bending, bevelling, beading or deep drawing are necessary before welding, the work-hardened material must be soft- or solution-annealed before welding. For the fabrication of plant subject to official supervision, such heat treatment is prescribed if cold forming of more than 5 to 15% is carried out by the fabricator. In this case, the specifications contained in the VdTÜV Material Sheets, in AD-Merkblatt HP 7/3 and in the manufacturers' material data sheets must be observed under all circumstances. Since serious errors are by no means infrequent, particularly with regard to the cleanliness of the heat-treated material, it should be emphasized that considerable attention must be paid to cleanliness when heat-treating all nickel materials. Surfaces in particular must be absolutely free of sulphur- and lead-bearing substances; even fingerprints can be detrimental. The annealing atmosphere should also be free of sulphur and be controlled so that it is slightly reducing or neutral. A slightly oxidizing atmosphere should be used if complete freedom from sulphur cannot be assured. However, the annealing atmosphere must not fluctuate between oxidizing and reducing conditions. Under appropriate reducing conditions, nickel alloys containing no chromium, titanium or aluminium can be readily bright-annealed, whereas if these reactive elements are present, heat-treatment atmospheres of dried hydrogen or completely cracked ammonia, with a dew point below -55 °C, are necessary for this purpose.

The cleanliness requirements imposed on the material to be heat-treated apply equally to welding. In particular, the weld area and the adjacent heat-affected zone must be kept metallurgically clean. If gross welding errors are disregarded for the time being, there can be no greater danger for the quality of a welded joint than the presence of surface impurities in the weld region. This also means that welders must wear clean gloves and use clean tools and equipment.

Weld-joint preparation should be carried out in accordance with Figure 3.1. It is in this context important to point out the different physical behaviour of nickel alloys in comparison to carbon steel, i.e. the lower thermal conductivity and the higher thermal expansion. This behaviour is taken into account by, amongst other things, a larger root gap spacing, whereas because of the viscous behaviour in the liquid state larger opening angles have to be used to counteract the marked shrinkage behaviour. It is advisable, because of the more or less pronounced tendency of all the materials described here to smear, to prepare the weld angles by cutting processes such as turning, milling or planing. These processes are usually carried out dry. If cutting fluids or lubricants have to be used, it is essential that these are sulphur-free and that careful cleaning is subsequently carried out in every case.

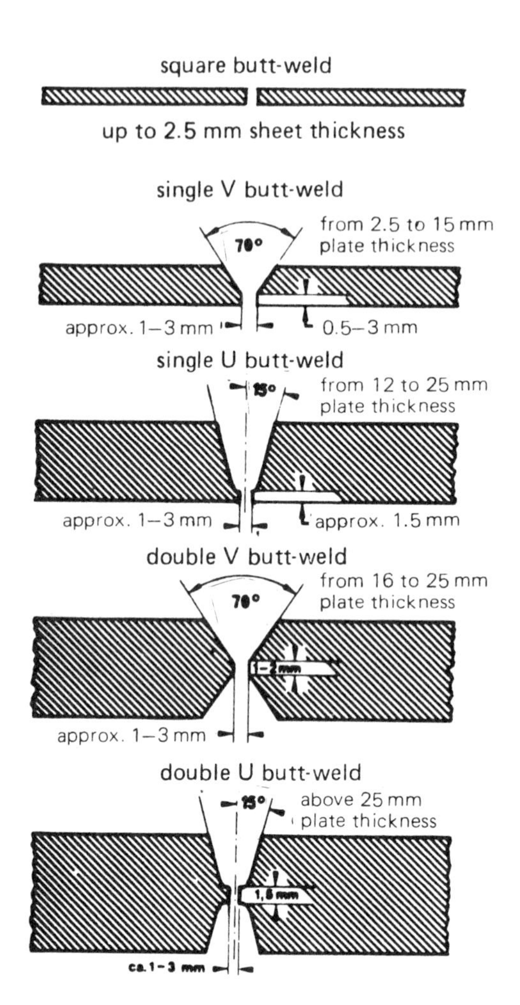

Figure 3.1:
Weld preparation for
nickel and nickel alloys

Machining of nickel alloys may be more time consuming than machining of the lower alloyed stainless steels. This has to be taken into account when calculating the cost of the preparatory work.

3.1.3 Welding processes

All the materials under consideration can be welded by the usual fusion welding processes [1]. These are:

GTAW, gas tungsten arc welding (141)
GTAW-HW, gas tungsten arc welding using preheated (hot) filler wire (141)

PAW, plasma arc welding (15)
GMAW, gas metal arc welding using inert shielding gas (131)
GMAW, gas metal arc welding using active shielding gas (135)
SMAW, shielded metal arc welding, - manual metal arc welding using a coated stick electrode (111)
SAW, submerged arc welding (121)
L, laser beam welding (751)
The code corresponding to ISO 4063 is indicated in parenthesis.

Gas tungsten arc welding (GTAW) can produce the best weld quality. It is possible to weld with low heat input, particularly if a filler metal is used, since the weld pool temperature can be favourably influenced using the melting welding rod or cold wire electrode. The GTAW process is used in the range of low to moderate wall thicknesses and in addition for the welding of root passes and backings for greater wall thicknesses. The current type is material-dependent and nickel alloys are welded with direct current (electrode on the negative pole).

Gas tungsten arc welding using a preheated or hot filler wire (GTAW) enables the same high weld quality as with the GTAW process. The process can be described as the GTAW arc being used to melt the base metal whilst the filler metal is continually transported via a wire feed system to the arc or the weld pool. This filler metal with a diameter range of 0.8 - 1.2 mm is connected through a contact tube to its own power supply. The filler metal is immersed in the weld pool and must retain this position during the welding process, i.e. there is constantly a short circuit between the filler metal and the work (weld pool). The correct inclination of the torch is important for the optimum operation of GTAW hot wire welding. The angle of inclination of the contact tube should be in a range of 20-40° to the material surface or the horizontal. The length of the free wire end must, for example, not exceed 15 mm with a 0.8 mm welding wire diameter, since the wire otherwise burns away by resistance heating before immersion in the weld pool. The GTAW torch is connected to the negative pole and the component to the positive pole as usual. A separate alternating current source with adjustable voltage is used to heat the hot wire. The voltage is between 5 and 12 V. To avoid oxidation of the heated wire, the contact tube can be equipped with a shielding gas supply. If a welding speed of 10 - 15 cm/min. can normally be achieved in GTAW welding with a cold wire filler, the range in GTAW hot wire welding is 20-30 cm/min. and sometimes significantly faster. In comparison to other processes, the GTAW hot wire process has an extremely positive effect on the welding results, for example less distortion, a narrow heat-affected zone, reduced risk of hot cracking and less dilution with the base metal.

In the GTAW process it may be advantageous with respect to higher speed and flatter and smoother weldments to make use of argon with addition of 2 - 3 % hydrogen as a shielding gas.

In plasma arc welding (PAW) the transferred arc burns between a tungsten electrode and the work through the water-cooled, tapered copper nozzle, the tungsten electrode being connected as the cathode and the nozzle as the anode. The plasma gas (synonymous with a ionized gas at high temperature) is blown into the annular space between the tungsten electrode and the nozzle. A second gas flow, the shielding gas, is needed in addition to the plasma gas, to protect the weld pool from atmospheric influences. The shielding gas is usually argon without or with addition of some hydrogen as in the GTAW process. The PAW process itself can be explained as follows: the plasma beam penetrates the material or the components to be joined by a square butt joint. The so-called keyhole forms, which moves along the joint edges of the components to be joined. The molten material flows evenly together behind the keyhole and forms the matching weld metal. The PAW process produces a high weld quality and can be used without problems in the wall thickness range 2-8 mm without weld preparation - therefore in square butt welds - with and without filler metal, though long years' experience has shown that it is advantageous to make use of a filler metal.

In gas metal arc welding using an inert shielding gas (GMAW) the heat source is the arc between the continuously fed melting filler metal (wire electrode) and the base metal under a shielding gas (argon or helium). Welding with a pulsed arc is an interesting alternative for welding nickel alloys and is growing in importance. The possibility of superimposing current pulses with adjustable frequencies on the welding current (base current) offers the advantage of being able to control the course of the welding process so that as well as the reliable welding of thin cross-sections it is also possible to weld with relatively low heat inputs.

In gas metal arc welding (GMAW) using an active shielding gas additions of hydrogen and carbon dioxide are made to an argon / helium shielding gas [2]. Gas compositions are e.g. Ar + 50 % He + 0.05 % CO_2 or Ar + 30 % He + 2 % H_2 + 0.05 % CO_2 in combination with a pulsed arc. Such gas mixtures may be helpful in i) improving flow, penetration and wetting ii) reducing any splashes and iii) increasing welding speed considerably. It has to be mentioned that a freely programmable pulsed current source is a mandatory prerequisite to obtain a stable process.

Shielded metal arc welding (SMAW), i.e. manual welding with coated stick electrodes continues to have a specific field of application because of its diverse possibilities of use. For a low equipment cost welding can be carried out in component regions that are difficult to reach and in constrained areas. Manual metal arc welding with coated electrodes has the advantage that sufficiently experienced welders are available for this process; even though special training may be necessary for special alloys. Correctly performed, this process produces the lowest heat or energy input of all fusion welding processes, particularly in multipass welding. It is usually carried out using direct current with the electrode connected to the positive pole.

In submerged arc welding (SAW) the continuously advancing wire electrode is melted under a blanket of flux powder. The powder melts to form a slag bubble that is filled with ionized gases in which the arc burns. Since the welding process takes place completely under cover the thermal efficiency is very high. This high thermal efficiency enables, for example, a heat input up to 7 times higher than in welding with rod electrodes and thus correspondingly higher fusion rates. It may be necessary to reduce the heat input if the base metal has a tendency to sensitization or to precipitation of intermetallic phases. In such cases the electrode diameter should be limited to 2 mm maximum. Submerged arc welding of nickel alloys has grown in importance due to the development of suitable flux powders in recent years.

In the laser beam welding process (L) the extremely high beam energy of the laser is utilized to melt the material. The joint faces stand parallel to each other as butt welds with an extremely narrow gap (< 0.20 mm). The contactless energy transfer in a defined atmosphere (inert) enables high speeds with extremely narrow welds to be used and with the highest weld quality being achieved. This results in low thermal stressing of the material being welded, little structural change in the weld region, low distortion and the maintenance of tight tolerances, to name only some of the advantageous process features. A disadvantage is that this excellent welding process can only be used as a fully mechanized process and thus from the point of view of chemical plant construction is mainly used for the production of longitudinally welded tubes.

In all of these welding processes protection of the weld pool plays a decisive role. Particular care should be taken to exclude as far as possible atmospheric influences that would cause oxidation and burning. The coatings, flux or gas must also not contain carbon-bearing constituents, otherwise an undesirable increase in the carbon content of the deposited metal could result. Furthermore, moisture must be kept as low as possible since it increases the tendency towards pore formation. Copper backing strips placed in the vicinity of the welding zone can help to dissipate heat and reduce distortion. Clamping equipment is recommended for aligning the components. Components allowed to cool in clamping devices after welding display the least distortion. Where clamps cannot be used, a close series of tack welds is recommended. A welding procedure sheet is necessary for complicated joints.

For multi-pass welding, in many cases several thin passes are better than only one or two passes, which may lead to reduced resistance to corrosion. String beads are preferable to weave beads. The weld should be thoroughly cleaned after each pass before the application of further passes. Where possible the root pass on V or U welds should be back-welded.

From the welding process point of view, there has recently been a significant shift from manual metal arc welding with coated stick electrodes to the gas-shielded processes. The reason for this change is that a high weld quality can be more easily achieved with the gas-shielded processes than with manual metal arc welding. The burn-off losses from the filler metal are extremely low in gas-shielded welding processes. This is a decisive aspect for components subjected to severe corrosion attack. An additional positive feature is the possibility of producing welds characterized by uniformly smooth and finely scaled surfaces in the capping pass and root, which as a result provide no possibility for formation of deposits and thus crevice corrosion (see Section 2.1).

3.1.4 Filler metals

High-alloy special steels must frequently be welded with a more highly alloyed composition in order to achieve a corrosion resistance in the weld metal which is equal to that of the parent material [3, 4]. In contrast, nickel and nickel alloys are usually welded using filler metals of similar composition. These are standardised in DIN 1736 part 1. Table 3.1 gives some welding filler metals with details of their main alloying elements. Others do not differ from the base materials mentioned in Tables 1.1 - 1.7. By way of example details are given below of their application in welded joints and cladding work. Welding rods, welding wire and wire electrodes are available for manual GTAW, PAW, pulsed-arc GMAW, SMAW as well as SAW. With coated electrodes the chemical composition of the molten weld metal is of decisive importance. The alloying elements may in this case originate either entirely or partly from the core wire as well as from the coating. Weld samples according to DIN 32525 Part 1 and AWS are melted down to determine the chemical composition and mechanical properties. Table 3.2 shows, as an example, the mechanical properties of the weld determined by means of such weld specimens.

Table 3.1: Welding consumables for nickel and nickel alloys in the form of wire and strip electrodes and filler wire and rods

Welding consumables			Main alloying constituents, nominal composition, wt. %					
Material	German Mat. No	AWS designation	Ni	Cr	Mo	Cu	Fe	others
Nickel S 9604 Alloy FM 61	2.4155	ERNi-1	96	-	-	-	-	3.3 Ti
Nicorros S 6530 Alloy FM 60	2.4377	ERNiCu-7	65	-	-	29	-	3.2 Mn; 2.5 Nb; 0.4 Ti
Cunifer S 7030 Alloy FM 67	2.0873	ERCuNi	30	-	-	68	-	0.4 Ti
Nimofer S 6928 Alloy FM B-2	2.4615	ERNiMo-7	64	0.7	28	-	1.7	
Nicrofer S 7020 Alloy FM 82	2.4806	ERNiCr-3	72	21	-	-	-	3.2 Mn; 2.5 Nb; 0.4 Ti
Nicrofer S 3127 Alloy FM 31	1.4562		31	27	6.5	1.3	31	0.20 N
Nicrofer S 5520 Alloy FM 617	2.4627	ERNiCrCoMo-4	54	22	9	-	0.5	12 Co; 0.5 Ti; 1.0 Al
Nicrofer S 5716 Alloy FM C-276	2.4886	ERNiCrMo-4	57	16	16	-	6	3.5 W; 0.5 Mn; 0.2 V
Nicrofer S 5923 Alloy FM 59	2.4607	ERNiCrMo-12	59	23	16	-	1	
Nicrofer S 6020 Alloy FM 625	2.4831	ERNiCrMo-3	64	22	9	-	1	3.4 Nb
Nicrofer S 6616 Alloy FM C-4	2.4611	ERNiCrMo-7	66	16	16	-	1	0.3 Ti

Table 3.2: Mechanical properties of weld metal in nickel and nickel alloys at room temperature and at -196 °C

Welding consumables			tensile strength	yield strength		elongation to fracture	impact strength ISO-V-notch	
Material	German Mat. No	AWS designation	Rm N/mm² -min-	$Rp_{0.2}$ N/mm² -min-	$Rp_{1.0}$ N/mm² -min-	A_5 % -min-	20° C J -min-	-196° C J -min-
Nickel S 9604 Alloy FM 61	2.4155	ERNi-1	410	200	220	30	100	80
Nicorros S 6530 Alloy FM 60	2.4377	ERNiCu-7	450	200	220	30	100	80
Cunifer S 7030 Alloy FM 67	2.0873	ERCuNi	360	120	140	25	80	50
Nimofer S 6928 Alloy FM B-2	2.4615	ERNiMo-7	760	360	390	30	100	80
Nicrofer S 7020 Alloy FM82	2.4806	ERNiCr-3	600	340	360	30	120	90
Nicrofer S 3127 Alloy FM 31	1.4562		700	340	360	30	70	55
Nicrofer S 5520 Alloy FM 617	2.4627	ERNiCrCoMo-4	700	400	420	25	70	55
Nicrofer S 5716 Alloy FM C-276	2.4886	ERNiCrMo-4	755	310	330	25	70	55
Nicrofer S 5923 Alloy FM 59	2.4607	ERNiCrMo-12	700	370	400	30	100	80
Nicrofer S 6020 Alloy FM 625	2.4831	ERNiCrMo-3	700	420	440	25	80	60
Nicrofer S 6616 Alloy FM C-4	2.4611	ERNiCrMo-7	650	300	340	25	90	80

The filler metals that can be considered for the high alloy special stainless steels are to some extent covered by DIN 8556 part 1 and can be used for the manual GTAW, PAW, pulsed-arc GMAW, SMAW and SAW.

3.1.5 Post-weld treatment

Any oxide layer produced during welding or heat treatment, for instance annealing colours or scale, must subsequently be removed so that the surface can attain a maximum degree of corrosion resistance (see Sections 1.5.1 and [3]). Treatment with a suitable mixture of acids (e.g.. nitric acid/hydrofluoric acid base) is recommended to attain satisfactory removal of the oxide. Pickling should be carried out carefully and under controlled conditions. Special companies should be used for the treatment of larger objects or components. In many cases brushing the welds with stainless steel wire brushes, if possible while still hot, is sufficient. The use of special pickling pastes has also proved effective. It should be pointed out that in each case a correct disposal of the pickling agent according to the law in force is necessary.

3.1.6 Practical examples

The requirements mentioned at the outset apply in particular to the welding of nickel alloys and specifically to nickel. Failure to satisfy these requirements results in the consequences shown in Fig. 3.2. Dirty joint edges, excessive heat input and too large a welding rod diameter have produced a substantial defect in the shape of cracks in the heat- affected zone of a workpiece of 8 mm thick LC-Nickel 99.2. As well as scrupulous cleanliness it is imperative to ensure that the heat input is continuously controlled and held as low as possible as appropriate to the material. This applies to LC-Nickel 99.2 - alloy 201 and also nickel alloys. The necessary heat energy required for each nickel alloy must naturally be specifically determined before the welding of the component using procedure tests or test pieces. Preheating of nickel and nickel alloys is not necessary.

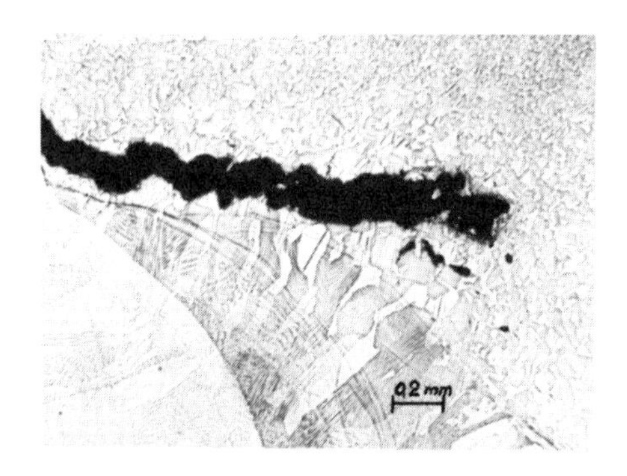

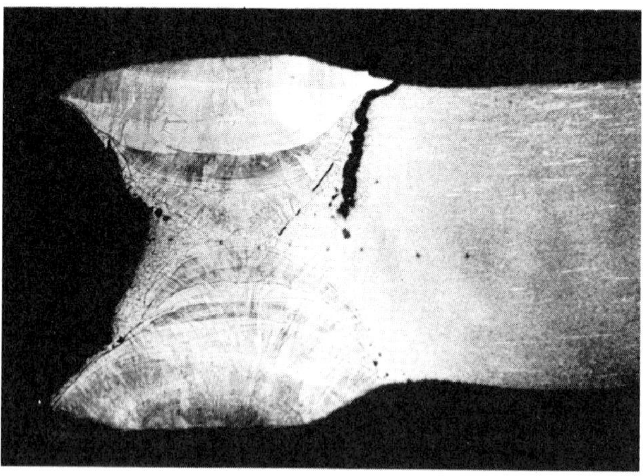

Figure 3.2: Cracking in SMAW of LC Ni 99.2, resulting from contaminated seam edges and too high a specific heat input (more than 15 kJ/cm)

Numerous investigations [4-11] have provided an extensive scientific basis for the welding practice and weldability of nickel alloys and high-alloy special stainless steels.

Manual GTAW of a nickel alloy is illustrated in detail as an example in Table 3.3 (a work record). It gives information on all details, e.g. on the filler metal, on suitable parameters and, lastly, on the result of the mechanical-technological evaluation. Fig. 3.3 shows the manual GTAW process being used in the vertical-up and horizontal positions on site. The scrubber is clearly being constructed by simultaneous welding on both sides, which guarantees a high weld quality.

Table 3.3: Work record of a manual GTAW test on alloy 59

1. Base material	**Nicrofer 5923 hMo / Alloy 59**	**Plate thickness**	**16 mm**
2. Welding electrode	**Nicrofer S 5923 / Alloy 59**	**Filler wire diameter**	**2.5 mm**

3. Composition, wt. %	**Cr**	**Ni**	**Mn**	**Si**	**Mo**	**Nb**	**Fe**	**Ti**	**C**	**Cu**	**V**
Base material	22.50	60.65	0.15	0.02	15.9		0.23		0.002	0.01	0.15
Filler wire	22.65	60.80	0.15	0.03	15.60		0.29		0.005	0.01	0.14

4. Welding parameters	**root**	**intermediate passes**	**cap passes**
type of current/polarity	= / -	= / -	= / -
current, A	110	130	10
voltage, V	10	12	12
welding speed, cm/min	6	12	13
interpass temperature, °C		130	130
heat input, kJ/cm	11	7,8	7,2
shielding gas	ArW3	ArW3	ArW3
welding position	W	W	W

single 70° V groove
root gap: 1.5 mm

5. Tensile test data

testing temperature	$R_{p0.2}$ N/mm²	$R_{p1.0}$ N/mm²	Rm N/mm²	A_5	position of fracture	remarks
RT		418	792	42	base material	
RT		410	770	46	base material	

6. Bending test, bending radius 1.5 wall thickness

	Results		
cap passes in tension	180° / without cracking	180° / without cracking	180° / without cracking
root in tension	180° / without cracking	180° / without cracking	180° / without cracking
side bend test	180° / without cracking	180° / without cracking	180° / without cracking

7. Impact test, ISO-V-notch

position	**test temperature**	**impact energy, J**
weld	**RT**	**166, 145, 155**
	- 196° C	**110, 132, 133, 134, 117**
weld / base metal	**RT**	**255, 258, 260**

Figure 3.3: Application of manual GTAW in the field, - welding a scrubber from plates of a nickel-base alloy in vertical-up and horizontal-vertical position simultaneously on both sides

A work record of GMAW using the active shielding gas Cronigon He 30 S (Ar + 30 % He + 2 % H_2 + 0.05 % CO_2) is shown in Table 3.4 and the welded component in Fig. 3.4.

Fig. 3.5 shows, as an example, the PAW process - also known as the plasma keyhole process - to illustrate the advantages of the gas shielded process mentioned previously. Table 3.5 gives an example of a corresponding work record.

A cross-section of a welded joint in Nicrofer 5923 hMo - alloy 59 (2.4605) produced using the laser technique is shown in Fig. 3.6. The typical weld profile of a laser weld, which can be used with nickel-base materials without any problems, can be recognized. Segregation in the weldment is extremely low as shown in Fig. 3.7.

Table 3.4: Work record of a manual GMAW test on alloy 59 using active shielding gas

1. Base material	Nicrofer 5923 hMo / Alloy 59	Plate thickness	16 mm
2. Welding electrode	Nicrofer S 5923 / Alloy 59	Electrode diameter	1.2 mm

3. Composition, wt. %	Cr	Ni	Mn	Si	Mo	Nb	Fe	Ti	C	Cu	V
Base material	22.50	60.65	0.15	0.02	15.9		0.23		0.002	0.01	0.15
Welding electrode	22.65	60.80	0.15	0.03	15.60		0.29		0.005	0.01	0.14

4. Welding parameters	intermediate passes	cap passes	
type of current / polarity	= / +	= / +	
current, A	175	175	
voltage, V	30	30	
welding speed, cm/min	45	45	
interpass temperature, ° C	130	130	
heat input, kJ/cm	7.0	7.0	single 70° V groove
wire speed, m/min	6.2	6.2	root gap: 1.5 mm
frequency, Hz	90	90	root: GTAW
shielding gas	Cronigon He 30 S (Ar + 30 % He + 2 % H_2 + 0.05 % CO_2)		

5. Tensile test on weld without adjacent base material

testing temperature	$R_{p0.2}$ N/mm²	$R_{p1.0}$	Rm	A_5	position of fracture	remarks
RT	378	442	732	56		
RT	372	435	724	58		
RT	381	444	726	56		

6. Bending test, bending radius 1.5 wall thickness

	Results		
cap passes in tension	180° / without cracking	180° / without cracking	180° / without cracking
root in tension	180° / without cracking	180° / without cracking	180° / without cracking
side bend test	180° / without cracking	180° / without cracking	180° / without cracking

7. Impact test, ISO-V-notch on weld without adjacent base metal

position	test temperature	impact energy, J
weld	- 196° C	176,170, 183, 162, 164, 190
	RT	209, 180, 214

Table 3.5: Work record of a manual PAW test on alloy C-4

1. Base material	Nicrofer 6616 hMo / Alloy C-4	Plate thickness	6 mm
2. Welding electrode	Nicrofer S 6616 / Alloy C-4	Filler wire diameter	1.2 mm

3. Composition, wt. %	Cr	Ni	Mn	Si	Mo	Nb	Fe	Ti	C	S	P
Base material	16.0	67.1	0.04	0.03	15.8		0.26		0.008	0.002	0.003
Filler wire	16.0	67.0	0.04	0.03	16.0		0.25		0.008	0.002	0.003

4. Welding parameters	root	
type of current/polarity	= / -	
current, A	25	
voltage, V	195	
welding speed, cm/min	25	
interpass temperature, °C	130	
heat input, kJ/cm	11,7	square butt
shielding gas	Argon*	edge distance 0 mm
plasma gas	Argon*	edge preparation: shear cut

* additions of H_2 up to 3 % may be advantageous

5. Tensile test data

testing temperature	$R_{p0.2}$ N/mm²	$R_{p1.0}$ N/mm²	Rm N/mm²	A_5	position of fracture	remarks
RT		295	748	42.5	weld	
RT		290	750	40.8	weld	
RT		284	756	41.0	weld	

6. Bending test, bending radius 1.5 wall thickness

	Results		
cap in tension	180° / without cracking	180° / without cracking	180° / without cracking
root in tension	180° / without cracking	180° / without cracking	180° / without cracking
side bend test	180° / without cracking	180° / without cracking	180° / without cracking

3.2 Special aspects of joint welding

3.2.1 Welding of clad materials

Clad materials are used in equipment and plant construction for the chemical and petrochemical industry, offshore applications, seawater desalination, tank construction and, last but not least, flue gas desulphurisation [12-14]. The clad materials are produced by roll-bond cladding [15] or explosive cladding. Usually a low-alloy structural steel suited to the requirements is used as the substrate and structural material, whereas the corrosion protection is provided by a comparatively thin cladding deposit of nickel-base alloys. The minimum thickness of the cladding layer should not be less than 2 mm. From the point of view of welding practice, even a cladding thickness of 3 mm is recommended since the problems in welding increase with decreasing cladding thickness.

Figure 3.4: Double louver damper for a flue gas duct having been made from alloy 59 using the GMAW process with an active shielding gas according to Table 3.4

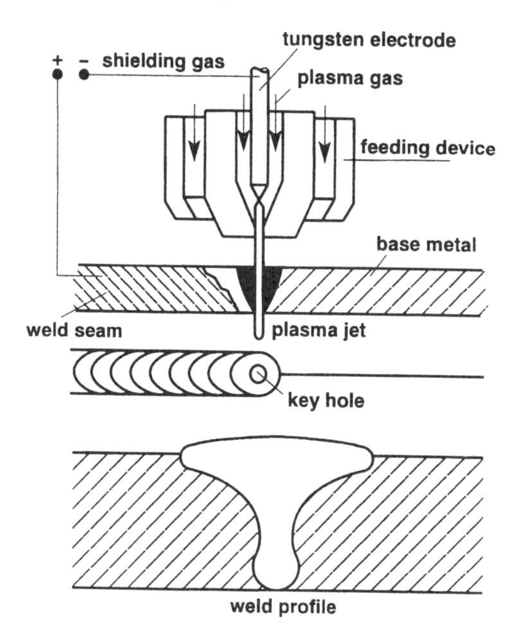

Figure 3.5:
The PAW process and the resulting weld profile

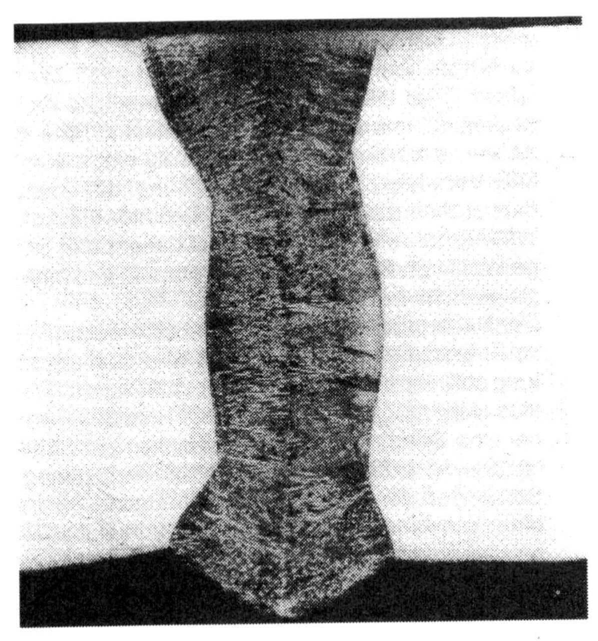

Figure 3.6: Laser weldment on 2 mm strip of alloy 59 for manufacturing of longitudinally welded heat exchanger tubes

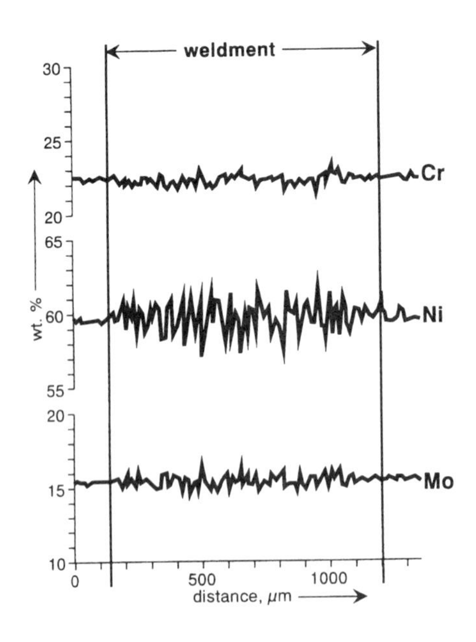

Figure 3.7:
Segregation of alloying constituents in a laser weldment of alloy 59 as evaluated by means of microprobe analysis

Before the welding of clad materials is undertaken, the selection of the welding process to be used must be clarified. The welding of the substrate is usually carried out using the SMAW process. Where possible, the SAW process or welding with powder filled wire is used because of the higher melting efficiency with the same high quality of the welded joint. Manual metal arc welding with coated electrodes (SMAW) is still frequently used for the cladding material; however, the gas-shielded welding processes are recommended, preferably the GTAW process and, with a large volume of welds, the fully mechanized GTAW process. The conventional method with cold wire filler and the significantly more attractive and economical version with hot wire filler are available. Both can be used very successfully.

Weld preparation is of considerable importance [10]. Strict adherence to DIN 8553 is therefore recommended. Weld preparation shall be carried out mechanically. If this is not possible then plasma cutting can be used, although treatment of the joint edges is necessary to remove the oxide. The resultant groove must be wide enough to have sufficient room for welding the cladding layer and to guarantee proper access to the root. On the other hand, the procedure followed should ensure that during subsequent welding the weld metal for the substrate cannot come into contact with the cladding layer. Fig. 3.8 and Fig. 3.9 show examples for the edge profiles to be used, Tables 3.6 and 3.7 give examples of the corresponding welding parameters.

Table 3.6: Parameters for welding of a scrubber wall made of roll clad plate, 2 mm alloy 59 on 16 mm carbon steel, vertical-up position

pass	welding process	current A	voltage V	welding speed cm/min	heat input kJ/cm	filler diam., mm	interpass temperature, °C	shielding gas	AWS designation
1-2	SAW					carbon steel			
3	SMAW	75	24	21	5.14	Alloy 59, coated electrode 3.25	< 100		
4	GTAW manual	170	15	14	10.93	Alloy 59, 2.4	< 100	Ar + 2 % H_2	ER NiCrMo-12
5	GTAW manual	160	15	14	10.29	Alloy 59, 2.4	< 100	Ar + 2 % H_2	ER NiCrMo-13
6-8	GTAW manual	160	15	14	10.29	Alloy 59, 2.4	< 100	Ar + 2 % H_2	ER NiCrMo-14
Visual observation:	excellent manual GTAW, fine scaled uniform bead, without excessive elevations, metallic clean								

The general prerequisite for achieving a welded joint on clad sheets is accurate positioning of the sheets or components to be welded in order to guarantee the uniformity of the cladding at the region to be welded. It is essential to have the edges to be welded well in parallel position. Welding in gravity position should be aimed at, since this should give the best results from the point of view of low heat input and dilution. In practice, welding must also be carried out in the vertical-up and horizontal positions. Satisfactory results under site conditions can only be achieved in this case by using well qualified personnel.

Table 3.7: Parameters for welding of a scrubber wall made of roll clad plate, 2 mm alloy 59 on 16 mm carbon steel, horizontal-vertical position

pass	welding process	current A	voltage V	welding speed cm/min	heat input kJ/cm	filler diam., mm	interpass temperature, °C	shielding gas, type of flux	AWS
1-5	SAW					carbon steel			
6	SAW	250	29	42	10.4	Alloy 59, 1.6	100	UTP 5005	
7-8	SMAW	80	24	17	6.8	Alloy 59, coated electrode 3.25	100		ER NiCrMo-12
9-11	GTAW fully mechanized	185	14	21	7.4	Alloy 59, 1.2	80	Ar + 2 % H_2	ER NiCrMo-13
Visual observation: fine scaled uniformly weaved bead without any grooves, metallic clean									

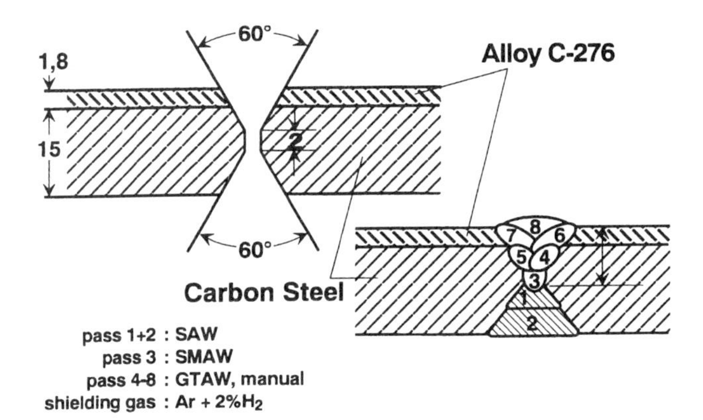

Figure 3.8: Weld preparation for welding of a scrubber wall from roll clad plates, vertical-up position, according to Table 3.6

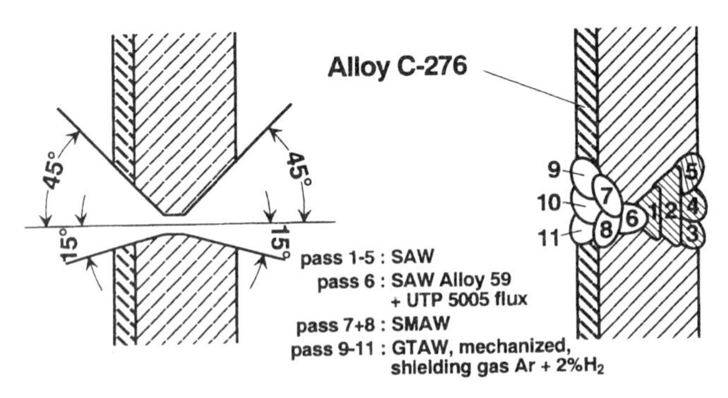

Figure 3.9: Weld preparation for welding of a scrubber wall from roll clad plates, horizontal-vertical position, according to Table 3.7

Following the usual method of carrying out two separate welding operations, the substrate is usually welded first and then the clad side following the given guidelines. The thickness of the cladding layer, the requirements of the user of the future plant and, last but not least, the composition of the media govern the procedure and the number of weld passes. The aim is to have an Fe content in the capping pass that largely matches the cladding material. Otherwise the corrosion resistance will be reduced. This has to be tested in cases of doubt [12] since there are applications that do allow a larger iron dilution, for example CuNi10Fe in seawater applications [16,17]. When using the stringer technique (several thin passes using the smallest possible wire and welding rod diameter) a rod diameter of 2.0 to 2.5 mm is recommended for manual GTAW and a wire diameter of 0.8 to 1.2 mm for mechanized GTAW using both the cold or the hot wire technique.

It has been found with nickel alloys that Fe contents corresponding to the unwelded clad metal in the deposited weld metal are only attained in the 3rd pass, with the consequence that the weld reinforcement on the clad side is 2 to 3 mm. If this reinforcement is not permissible, the first or second pass must be heavily ground or the groove on the clad side must be more deeply worked into the substrate.

Fig. 3.10 shows, as an example, the results of a microprobe analysis of GTAW-HW welded roll clad alloy 59 in the region of the capping pass. It can be seen that the

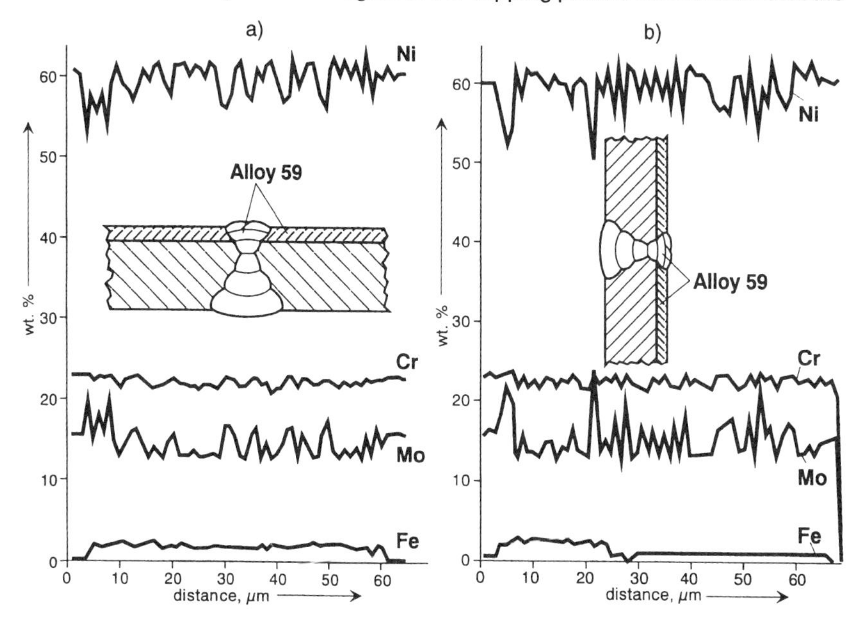

Figure 3.10: Segregation of alloying constituents in the capping pass of a GTAW-HW weldment on alloy 59 roll clad to carbon steel
a) vertical-up position b) horizontal-vertical position

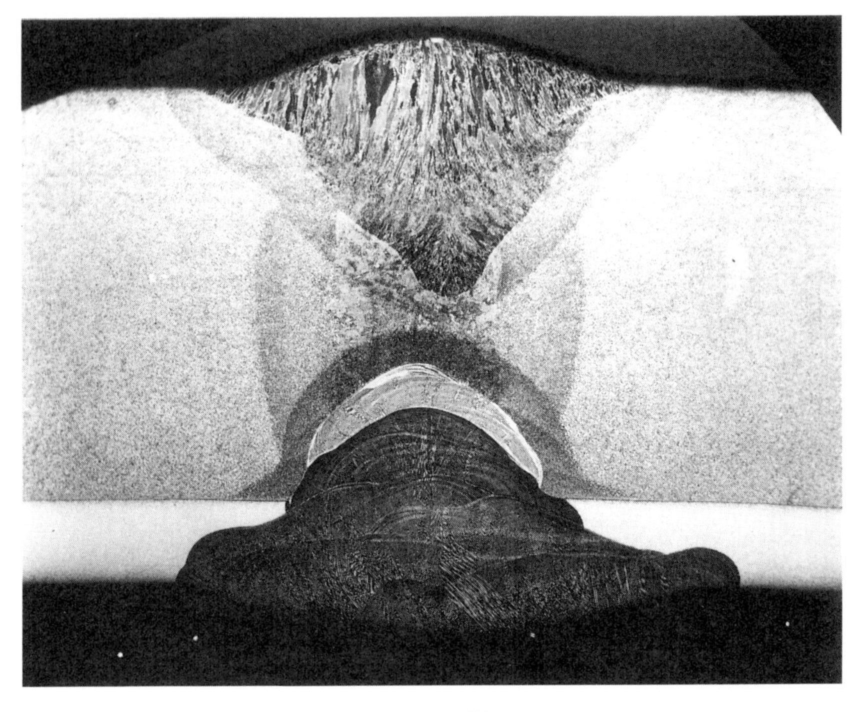

Figure 3.11: Micrograph corresponding to Figure 3.10a

iron content does not exceed 2.5 wt.%. The micrograph corresponding to Fig. 3.10a is shown in Fig. 3.11.

3.2.2 Wallpaper cladding by welding

Besides the processing of rollbond-clad or explosively clad sheets and weld-clad components which is discussed in Section 3.3, there has recently been an increasing interest in the so-called "wallpapering" process. In this thin claddings of highly corrosion-resistant materials are applied to the regions most endangered by corrosion. This is especially advantageous if subsequent cladding or covering is required, e.g. in the case of repair or refurbishment of very large-volume components such as may be required for flue gas desulphurisation in fossil fired power stations.

There are various techniques available for the application of such claddings that are all practice-proven [18, 19]. The welding processes used also play an important role here. It cannot be ignored that a single-pass, impervious, high-quality and fully reproducible weld has to be achieved in all positions; usually these are fillet welds. It is obvious that in this case serious consideration has to be given to the use of welding processes that can be mechanized and that the use of filler metals is absolutely necessary in every case. The techniques described below using welding processes that can be fully mechanised - mainly the GTAW process - can all be reliably controlled.

In manual GTAW or pulsed-arc GMAW welding with overlap and without straps, the cladding sheets or strips are fixed to the wall of the equipment which has been appropriately prepared in the weld region on the side in contact with the medium. They are bonded to the substrate at their edges by fillet welds using filler metals - stitch welds are usually adequate. The next sheet or strip is fixed with a fillet weld, a so-called single-pass seal weld, whereby a minimum overlap of 30 mm over the already fixed material should be achieved. This is continued as shown in Fig. 3.12 until the entire area is covered in a shingle-like manner. The welds (1) and (2) have different functions. Weld (1) is for fixing; weld (2) must be impervious and fulfill the corrosion requirements.

Manual GTAW or pulsed-arc GMAW welding with cover strips is another practical method. The cladding sheets or strips are first welded onto the equipment walls in contact with the medium by stitch welding. Then the resultant gaps between the applied sheets or strips are generously covered with strips of the same material and of at least the same thickness and fixed by fillet welds in the form of single-pass seal welds. This process is schematically illustrated in Fig. 3.13.

In manual GTAW or pulsed-arc GMAW welding on straps, instead of the cover strips straps of the corrosion-resistant material are first fixed to the substrate. The sheets or strips of the selected corrosion-resistant material are then attached to these straps by fillet welds as shown in Fig. 3.14.

As a matter of principle the fillet welds which are exposed to the corrosive medium must be made by means of the GTAW process, if possible not manually, but fully mechanized in order to assure the same high weld quality in every place. If the GTAW-HW process is available welding speeds of 0.8 to 1 m/min may be achieved.

The techniques described above are always expensive if large sheets (e.g. 8000 x 2500 mm) are not available. If, however, large sheets can be used, point by point fixing across the surface is necessary. Experience has shown that at least 5 such fixings per m^2 are required. Plug welding with covers is used for this type of fixing, as shown in Fig. 3.15.

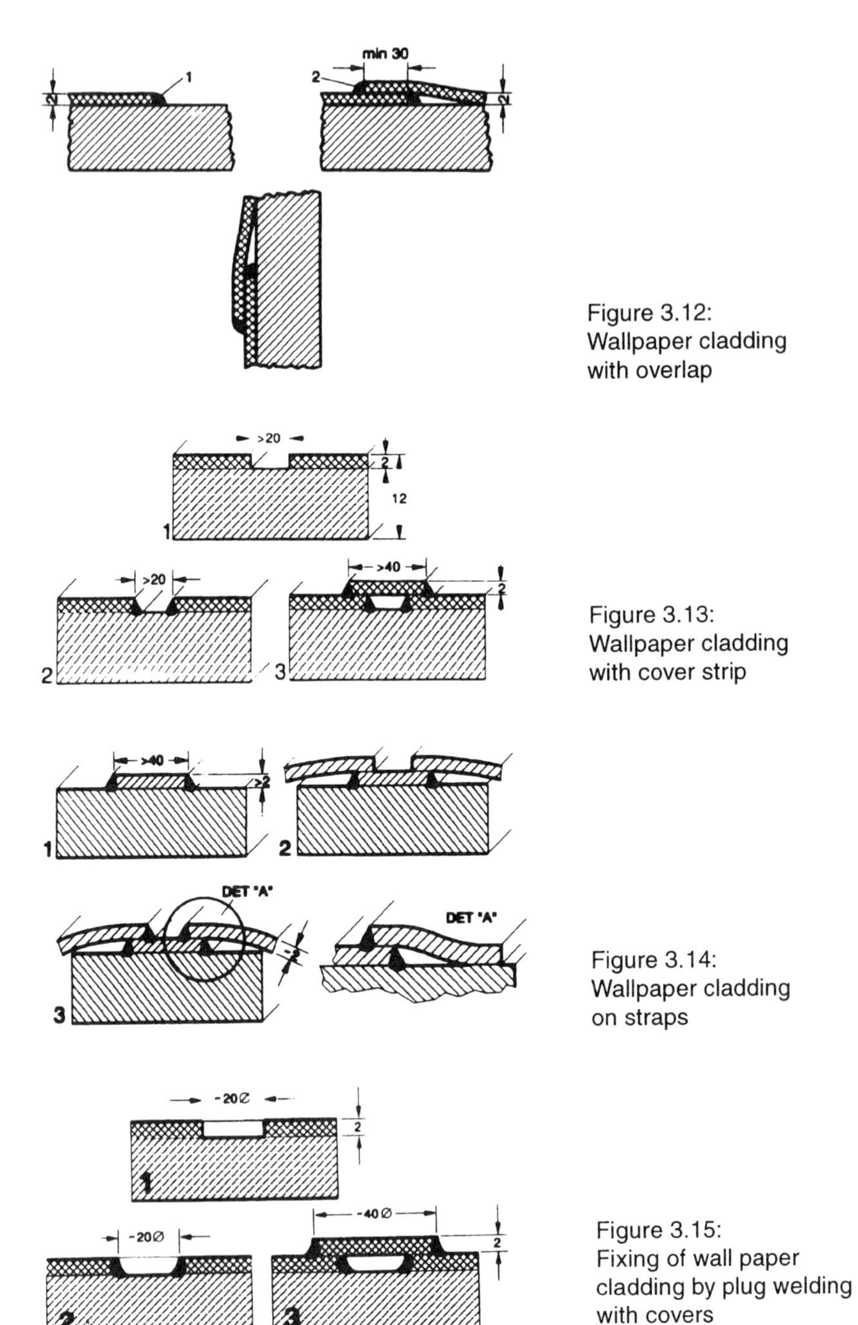

Figure 3.12: Wallpaper cladding with overlap

Figure 3.13: Wallpaper cladding with cover strip

Figure 3.14: Wallpaper cladding on straps

Figure 3.15: Fixing of wall paper cladding by plug welding with covers

3.3 Overlay welding

3.3.1 General

Cladding by means of overlay welding, using wire and strip electrodes, has firmly established itself in the manufacture of chemical equipment. Cladding by means of overlay welding is used where the fabrication of only a small number of items, or of relatively small items of compact shape, is required. Typical examples are thick-walled tube plates for steam generators, pressure vessels for nuclear reactors, or tubesheets for heat exchangers, as frequently used in chemical plant construction.

Requirements concerning staff training and on the cleanliness of the materials to be worked, of the workplace itself, and of all equipment and tools are identical to those described for welded joints (see Section 3.1.2.) The filler metals are also identical (see Section 3.1.4) and post-weld treatment (Section 3.1.5), where necessary, must be carried out in the same way. Stringent requirements are placed on the welding processes that can be considered for this purpose. Preference is given to processes with a low degree of penetration, that is, with only slight dilution with the substrate, and with as high a deposition rate as possible so as to produce a uniformly shaped, finely rippled surface with flat flanks. The welding processes dealt with in this chapter essentially meet these requirements and have thus become established in practice, although with varying degrees of success.

3.3.2 Welding processes and examples of application

3.3.2.1 Submerged-arc strip overlay welding

Submerged-arc strip overlay welding is still used for a large proportion of weld-cladding work. The strip electrode sizes are usually 30 x 0.5 mm and 60 x 0.5 mm. The quality of the welding and the deposited weld metal is influenced to a significant extent by the welding flux used. This influences the consumption and deposition rates and determines the slag separation, the surface quality and the flank angle. The most appropriate strip/flux combination and the corresponding parameters should therefore always be determined by preliminary tests.

Although the penetration in submerged-arc cladding with strip electrodes made of nickel alloys is relatively low, an iron dilution of the order of 15-20% must be expected in the first pass. This means that with a weld pass thickness of 3-5 mm, a three-pass cladding normally has to be applied to achieve a sufficiently low dilution in the final pass [20]. Only then can adequate corrosion resistance be expected.

Table 3.8 shows in the first row data for a submerged-arc cladding operation using strip electrodes of Nicrofer B 6020 - alloy 625 [20]. The necessity for three-pass cladding is indicated. It arises from the fact that an iron content of max. 5 wt.% in the capping pass has to be achieved.

Table 3.8: Overlay welding with alloy 625 [20]

No.	Process	Electrode (mm)	Voltage (V)	Current (A)	Speed (mm/min)	Deposition Rate (mm²/min)	Layers required
1	Submerged arc	0.5 x 60	28	750	120	7,200	3
2	Electroslag	0.5 x 60	24	850	150	9,000	2
3	Electroslag	0.5 x 90	24	1500	155	13,950	2
4	Gas metal pulsarc	1.6 dia. x 20 osc.	28	200	185	3,700	2

3.3.2.2 Electroslag overlay welding

Developments in processes and powders have been such that nickel alloys can be processed by electroslag cladding using strip electrodes or welding strip with, for example, dimensions of 60 x 0.5 mm. Electroslag is a high-output welding process. The operating principle resembles the submerged-arc cladding process; however, the basis of this welding process is not the energy of the arc but the heat development as the electric current passes through the molten slag.

To obtain this, an arc is first struck on an electroslag powder fed to the welding zone in a similar manner to the submerged arc process. If sufficient slag is melted by the arc, the initial arc process changes into an arcless process. Owing to the increasing slag volume, the larger contact area with the welding strip and the increasing slag temperature, the resistance of the slag is reduced to such an extent that it falls significantly below the resistance of the arc and the welding current then only flows through the slag. As a consequence the arc is extinguished. Although the liquid slag has a good electrical conductivity, its residual resistance is sufficient for temperatures as high as 2000 °C to be achieved due to the joulean heat, thus providing the necessary heat for melting the substrate and the welding strip.

The electroslag process has considerable advantages over the submerged arc process. In cladding it is always a question of achieving the absolute minimum iron dilution with the substrate. Depending on the alloy type, specific maximum iron contents have to be maintained in the capping pass. Whereas, as shown in Fig. 3.16, in electroslag cladding the desired weld metal is achieved by the second pass, this cannot be achieved in less than three passes in submerged-arc strip cladding by welding. In practice this means that the electroslag process can give not only a saving in material but also a reduction in welding time of a significant amount (30%). However, the iron dilution achieved in the first pass is dependent on the welding parameters, as is clearly shown in Fig. 3.17 [20].

From the equipment point of view, the transformers suitable for the submerged arc process can be used, providing their power is adequate. The use of electromagnets located left and right of the strip head that balance the magnetic fields produced in the weld pool by the current flow has proved advantageous with regard to achieving favourable wetting conditions and uniformly formed bead flanks.

As can be seen from Table 3.8, higher current densities are used with the electroslag process than those normally associated with the submerged arc process. Table 3.9 shows, as an example, welding data for electro-slag overlay welding with the alloy Nicrofer B 6616 - alloy C-4.

Table 3.9: Overlay welding with alloy C-4 [20]

No.	Process	Electrode (mm)	Powder Type	Voltage (V)	Current (A)	Speed (mm/min)	Deposition Rate (mm^2/min)	Layers required
1	Electroslag	0.5 x 60	B	22	800	150	9,000	3
2	Electroslag	0.5 x 60	B	24	800	130	7,800	2
3	Electroslag	0.5 x 60	C	24	925	170	10,200	2
4	Gas metal pulsarc	1.6 dia. x 20 osc.	-	28	200	160	3,200	3
5	Gas metal pulsarc*	1.6 dia. x 20 osc.	-	24	165	160	3,200	2

* additional feeding with 1.2 mm filler wire in the 1st layer

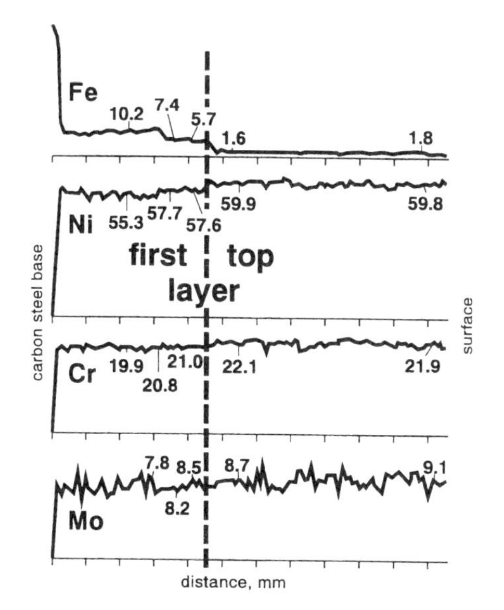

Figure 3.16: Alloy analysis (wt. %) across an overlay weld with alloy 625 [20]

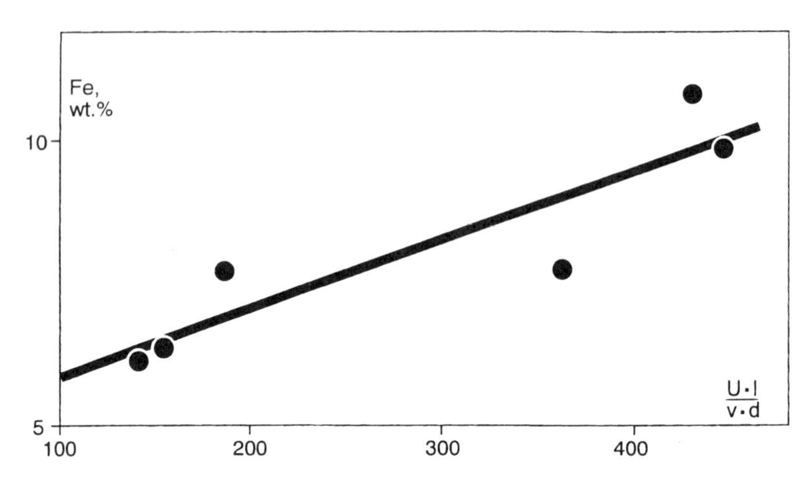

Figure 3.17: Alloy C-4 electroslag overlay welding, - iron content of the 1st layer as function of electrical power and reverse function of welding speed and layer thickness [20]

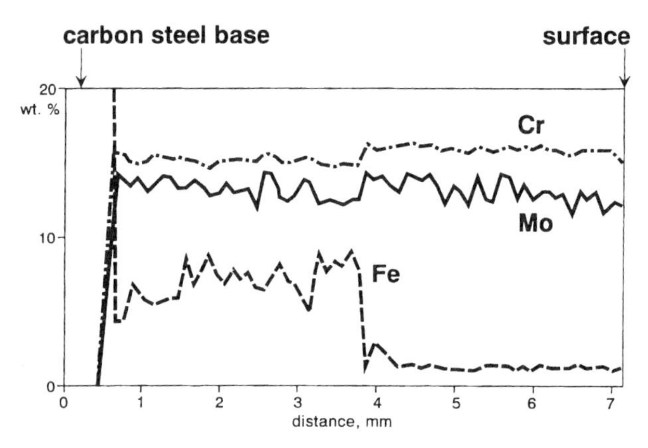

Figure 3.18: Alloy analysis across a gas metal pulsarc overlay weld of alloy C-4 [20]

3.3.2.3 Pulsed-arc GMAW overlay welding

In pulsed arc GMAW overlay welding an iron dilution of 12-20% can be expected in the first pass, whereas the cladding thickness varies in the range 3 to 5mm. In order to reduce the heat input and penetration, pulsed-arc GMAW cladding with a cold wire was developed. Using this process it is possible to achieve cooling of the weld pool through the cold wire and simultaneously a reduction in the penetration. The iron dilution can consequently be reduced to 5-15% in the first pass. It should however be pointed out that, particularly for nickel alloys, suitable parameters must be determined using preliminary tests before commencing work. An important process variation is the pulse technique. Control of the droplet transfer gives a more favourable heat input with a lower dilution with the substrate.

According to research with nickel alloys to date, the iron dilution in the first pass is below 10%, whereas the pass thickness is about 3mm. Considered overall, pulsed-arc GMAW cladding is an interesting alternative to electroslag welding where smaller areas have to be clad by welding. It does not depend on a flux powder. Two passes are sufficient, as Fig. 3.18 shows as an example. The cladding output that can be achieved is however only half and frequently even only a third of that of electroslag cladding. Table 3.9 illustrates this using the results of cladding with the nickel-chromium-molybdenum alloy Nicrofer 6616 hMo - alloy C-4.

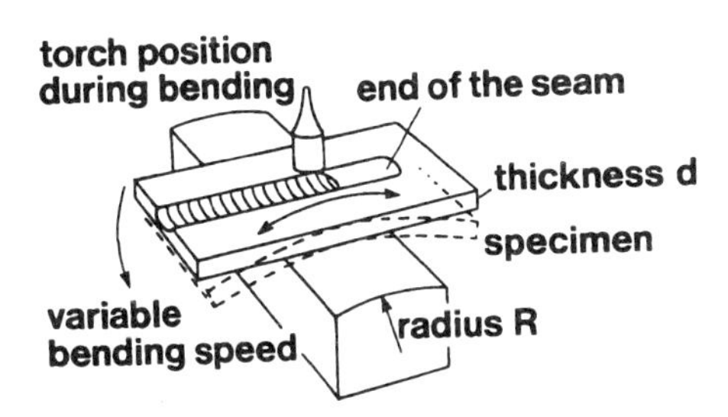

Figure 3.19:
Modified varestraint test for evaluation of hot-cracking behaviour [21]

3.3.2.4 Manual GTAW-HW overlay welding

The process description previously outlined in Section 3.1.3 applies here. This process will primarily remain limited to a special field of application, i.e. small, complicated components, and in this case for high-quality internal and external cladding.

3.4 Weldability testing methods

As a criterion of weldability, testing of the hot cracking behaviour in the modified Varestraint test (MVT) of the Bundesanstalt für Materialforschung und -prüfung (BAM) has proved to be of increasing importance for the materials considered here (6,21). In this a specimen is melted with a GTAW torch under defined conditions over a specific length as shown in Fig. 3.19 and mechanically bent during this process (6). The total length of the cracks visible on the surface at a magnification of 25 is then determined as a function of the applied bending strain and serves as a measure of the sensitivity to hot cracking. The results are shown in Fig. 3.20. It can be seen that in this test the alloy Nicrofer 4221 - alloy 825 exhibits relatively high sensitivity to hot cracking, whereas that of the alloy Nicrofer 6020 hMo - alloy 625 is significantly lower, followed by the alloys Nicrofer 5716 hMoW - alloy C-276, Nicrofer 5923 hMo - alloy 59 and Nicrofer 6616 hMo - alloy C-4. The relatively high position of the alloy Nicrofer 4221 - alloy 825 agrees with practical experience: this material cannot be welded with a matching filler, whilst this is possible with the alloys below alloy 926. As an experienced equipment fabricator would expect, the test result for Nicrofer 6616 hMo - alloy C-4 again indicates the excellent fabricating qualities of this material. The alloy Nicrofer 5923 hMo - alloy 59 is not far away in a similar range but has at the same time significantly superior corrosion resistance [22].

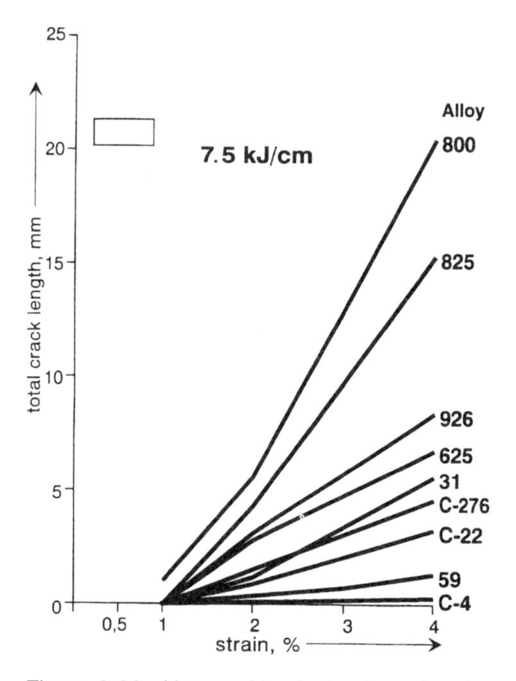

Figure 3.20: Hot cracking behaviour of various nickel alloys and high-alloy special stainless steels as evaluated by means of the MVT-test

3.5 References Chapter 3

[1] J. Ruge: Handbuch der Schweißtechnik, 2. Aufl., Springer-Verlag, Berlin-Heidelberg-New York, Bd. 1 / 2 (1980)

[2] H. Geipl, L. Zell, T. Hoffmann: Neue Entwicklungen beim Metall-Schutzgasschweißen von Nickelwerkstoffen, DVS-Bericht 172 (1996) 78 - 84

[3] M. B. Rockel, W. Herda: Die Oberflächenbehandlung von Schweißverbindungen hochkorrosionsbeständiger 6 % Mo-Stähle und Nickel-Basis-Legierungen, Werkstoffe und Korrosion 43 (1992) 354 - 357

[4] R. Kirchheiner, F. Hofmann, T. Hoffmann: Corrosion Properties of Welded Stainless Steels and Nickel Alloys in a FGD Absorber Tower, CORROSION 85, Paper No. 51, The National Association of Corrosion Engineers, Houston, Texas, 1985

[5] M. Renner, G. Rudolph: Einfluß der Streckenenergie auf das Korrosionsverhalten von lichtbogenhandgeschweißtem Nicrofer 3127 LC / Alloy 28, METALL 40 (1986) 1012 - 1018

[6] T. Hoffmann, M. B. Rockel, K. Wilken: Schweißeignung hochlegierter Stähle in der chemischen Industrie und im Apparatebau, DVS-Bericht 105 (1986) 112 - 118

[7] W. Hegner, T. Hoffmann, F. Hofmann: Unterpulver-Engspaltschweißen des Hochtemperaturwerkstoffes NiCr23Co12Mo, DVS-Bericht105 (1986)

[8] T. Hoffmann, M. Renner, G. Rudolph: Gefüge und Eigenschaften von Schweißgut aus hochlegierten korrosionsbeständigen Nickelwerkstoffen, Schweißen & Schneiden 38 (1986) 551 - 557

[9] R. Kirchheiner, F. Hofmann, T. Hoffmann: A Silicon Alloyed Stainless Steel for Highly Oxidizing Conditions, CORROSION 86, Paper No. 120, The National Association of Corrosion Engineers, Houston, Texas, 1986; Materials Performance 26 (1987) No. 1, 49 - 56

[10] G. Rudolph, T. Hoffmann: Verbesserung der Schweißtechnik und Schweißeignung von Nickellegierungen, METALL 41 (1987) 988 - 996, 1210 - 1220

[11] M. B. Rockel, T. Hoffmann: Korrosionsbeständigkeit hochlegierter Nickelwerkstoffe und Sonderedelstähle im geschweißten Zustand, DVS-Bericht 172 (1996) 16 - 26

[12] R. Kirchheiner, T. Hoffmann, F. Hofmann: Using Explosion Clad Plates of Alloy 625 in Flue Gas Desulfurization Plants, CORROSION 87, Paper No. 257, The National Association of Corrosion Engineers, Houston, Texas, 1987

[13] M. Richter, R. Kirchheiner, K. Waggershauser, H. Lorenz: Anwendung neuer Nickel-Basiswerkstoffe - Erfahrungen bei der schweißtechnischen Verarbeitung in Rauchgasentschwefelungsanlagen, - Kraftwerk Boxberg, DVS-Bericht 164 (1995), 48 - 64

[14] R. Kirchheiner, W. Herda, N. Schupp: Reliable fabrication with Alloy 59 hot roll-clad plates in a major FGD-project, Paper No 453, CORROSION 96, NACE International, Houston, Texas, 1996

[15] R. Pleschko, R. Schimböck, E. M. Horn, P. Mattern, M. Renner, W. Heimann: Walzplattieren mit dem hochkorrosionsbeständigen nichtrostenden austenitischen Stahl X1NiCrMoCuN25206, W.-No. 1.4529, Werkstoffe und Korrosion 41 (1990) 563-570

[16] T. Hoffmann, G. Rudolph: Welding of Cupro-Nickel and Cupro-Nickel Clad Steel, ATB Metallurgie XXV (1985) No. 1, S. 69 - 79

[17] M. B. Rockel, G. Rudolph: Korrosionsverhalten von Schweißverbindungen des Verbundwerkstoffs CuNiFe/Stahl, Werkstoffe und Korrosion 36 (1985) 348 - 357 (Teil 1: Kurzzeit- und Meerwasserloop-Versuche), 358 - 367 (Teil 2: Korrosionsverhalten im Meerwasserprüfstand auf Helgoland)

[18] T. Hoffmann, D. C. Agarwal: "Wallpaper" installation guidelines and other fabrication procedures for FGD maintenance, repair and new construction with VDM high performance nickel alloys, VDM report No. 17 (1991)
[19] G. Carniato, G. K. Grossmann: New approach to welding fabrication of corrosion-resistant Ni-Cr-Mo alloy sheets for lining installation in flue gas cleaning equipment, as exemplified in an Italian power station, Proc. Power-Gen 93, Singapore 1993
[20] U. Heubner, T. Hoffmann: Overlay Welding of Corrosion Resistant Nickel Superalloys, Weldability of Materials, ASM, Materials Park, Ohio, 1990, S. 175 - 182
[21] U. Brill, T. Hoffmann, K. Wilken: Solidification Cracking: Super Stainless Steels and Nickel-Base Alloys, Weldability of Materials, ASM International, Materials Park, Ohio, 1990, S. 99 - 105
[22] R. Kirchheiner, M. Köhler, U. Heubner: Nicrofer 5923 hMo, ein neuer hochkorrosionsbeständiger Werkstoff für die Chemische Industrie, die Umwelttechnik und verwandte Anwendungen, Werkstoffe und Korrosion 43 (1992) 388 - 395

4 High-temperature materials

U. Brill and J. Klöwer

4.1 Introduction

The use of high-temperature materials is especially important in power station construction, heating systems engineering, furnace industry, chemical and petrochemical industry, waste incineration plants, coal gasification plants and for flying gas turbines in civil and military air crafts and helicopters. Particularly in recent years, the development of new processes and the drive to improve the economics of existing processes have increased the requirements significantly so that it is necessary to change from well proven materials to new alloys. Hitherto, heat resistant ferritic steels sufficied in conventional power station constructions for temperatures up to 550°C, newly developed ferritic/martensitic steels provide sufficient strength up to about 600 - 620°C. In new processes, e. g. fluidised-bed combustion of coal, process temperatures up to 900°C occur. However, this is not the upper limit, since in combustion engines, e. g. gas turbines, material temperatures up to 1100°C are reached locally. Similar development trends can also be identified in the petrochemical industry and in heat treatment and furnace engineering. The advance to ever higher material temperatures now not only has the consequence of having to use materials with enhanced high-temperature strength properties; considerable attention now also has to be given to their chemical stability in corrosive media. If temperatures exceed 550°C, the mechanical properties of the heat-resistant ferritic steels are drastically reduced, especially the creep strength. The low creep strength of the ferritic materials at high temperatures is largely attributable to the relatively high mobility of the atoms in the body centred cubic lattice. As Fig. 4.1 clearly shows that using materials with a face centred cubic lattice provides a significantly higher creep strength [1]. The 3 materials on the left in Fig. 4.1 are body centred cubic, while those on the right are face centred cubic. The greater high-temperature strength is largely attributable to the more densely packed face centred cubic lattice with the atoms having lower mobility. Classic examples of this are the nickel-alloyed austenitic stainless steels with approximately 16 - 25 wt% chromium and 8 - 20 wt% nickel. A further increase in the high-temperature strength properties of these austenitic steels can be obtained by solid solution hardening and certain other mechanisms. Elements other than the alloying components iron, chromium and nickel are necessary for this purpose, for instance Mo, W, Co, Ti, Al, Nb. The austenitic stainless steels and also the iron-based alloys have the disadvantage of a very limited solubility for these elements. This can be compensated by increasing the

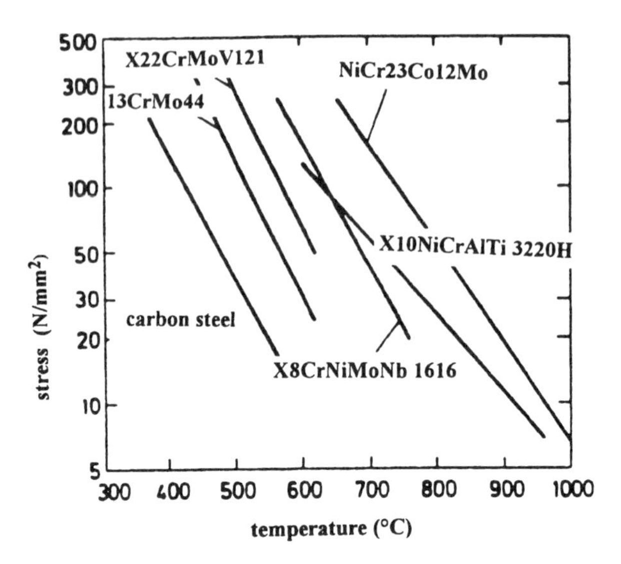

Figure 4.1: 100,000h-creep strength of some steels and Ni-base alloys

nickel content and hence producing extremely highly alloyed materials containing practically no iron. High-temperature materials are thus frequently based on the element nickel, since this produces a face centred cubic lattice and a high solubility for those alloying elements that promote high-temperature strength properties and corrosion resistance.

4.2 High-temperature, high-strength materials

4.2.1 Methods of increasing high-temperature mechanical properties

A total of six main parameters can be used to increase the high-temperature mechanical properties of a face centred cubic austenitic matrix: solid solution hardening, precipitation hardening, carbide and nitride strengthening, microstructural design and consideration of trace elements. Only precise harmonisation of these parameters will produce the material required for a particular application. In some instances strength characteristics are more in demand, whilst in others safety considerations may place the emphasis on toughness. Other applications may place greater demands on the materials workability or weldability and, last but not least, great attention is often paid to corrosion resistance. In the course of time, many material variants have attempted to meet all these individual requirements to a greater or lesser degree. However, in nearly all cases it is necessary to accept a compromise solution.

4.2.2 Solid solution hardening

The term solid solution hardening signifies the attainment of an increase in matrix strength by the addition of an element which is soluble in the matrix, but exhibits a larger atomic radius. This is achieved mainly by distortion of the atomic lattice, inhibiting dislocation movement. However, the movement of dislocations is necessary for deformation processes, which include creep.

Solid solution hardening also has the effect of decreasing the stacking fault energy in the crystal lattice, leading primarily to inhibition of dislocation cross slip, which is the main promoter of deformation in imperfect crystals at elevated temperatures. At temperatures > 0.6 T_m, (typically above approx. 820°C for these alloys), high-temperature mechanical properties, especially time-dependent creep, are defined mainly by diffusion. Molybdenum and tungsten are particularly suitable for solid solution hardening. However, the element molybdenum is preferred due to its considerably lower atomic weight.

4.2.3 Precipitation hardening

A considerable increase in the creep strength of alloys for high-temperature applications can be obtained by precipitation hardening. In the case of nickel-based and iron-based alloys, this can be achieved using the elements Ti, Al and Nb. As already shown by way of example in Section 1.3 and Fig. 1.11, these elements are only soluble to a limited extent in the systems, with the result that, by means of suitable heat treatment, finely distributed precipitates can be generated in the matrix from a supersaturated solid solution. The precipitates resulting from such temperature-dependent decrease in solubility, generally Ni_3 (Ti,Al) γ'-phase and Ni_3Nb γ''- -phase, inhibit the movement of dislocations. Movement of a dislocation in the matrix can then only take place by cutting through or by bypassing the precipitated particles. As Fig. 4.2 shows, the best lattice state, that is to say the greatest interference with movement of dislocations in a matrix with finely distributed precipitates, is achieved when such precipitates have a diameter of the order of 20 - 50 nm [2]. Where the particles are too small, cutting through the particles is energetically more favourable for dislocations. In the case of coarser particles at a lower degree of dispersion, i. e. greater particle spacing, bypassing the particles involves the least energy expenditure. Both cases cause a drop in the mechanical properties of the material.

The use of precipitation hardening in high-temperature alloys is limited from the point of view of operating temperatures, since high-temperature mechanical properties are impaired by overaging and particle dissolution. This can be counteracted by using cobalt as an alloying element. As shown in Fig. 4.3 this raises the solubility limits for Ti and Al, and thus the upper limit for the service temperature [3]. A further restriction on applications for precipitation hardened high-temperature materials is their low ductility and limited weldability.

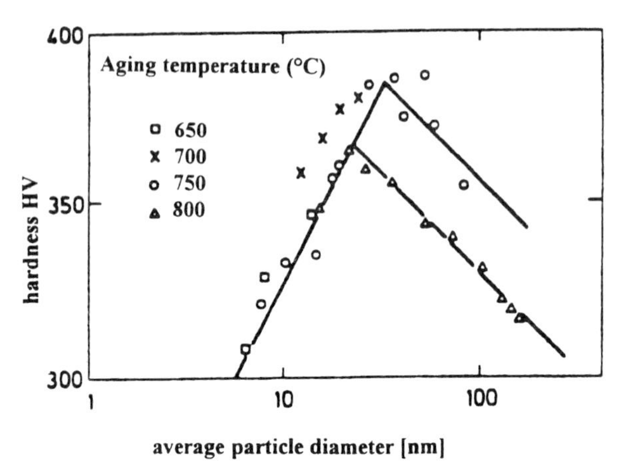

Figure 4.2: Hardness of a Ni-base alloy comprising 22 wt% Cr, 2.8 wt% Ti and 3.1 wt% Al in dependence on the diameter of precipitated particles, after Mitchell [2]

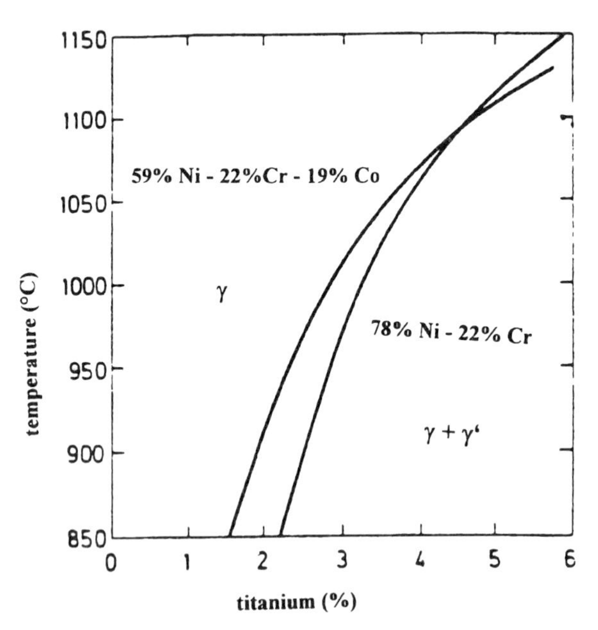

Figure 4.3: Solubility of titanium in a Ni-base alloy with 22 wt% Cr and dependence of solubility on addition of 19 wt% Co. The Ti/ Al ratio was kept constant with 1:1, after Heslop [3]

4.2.4 Carbide hardening

Increasing the high-temperature strength by carbide hardening is of great significance, particularly for high-temperature austenitic stainless steels. This is because solid solution hardening of these alloys, using elements such as Mo and W, is only possible to a limited extent, due to their limited solubility in the iron matrix. In these alloys, the precipitation of titanium, niobium and vanadium carbonitrides is important. In addition, strength-increasing stacking faults are produced, though these occur at the expense of ductility. An increase in the nickel content of austenitic alloys decreases the carbon solubility, which, for a given carbon content, leads to increased carbide precipitation. This occurs mainly at the grain boundaries for energy reasons.

Correctly applied, this phenomenon can be particularly useful for increasing the creep strength of these alloys, since under loads experienced in service, creep processes at high temperatures mainly take place at the grain boundaries. As shown for example in Fig. 4.4 the creep strength of the iron-nickel-chromium-based high-temperature alloy 800 H is a function of its carbon content, and thus of the amount of carbides which, precipitated preferentially at the grain boundaries, inhibiting grain boundary slip. Creep processes are therefore transferred from the grain boundary to the interior of the grain. Small, globular, noncohesive carbides are best suited for stabilising the grain boundaries. These are mainly primary precipitates of MC and M_6C carbides based on Mo, W, Ti and Nb. Since most high-temperature alloys have high chromium contents for corrosion related

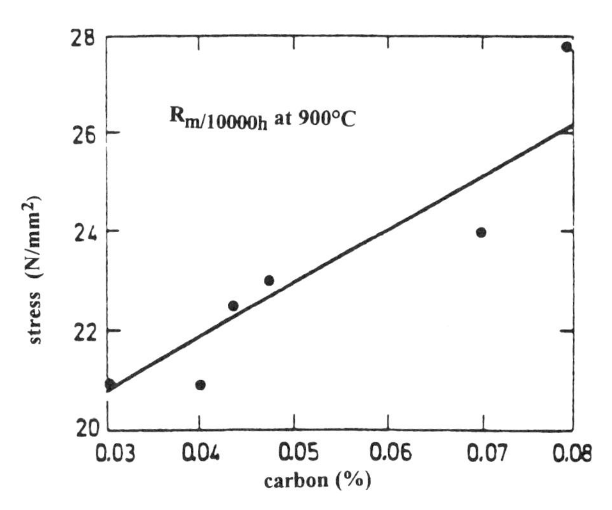

Figure 4.4: 10,000 h creep strength of Nicrofer 3220 H-alloy 800 H (1.4958) as a function of the carbon content of the alloy

reasons (approx. 16 - 30 wt%), the formation of $M_{23}C_6$ carbides is unavoidable. With prolonged exposure to high-temperatures the MC and M_6C carbides also tend to transform into $M_{23}C_6$ according to equations 1 and 2:

$$MC + \gamma \rightarrow M_{23}C_6 + \gamma' \quad \text{(eq. 1)}$$

$$M_6C + M' \rightarrow M_{23}C_6 + M'' \quad \text{(eq. 2)}$$

As can be seen in Fig. 4.5, this type of carbide tends towards the formation of cohesive grain boundary precipitates. Since carbides tend to be brittle because of their structure, these cohesive grain boundary precipitates represent a preferred path for crack growth and this results in a decrease in creep resistance.

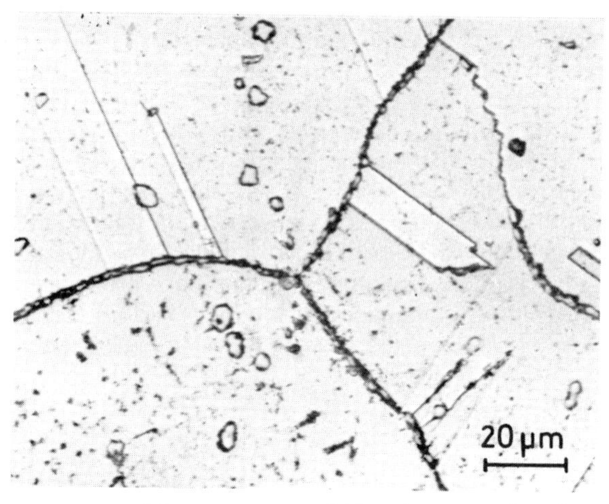

Figure 4.5: Microstructure of Nicrofer 5520 Co-alloy 617 (2.4663) creep specimen after 7,209 h at 850°C and δ = 54 N/mm² [1]

In contrast, with noncohesive globular precipitated carbides stresses can be reduced by slight grain boundary slip and thus the formation of microcracks can be prevented or hindered. As shown clearly in Fig. 4.6, at high temperatures $M_{23}C_6$ carbides also tend, on the other hand, to agglomeration, i. e. to the formation of fewer larger carbides on grain boundaries. These represent initiation points for crack formation. In addition the slip of grain boundaries is favoured. Both produce a decrease in creep strength.

In solid solution hardened high-temperature alloys, the carbide forming elements Mo and W also tend to precipitate as carbides at the grain boundaries after long-term exposure to high temperatures. This causes the grain boundary zones to become depleted in these elements. This has the disadvantage that solid solution hardening decreases in this area, but also the advantage that stresses oc-

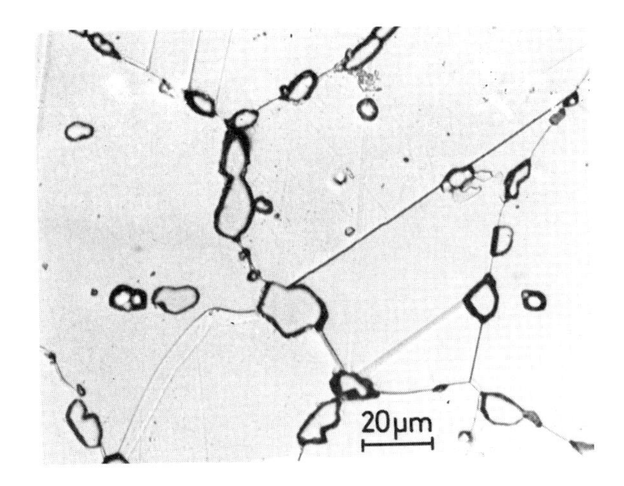

Figure 4.6: Microstructure of a Nicrofer 5520 Co-alloy 617 (2.4663) creep specimen after 10,000 h at 1000°C and δ = 6 N/mm² [1]

curring are dissipated via these "softer" zones and not via the carbide-decorated grain boundaries. The result is a beneficial influence on creep rupture strength [3]. The increasing demand for alloys suitable for temperatures up to 1200°C was the reason for working on other strengthening mechanisms rather than solid solution strengthening of Ni-base alloys.

Because of the high mobility of the atoms at high temperatures solid solution strengthening becomes ineffective, and led to the idea of applying the mechanism of carbide strengthening also to Ni-base alloys.

The idea to use the solubility drop of carbon from the liquid to the solid state in Ni-base alloys for carbide strengthening of the alloy, was first practised in 1992 with the commercial alloy Nicrofer 6025 HT -alloy 602 CA (2.4633). The analysis of this alloy is given in Tab. 1.7. The high chromium and high carbon content of this alloy resulted in the precipitation of primary chromium carbides of the type $M_{23}C_6$. Fig.4.7 shows the typical microstructure of the alloy in the 'solution annealed condition'. These primary precipitated chromium carbides are not soluble up to temperatures of 1250°C thus resulting in excellent creep strength in the temperature regime of 1000 - 1200°C.

Fig. 4.8 shows the superior creep strength of Nicrofer 6025 HT -alloy 602 CA-(2.4633) in comparison to other high-temperature superalloys.

Since 1992 Nicrofer 6025 HT-alloy 602 CA (2.4633) has found wide interest in furnace and heat treatment industry as well as in chemical, petrochemical and metallurgical process industry.

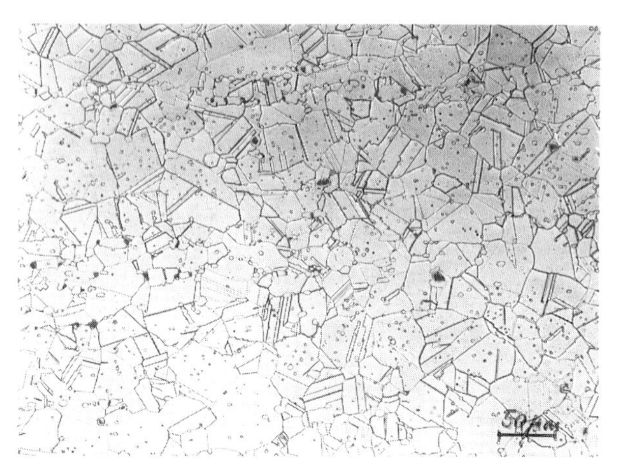

Figure 4.7: Typical microstructure of Nicrofer 6025 HT - alloy 602 CA (2.4633) in the solution annealed condition

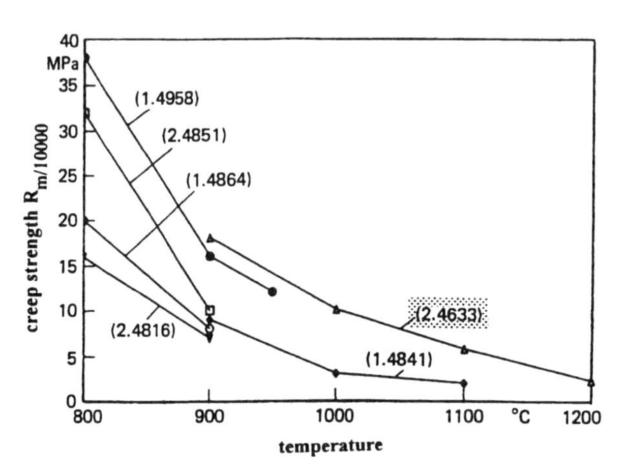

Figure 4.8: Comparison of the 10,000h creep strength of some high-temperature alloys

A further increase in carbon up to 0.2 - 0.4 wt% and balanced amounts of chromium, titanium, zirconium and boron led to an alloy with about 5 vol.-% of primary precipitated chromium carbides of the type M_7C_3 in the matrix.

This alloy Nicrofer 6125 GT -alloy 603 GT (2.4647) was first commercially available in 1996. The composition of the alloy is given in Tab. 1.7. Because of the further increased creep strength and excellent oxidation resistance due to an aluminium content up to 3.0 wt% this alloy is a candidate for stationary gas turbine applications like combustion chambers, honey combs, heat shields etc.

The 10.000 h - creep strength of Nicrofer 6125 GT -alloy 603 GT (2.4647) in the temperature range of 900 - 1200°C is given in comparison to the alloys Nicrofer 4626 MoW -alloy 333 (2.4808), Nicrofer 4722 Co -alloy X (2.4665) and Nicrofer 5520 Co -alloy 617 (2.4663) in Fig. 4.9, [4 - 6].

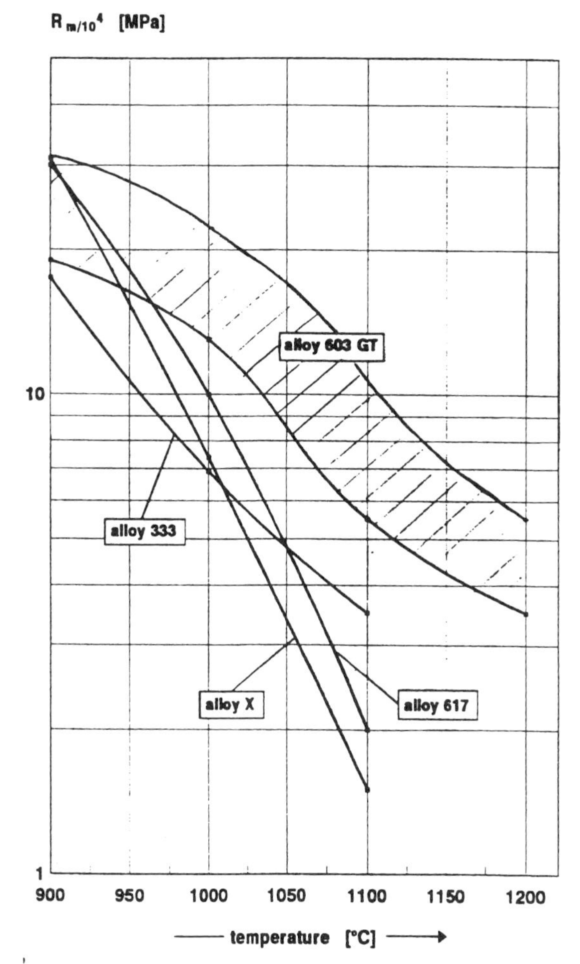

Figure 4.9: Comparison of the creep rupture strength $R_{m/10^4}$ of some high temperature alloys

4.2.5 Nitridation strengthening

As described above, carbide strengthening or particle strengthening is very effective at high temperatures. To even increase the effectiveness of these particles the following requirements have to be met by new particle strengthened superalloys:

- high thermal stability of the precipitated particles at least up to 1000°C
- high hardening rate due to a high solubility at melting temperature
- no significant decrease in incipient melting temperature
- precipitation kinetics and morphology must be adjustable
- precipitation strengthening should not result in a decrease of oxidation or corrosion resistance due to an excessive reaction with chromium
- low particle density for reduced component weight
- small particle size for homogeneous and effective particle distributions

According to Krupp VDM inventions these demands can be fulfilled by using chromium nitrides for strengthening a Ni-base matrix.

The high nitrogen content necessary for providing a sufficient high volume of precipitated nitrides can be achieved by the DESU-technology (pressurised electroslag remelting).
A typical analysis of alloy Cronix 70 NS is given in Tab. 1.7.

This alloy is age-hardenable in the temperature range of about 750 - 1150°C due to the precipitation of π-phase, which is a $Cr_{13}Ni_7N_4$-nitride. Because of the stability of the precipitates up to appr. 1150°C the precipitation of π-phase is even more useful than γ'-, γ''- or δ-phase precipitations used today for example in Nicrofer 5219 Nb -alloy 718 (2.4668) (Fig. 4.10). The stability of π-phase provides excellent creep and fatigue strength in the temperature range of 750 - 1000°C, which makes the alloy a favourite candidate alloy for the combustion chamber of flying gas turbines.

4.2.6 Microstructural design (grain structure)

It is well known that a fine-grained structure produces an improvement in the mechanical properties of metals. However, this is not true for the high-temperature strength properties under consideration here, which are essentially determined by the creep strength of these alloys. The latter is also affected by deformation processes at the grain boundaries, i. e. by their slip. Coarse grains present considerably less potential for slip processes than fine grains. This can be seen in Fig. 4.11, which shows the relationship between the creep rupture strength of Nicrofer 5520 Co -alloy 617 (2.4663) and the grain size. A substantial increase in creep rupture strength, particularly at high temperatures, can be attained by controlled increases in grain size. Duplex structures, i. e. the simultaneous pres-

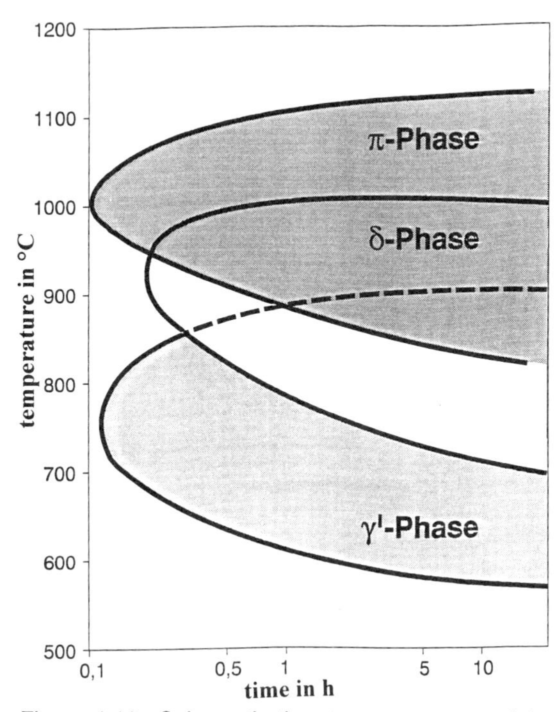

Figure 4.10: Schematic time-temperature precipitation diagram of the precipitation of the π-phase in Cronix 70 NS and γ' and δ -phase in Nicrofer 5219 Nb - alloy 718 (2.4668)

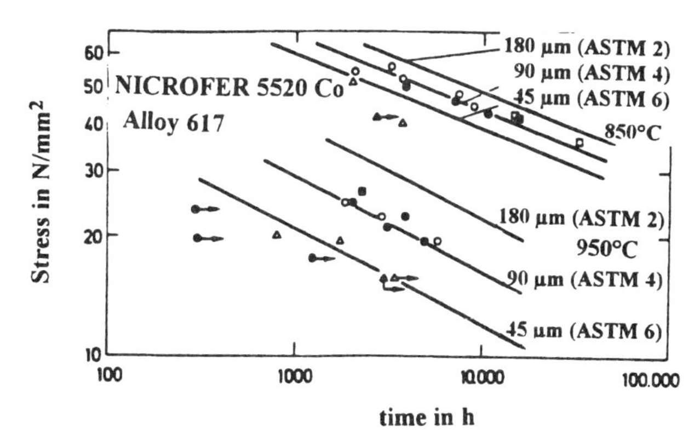

Figure 4.11: Creep rupture strength of Nicrofer 5520 Co -alloy 617 (2.4663) at 850°C for different grain sizes

ence of coarse and fine grains, should be avoided, since zones of different creep behaviour would then be present, leading to an overall decrease in creep resistance.

Regarding the grain size, carbide precipitation and distribution should always be considered as a variable affecting creep strength. With a fine-grained alloy, the density of carbide precipitation is low, considerably facilitating slippage processes at the grain boundaries. In contrast, with coarse grain size, there is the danger of cohesive, decoration of the grain boundaries with carbides and the corresponding disadvantage of rapid crack propagation. It also leads to loss of ductility and low temperature embrittlement.

Practical experience indicates that selection of an average grain size of between approximately 100 and 200 µm provides the best conditions for creep resistance in high temperature materials.

4.2.7 Trace elements

The trace elements under consideration occur either as unavoidable impurities (see Section 1.5.2) or as deliberate alloying additions and have in general only a limited solubility in the matrix. They therefore tend to segregate preferentially at the grain boundaries and affect their stability under creep loading. This can have both positive and negative consequence for the creep strength.

Little is known regarding the actions of the trace elements Pb, Bi, Sb, Sn, As, Te, Ga, S, P, Ag and their negative influence on creep strength. It can be assumed, however, that the decohesion process is simplified by the build-up of these elements in the distorted grain boundary lattice regions. At least a significant adverse influence of these elements on the creep rupture strength determined by creep processes has been empirically established (see Fig. 1.18). The precise maximum permissible contents for these elements are however, not known and can also vary from alloy to alloy. For instance, in the case of the widely used alloy Nicrofer 3220H - alloy 800H (1.4958) upper limits for the elements Pb and Bi of 7 and 1 g/t have proven favourable for creep rupture strength. However, it is a question of economics whether it is worth accepting the considerably higher material costs associated with a further reduction of these unwanted impurities for only a slight increase in the creep rupture strength. It should also not be forgotten that, in statistical terms, slight improvements are in practice submerged in a scatter band [7].

In addition to the elements recognised as detrimental, there are a number of helpful trace elements. These are the elements Zr, B and Mg [3]. In a similar manner to the impurities discussed earlier, these elements segregate at the grain boundaries. At these boundaries, they either fix detrimental impurities, such as sulphur, or are integrated into the distorted lattice. The latter phenomenon produces a strengthening of the grain boundaries, since the diffusion processes responsible for grain boundary slippage are inhibited or retarded. Furthermore, boron and zirconium inhibit the agglomeration of $M_{23}C_6$ carbides at the grain boundaries, and thus reduce the risk of microcrack formation. Boron also reduces the solubility of carbon at the grain boundaries. This means that the amount of carbide at the grain boundaries increases, but the formation both of coarse carbides and of continuous carbide layers is impeded. Both would be detrimental to creep rupture strength. Furthermore, zirconium inhibits grain growth at elevated temperatures. However, these elements should be used with great caution, since any excess can be extremely harmful. For instance, boron contents of up to 50 ppm have proved to be extremely favourable, whilst even a slight amount in excess of this limit can procure the reverse effect. There is then, for example, a danger of formation of low melting point borides, leading to corresponding embrittlement of the grain boundaries.

4.2.8 Examples of materials

The group of high - temperature, high-strength alloys, which are listed individually in Table 1.7, can be divided into predominantly solid solution hardening materials and precipitation hardening nickel-base alloys, as discussed in Section 1.2.7. Carbide hardening is also important for the group of predominantly solid solution hardening materials. Table 4.1 compares the creep rupture strengths of some typical materials from both groups [8-10] . It can be seen that the high creep rupture strength of the precipitation hardening materials is limited to the medium temperature range up to approximately 700 - 800°C. If this temperature limit is exceeded, these materials rapidly lose their outstanding mechanical properties by overageing and dissolution of the strength increasing precipitates (see Section 1.4 and Fig. 1.17). It is clear from Table 4.1, however, that the creep rupture strengths of the solid solution/carbide strengthened high-temperature, high-strength alloys connect smoothly with the working temperature range of the precipitation hardened materials at higher temperatures.

When selecting a suitable material for a practical application, however, not only creep rupture strength but (besides weldability) also corrosion resistance play an important role. Corrosion damage has only a relatively limited effect on the creep rupture strengths, which are usually determined by means of large diameter creep specimens. They cannot therefore be applied without further consideration to thin walled components. If components of small cross sectional area are subjected to high temperatures, the influence of corrosion damage should never be ignored. On the other hand, the influence of the surrounding atmosphere can also have a positive effect on the creep rupture strength. The creep rupture strength of the material Nicrofer 3220 H - alloy 800 H (1.4958) is significantly increased in a carburising atmosphere without the creep ductility falling to values lower than those applying to creep conditions in air [11].

Table 4.1: Comparison of the creep rupture strength of solid solution/ carbide strengthened and precipitation strengthened alloys

	Solid solution and carbide strengthened alloys			
	Rm/ 10,000 in MPa			
Alloy	**800°C**	**900°C**	**1000°C**	**1100°C**
Nicrofer 6025 HT -alloy 602CA (2.4633)	31	17	10	5.8
Nicrofer 3220 H -alloy 800H (1.4958)	37	17	-	-
Nicrofer 4722 Co-alloy X (2.4665)	59	31	7	-
Nicrofer 5520 Co-alloy 617 (2.4663)	65	30	10	4.0
	Precipitation - strengthened alloys			
	500°C	**600°C**	**700°C**	**800°C**
Nicrofer 5219 Nb -alloy 718 (2.4668)	700	560	240	80
Nicrofer 7520 Ti -alloy 80A (2.4952)	-	440	200	60
Nicrofer 5120 CoTi -alloy C 263-(2.4650)	650	500	330	150

When the number of parameters that can influence the creep strength (see Sections 4.2.1 to 4.2.7) are considered, it is not surprising that there is quite a wide variation in the data determined and published for the various materials [7]. In quality control of the materials for plant subject to supervision, for example pressure vessel construction, the available creep data must be carefully analysed and evaluated. In addition numerous safety factors have been introduced [12]. Interpolation and extrapolation of experimentally obtained creep strengths is usually possible with sufficient accuracy for the materials considered here by using the Larson-Miller parameter [13].

With the materials in common use, the severe reduction in the oxidation resistance at very high temperatures does not, as a rule, permit any creep application above 1000°C for the alloys Nicrofer 3220 H -alloy 800 H (1.4958) and Nicrofer 3718 So -alloy DS (1.4862), above 1150°C for Nicrofer 6023 H -alloy 601 H (2.4851) and above 1200°C for Nicrofer 6025 HT -alloy 602 CA (2.4633) (Fig. 4.12).

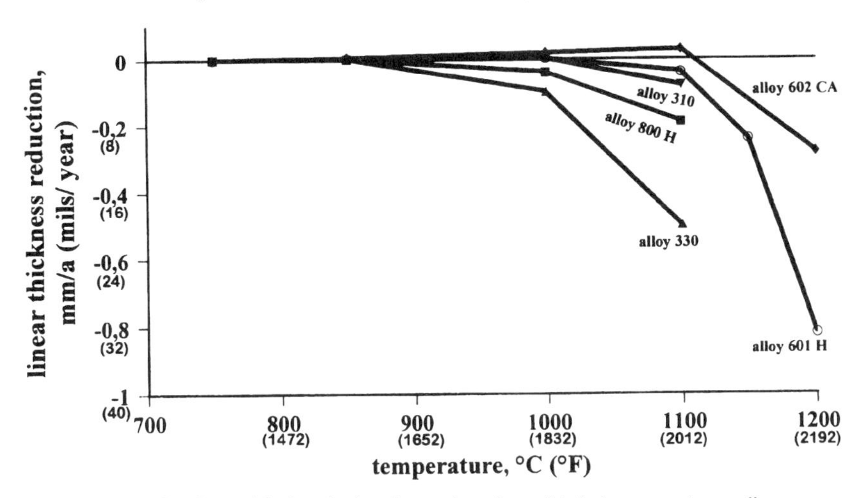

Figure 4.12: Cyclic oxidation behaviour of various high temperature alloys

4.3 Heat resistant materials

4.3.1 Introduction

Metal oxides, sulphides, carbides and other metallic compounds are thermodynamically more stable than the corresponding metals. That means that they will form at high temperatures under any atmosphere other than vacuum, either as outer layers on the metal surface or as internal corrosion products. Like aqueous corrosion, high temperature corrosion can lead to metal loss, to material wastage by pitting or crevice corrosion and to the formation of cracks along attacked grain boundaries. Internal corrosion (e.g. internal carbides, internal nitrides) may furthermore reduce the ductility and/or the mechanical strength of the material. Under certain conditions even liquid phases may be formed or a change of the microstructure may take place.

The reason that some metals can be used at higher temperatures despite the process of high temperature corrosion is that they form dense, slow growing oxide scales which cannot be penetrated and which therefore protect the subjacent metal against further attack from the gas phase.

To be protective, the following properties of an oxide scale are essential:

- They have to be dense, e.g. without cracks or other faults which would facilitate the penetration into the metal and enable internal corrosion.
- They must be slow-growing even at elevated temperatures.
- They must not evaporate.
- The should form even under conditions with very low oxygen partial pressure.
- There should be no competitive reaction to the formation of the protective oxide scale (at certain gas conditions the formation of nonprotective nitrides, chlorides or sulphides becomes more stable than the formation of protective oxides).
- They must have a certain ductility to prevent spalling.
- The should be resistant to dissolution by deposits of metals or salts.

The scales of chromium oxide meet these requirement under many process conditions. Consequently most high temperature alloys contain high amounts of chromium to enable the formation of dense protective chromia scales. Some typical heat resistant materials are listed in Table 1.6 in chapter 1. It can be seen that most of the common materials have chromium concentrations between 16 and 30 wt%. Chromia-forming Fe-Ni-Cr alloys, Ni-Fe-Cr and Ni-Cr alloys can be used at temperatures up to 1000 °C in a variety of corrosive gases.

There are, however, limitations of chromia formers. At temperatures above 1000 °C or under certain gas conditions, scales of aluminium oxide (alumina) are more protective than chromia scales. Alumina, especially α-Al_2O_3, shows low growth rates even at temperatures above 1000 °C [14]. Unlike chromia, alumina does not evaporate at temperatures above 1000°C and it is even stable in oxygen-deficient atmospheres.
Table 4.2 shows the equilibrium oxygen partial pressure of several metal oxides, i.e. the minimum oxygen partial pressure which is required for oxide formation. A comparison with Table 4.3, in which typical oxygen partial pressures of industrial processes are listed, clearly demonstrates, that in coal gasification processes and petrochemical plants stable oxide scales can only be expected in alumina formers. Chromia becomes unstable at the low oxygen partial pressures encountered in such processes. Consequently materials for use at very high temperatures or under oxygen deficient atmospheres must contain aluminium. Examples are Nicrofer 6023 H -alloy 601 (2.4851) (1.4 wt%Al), Nicrofer 6025 HT - alloy 602 CA (2.4633) (2.3 wt% Al) or the new Nicrofer 6009 -alloy 10Al (10 wt% Al).

If aqueous corrosion, corrosion by halogen-containing hot gases, molten salt attack or chloride-containing deposits are expected, even corrosion-resistant nickel-chromium-molybdenum alloys may be considered for application. When using these alloys, the upper temperature limits for pressure vessels according to ASME and VdTÜV must be noticed.

Table 4.2: Equilibrium partial pressures of several metal oxides. At oxygen partial pressures below those given in the table, the respective oxides are thermodynamically unstable (calculated with thermodynamical data after Barin [15])

Oxide	PO_2 in bar
2/3 Al_2O_3	1.6×10^{-34}
2/3 Cr_2O_3	3.4×10^{-22}
2 Ni O	9×10^{-11}
1/2 Fe_3O_4	4.9×10^{-15}
2/3 Fe_2O_3	9.5×10^{-14}

Table 4.3: Typical oxygen partial pressures of high temperature processes

Process	Typical oxygen partial pressure in bar	Ref.
Petrochemical plants	$10^{-25} - 10^{-30}$	[16]
Gas cooled reactors	$10^{-25} - 10^{-30}$	[16]
Fluidised bed combustion	$10^{0} - 10^{-5}$	[16]
Coal gasification	$\approx 10^{-25}$	[17]

4.3.2 Oxidation in air

All materials suffer oxidation during exposure in air with the corrosion rate depending on the protectiveness and the adherence of the chromia or alumina scale respectively. The oxidation behaviour of high temperature materials has been tested for a variety of materials under a variety of conditions. It is generally accepted that the oxidation resistance of materials increases with increasing chromium content [e.g. 1,18]. At the same time the oxidation resistance is also increased with increasing nickel content at least up to a concentration of 30 wt% nickel [19]. Conventional stainless steels like alloy 316 with 18 wt% chromium and 12 wt% nickel can only be used at temperatures up to about 850°C, since severe metal loss due to spalling of oxide scales takes place above that temperature. Fig 4.13 shows the mass changes measured in cyclic laboratory oxi-

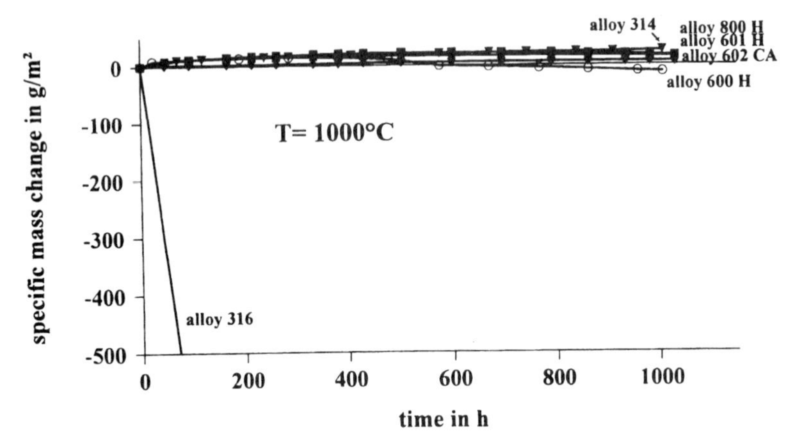

Figure 4.13: Specific mass change of heat-resistant materials and stainless steels during cyclic exposure in dry air (24-hours cycle) as a function of exposure time at 1000 °C. Mass losses indicate spalling or evaporation of metal, mass gains indicate the formation of internal or outer oxide scales (-100 g/m^2 mass loss and 50 g/m^2 mass gain are approximately corresponding to a metal loss of about 0.1 mm/year)

dation tests for several alloys at 1000 °C. Mass losses indicate spalling or evaporation of corrosion products, while mass gains indicate the formation of outer or internal corrosion scales. It can be seen that alloy 316 suffers significant metal loss already after very short exposure times while the heat resistant stainless steel alloy 314 (25 wt% Cr), Fe-Ni-Cr-alloys like Nicrofer 3220 H -alloy 800H (1.4958) and nickel-base alloys show only very low corrosion rates after 1000 hours of exposure. Generally all Fe-Ni-Cr-alloys, Ni-Cr -alloys and Ni-Fe-Cr-Mo alloys with more than 15 wt% chromium can be used at temperatures up to 1000°C. This does also apply to the high temperature/high strength - nickel-base alloys described in chapter 4.2 with the exception of niobium-containing alloys like Nicrofer 3228 NbCe -AC66 (1.4877) and Nicrofer 6020 hMo -alloy 625 (2.4856), which should only be considered for temperatures below 1000 °C.

If the temperature is increased to more than 1000 °C, only heat resistant materials with chromium concentrations of at least 20 wt% can be used. The time dependence of the oxidation behaviour at 1100 °C is shown in Fig. 4.14. A mass loss of less than 100 g/m^2, corresponding to a metal loss of less than 0.1 mm/year was calculated for the alumina formers Nicrofer 6023H -alloy 601H (2.4851) and Nicrofer 6025HT -602 CA (2.4633). Ni-Cr heating element alloys like Cronix 70 and Cronix 80 can also be applied at this temperature. These alloys contain, like Nicrofer 6025HT, additions of reactive elements like yttrium (602 CA), cerium or lanthanum (Cronix 70 and 80), which are known to prevent spalling of oxide scales [20-23].

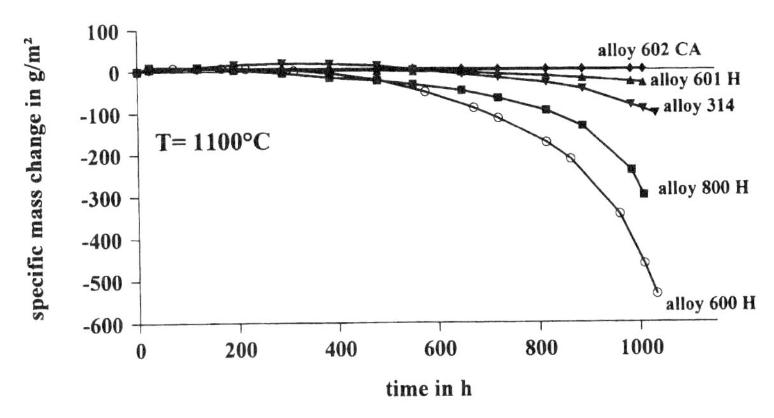

Figure 4.14: Specific mass change of heat-resistant materials and stainless steels during cyclic exposure in dry air (24-hours cycle) as a function of exposure time at 1100 °C. Mass losses indicate spalling or evaporation of metal, mass gains indicate the formation of internal or outer oxide scales (-100 g/m^2 mass loss and 50 g/m^2 mass gain are approximately corresponding to a metal loss of about 0.1 mm/ year)

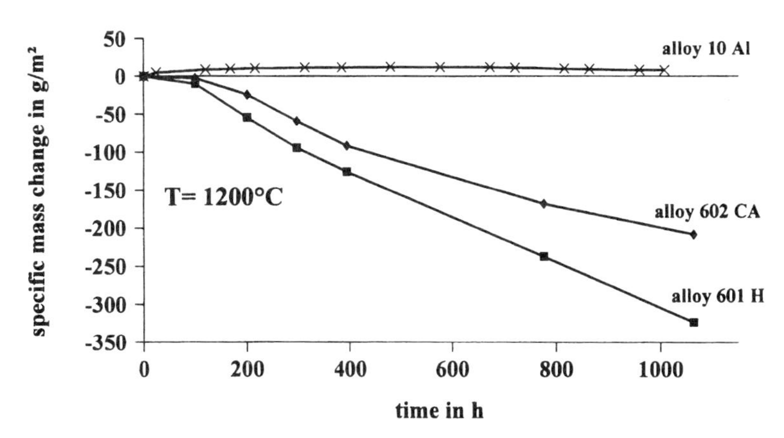

Figure 4.15: Specific mass change of heat-resistant materials and stainless steels during cyclic exposure in dry air (24-hours cycle) as a function of exposure time at 1200 °C. Mass losses indicate spalling or evaporation of metal, mass gains indicate the formation of internal or outer oxide scales (-100 g/m^2 mass loss and 50 g/m^2 mass gain are approximately corresponding to a metal loss of about 0.1 mm/ year)

Only alumina forming alloys like Nicrofer 6025HT, alloy 10Al or even alumina-forming Fe-Cr-Al-alloys can be used at 1200 °C in air. Fig. 4.15 shows the oxidation behaviour of the alumina-forming materials Nicrofer 6025 HT -alloy 602 CA (2.4633) and alloy 10Al in comparison to that of the chromium-forming alloy Nicrofer 6023H -601 H (2.4851) after 1000 hours of exposure in air. It can be seen that that the metal loss of alloy 602CA is still less than 200 g/m^2 corresponding to a metal loss of less than 0.2 mm/year.

Alumina-forming materials like Nicrofer 6025HT -alloy 602CA (2.4633) show also excellent resistance to combined attack by erosion and corrosion. This is especially important in industrial furnaces, where the rotating kiln or furnace rolls suffer abrasion at high temperatures. Fig. 4.16 shows the metal loss of several high temperature materials after 1000 hours exposure under combined high temperature corrosion and erosion. It can be seen that Nicrofer 6025HT -alloy 602 CA-(2.4633) suffers the smallest metal loss of all alloys investigated [24,25].

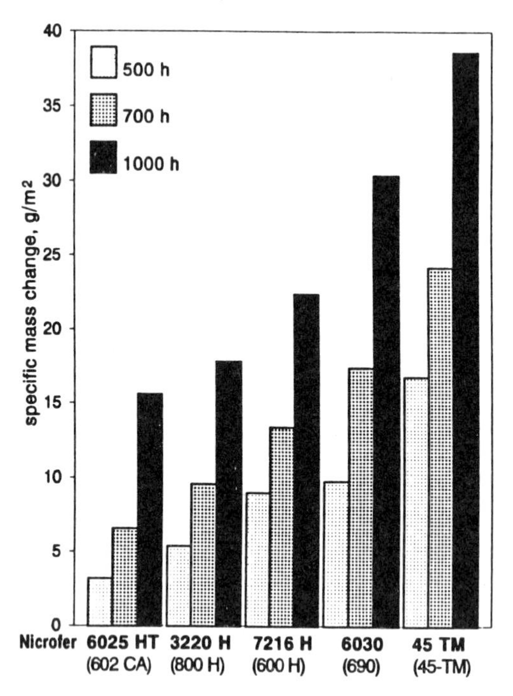

Figure 4.16: Metal loss of heat resistant alloys under combined high temperature oxidation and abrasion at 1000 °C. Relative velocity between erosive particles (SiO_2) and metal: 0.23 m/s

4.3.3 Carburisation

High concentrations of carbon are encountered by metallic materials in a variety of processes in chemical and petrochemical industry, in metallurgical plants and in industrial heat treating furnaces. Two different cases have to be considered:

CASE I: The carbon activity a_c is below one and corrosion attack occurs by internal carburisation.
CASE II: The carbon activity a_c is above one, i.e. the gas is oversaturated with carbon and metal wastage occurs by so-called 'metal dusting'.

a) CASE I: Internal carburisation

The material wastage occurs by penetration of carbon into the metal and subsequent formation of internal carbides, provided the carbon activity is below 1. Internal carburisation can result in the following consequences:

1. a loss of ductility due to the internal carbides
2. reduction of the melting point of the alloy by up to 350 °C [26].
3. changes of microstructure due to extensive formation of chromium carbides. In the case of stainless steels and Fe-Ni-Cr-alloys a transition from austenitic to ferritic behaviour has been observed.

The rate of internal carburisation is governed by the solubility of carbon, the rate of carbon diffusion into the metal and the rate of penetration into the metal [27]. Internal carburisation occurs mainly at high temperatures (above 1000 °C), where diffusion of carbon into the metal is fast and/or at low oxygen partial pressures, where protective oxide scales cannot be formed and hence the penetration of carbon into the material is possible. To obtain high resistance to internal carburisation, materials with low solubility and low diffusivity of carbon should be chosen. Since both the solubility of carbon [27] and the diffusivity of carbon [28] in nickel-base alloys are significantly lower than in iron-base alloys, nickel-base alloys generally are less susceptible to internal carburisation [29]. It can be seen from Fig. 4.17, that the amount of carbon pick-up during exposure in carburising CH_4/H_2 atmosphere is a clear function of the iron-concentration of the respective alloy. Nickel-base-alloys like Nicrofer 5520Co -alloy 617 (2.4663), Nicrofer 7216 H -alloy 600H (2.4816) and Nicrofer 6025HT -alloy 602CA (2.4633) show a much higher resistance to internal carburisation than the Fe-Ni-Cr alloys like Nicrofer 3220 H -alloy 800H (1.4958).

A very effective element to prevent carburisation is silicon. Silicon decreases both the solubility of carbon in iron-(nickel)-chromium alloys and the carbon diffusivity [27]. Consequently materials used for application in petrochemical industry contain silicon. An alloy widely used under carburising conditions is Nicrofer 3718 So -alloy DS (1.4862) with 2.2 % Si. This alloy can also be used for alter-

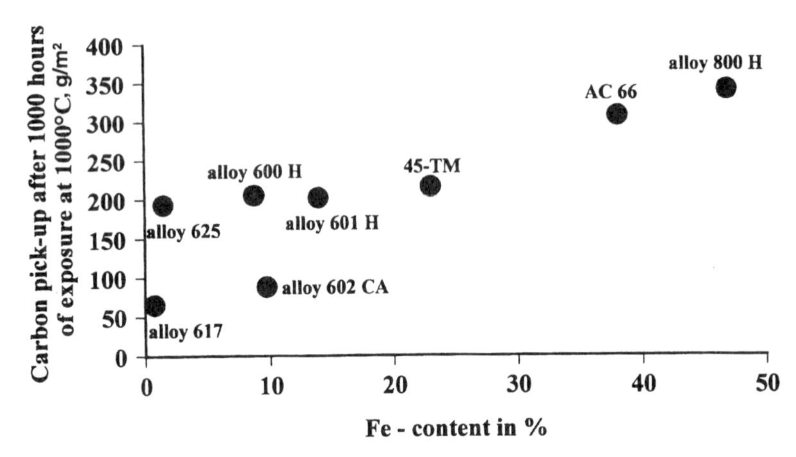

Figure 4.17: Carbon pick-up after 1000 hours of cyclic exposure in carburising CH_4/H_2 gas (carbon activity: a_c=0.8) as a function of the iron content of some heat-resistant and high temperature materials. T=1000 °C

nating carburising/oxidising conditions. Nicrofer 45-TM - alloy 45-TM (2.4889) provides an even better carburisation resistance than Nicrofer 3718 So -alloy DS (1.4862), because of its silicon concentration of 2.7 wt%. Nicrofer 45-TM - alloy 45-TM (2.4889) shows resistance to internal carburisation which is close to that of the nickel-base-alloy Nicrofer 7216H -alloy 600H (2.4816).

An increase in chromium will only increase the carburisation resistance, if the oxygen partial pressure is high enough to enable the formation of chromia. A comparison of Table 4.2 and Table 4.3 shows, that this is not the case for common petrochemical processes. Alumina scales, on the other hand, are protective even at very low oxygen partial pressures. Provided the scale remains intact, alumina formers do not suffer internal carburisation, since alumina is impervious to carbon [30]. Fig. 4.18 shows the impact toughness of two chromia-forming alloys Nicrofer 7216 H -alloy 600H (2.4816) and Nicrofer 3220 H -alloy 800H (1.4958) and one alumina-former Nicrofer 6025 HT -alloy 602CA (2.4633) after exposure in carburising CH_4/H_2 gas at 1000 °C as a function of time. It can be seen that the chromia-forming alloys suffer a loss of ductility while the ductility of the alumina-forming Nicrofer 6025HT -alloy 602CA (2.4633) remains unchanged even after 1000 hours of exposure.

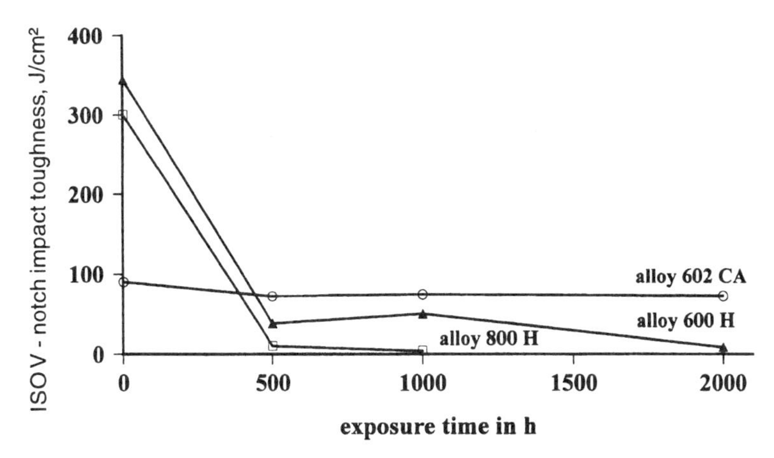

Figure 4.18: ISO V - notch impact toughness of some high temperture alloys after cyclic exposure in carburising CH_4/H_2 gas (carbon activity: a_c=0.8) as a function of exposure time at 1000 °C

b) CASE II: Metal Dusting

Severe material failures caused by so-called 'metal dusting' have been reported during recent years. The reason for these failures were strongly carburising CO-H_2 gas mixtures such as encountered in chemical plants for the synthesis of hydrocarbons, methanol, ammonia etc. as well as in plants for the reduction of iron ores.

Metal dusting is a rapid material wastage, which can occur under any process condition where the carbon activity of the gas atmosphere a_c becomes greater than one [31-33]. It is characterised by the disintegration of iron-, nickel- or cobalt-base materials to a dust which contains metal particles, carbon and sometimes also carbides and oxides. The mechanisms of metal dusting have been elucidated in the recent years for both iron-base alloys [31-33] and nickel-base alloys [34,35]. Exposed to the supersaturated gas, materials may dissolve high amounts of carbon. If graphite formation starts and deposition of graphite on the metal surface occurs, the carbon activity at the metal surface drops to a_c = 1 and the metal matrix, which now becomes supersaturated with carbon, decomposes. In the case of iron-base alloys, metastable carbides (Fe_3C) are formed as an intermediate phase which decomposes to graphite and metal particles. In nickel-base alloys, where metastable carbides do not exist, it is the growth of graphite in (and into) the supersaturated solid solution which destroys the metal. A dust of graphite and metal particles is formed as a reaction product with the metal particles serving as catalysts for further carbon deposition. Once started, the reaction can produce vast amounts of 'coke'.

Materials can be protected against metal dusting by adding sulphur-containing compounds (often H_2S) to the process gas. It is widely known that sulphur "poisons" the metal surfaces and prevents both carbon ingress and graphite nucleation [36]. However, in many catalytic processes sulphur-bearing substances cannot be injected since they deactivate the catalysts. To prevent material wastage under such conditions, an appropriate material with high resistance to metal dusting has to be selected.

Iron-base alloys are generally susceptible to metal dusting because of their tendency to form metastable iron carbides. The resistance to metal dusting increases with increasing nickel concentration, as can be seen from Fig. 4.19. Some iron-nickel-chromium alloys, however show a lower resistance to metal dusting than ferritic steels.

The carburisation behaviour of nine commercial nickel-base alloys and four iron-nickel-chromium alloys was tested at 650 °C in a strongly carburising H_2-CO-H_2O - gas with a carbon activity of $a_c >> 1$ [34, 35]. Table 4.4 shows the total exposure times and the final corrosion rates. The iron-nickel-chromium alloys suffered severe metal dusting after a very short test period. Nickel base alloys were generally less susceptible to metal dusting than iron-base alloys. However, their corrosion behaviour was found to depend sensitively on the chromium concentration of the respective alloys. Alloys like alloy 600H, with a chromium concentration of only 16 %, suffered wastage rates which were similar to those of

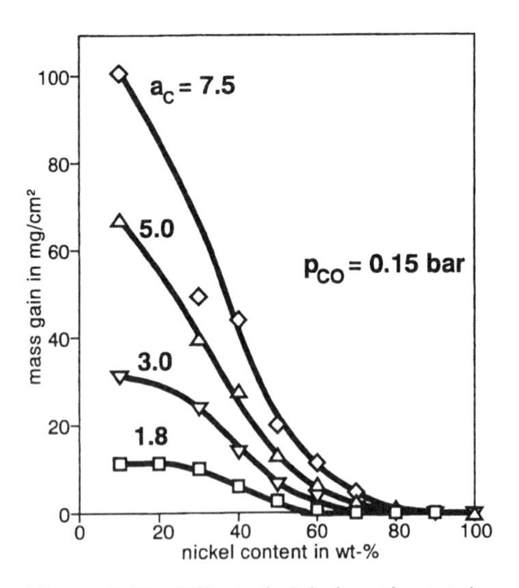

Figure 4.19: Effect of nickel on the total mass gain by carbon deposition of Fe-Ni alloys acc. to [34]

Table 4.4: Total exposure times and final wastage rates after exposure in strongly carburising H_2 - CO - H_2O gas at 650 °C (a_c = 14) [35]

Alloy-Designation	Chromium content in %	Surface	Total exposure time in hours	Final metal wastage rate in mg/cm²h
800 H	20	ground	95	2.1×10^{-1}
HK-40	25		190	4.0×10^{-2}
HP-40	25		190	3.8×10^{-2}
DS	18	ground	1988	4.3×10^{-3}
600 H	16	ground	5000	3.3×10^{-2}
601	23	black	6697	7.3×10^{-3}
601	23	polished	1988	4.9×10^{-3}
601	23	ground	10000	5.8×10^{-4}
C-4	16	ground	10000	1.1×10^{-3}
45-TM	27	black	10000	1.0×10^{-5}
602 CA	25	black	10000	1.1×10^{-5}
617*	22	ground	7000	3.7×10^{-6}
690	28	ground	10000	2.0×10^{-6}

* alloy 617: evidence of metal dusting

the more resistant iron-base alloys. Fig. 4.20 shows the metal wastage rate of some nickel base alloys with 16 % Cr (alloy 600H) 23% Cr (alloy 601) and 25% Cr (alloy 602CA). The nickel-base alloys with chromium concentrations of 25% and above like Nicrofer 6025HT -602CA (2.4633), on the other hand, showed no significant evidence of metal dusting even after 10000 hours of exposure. It was found that these alloys are protected against metal dusting by the formation of a dense, self-healing chromia scale, which prevents the penetration of carbon into the base metal. It can easily be seen that the metal wastage rate decreases with increasing chromium concentration.

The wastage rate of alloy 601 was found to depend sensitively on the surface preparation, as can be seen from Fig. 4.21. The lowest wastage rates were measured for the sample in the as-ground condition, for which the metal loss was about two orders of magnitude lower than for the specimen in the black condition.

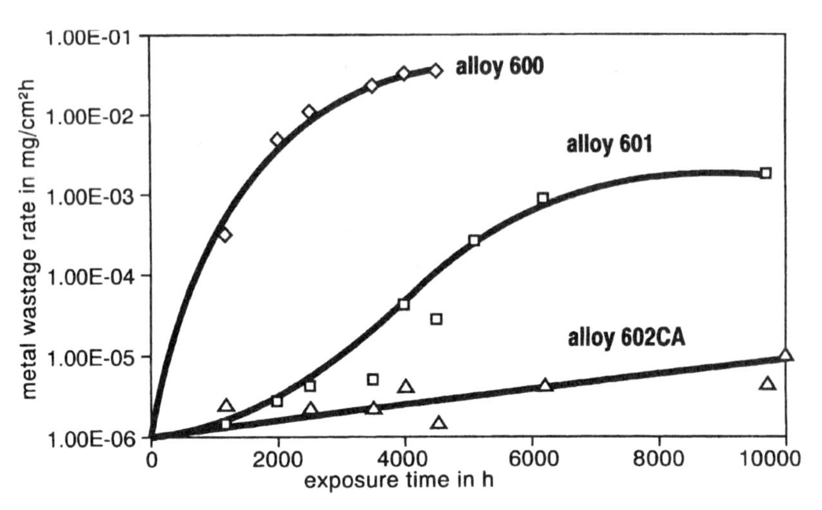

Figure 4.20: Metal wastage rate of some nickel base alloys after exposure in strongly carburising H_2-CO-H_2O gas at 650 °C [35]

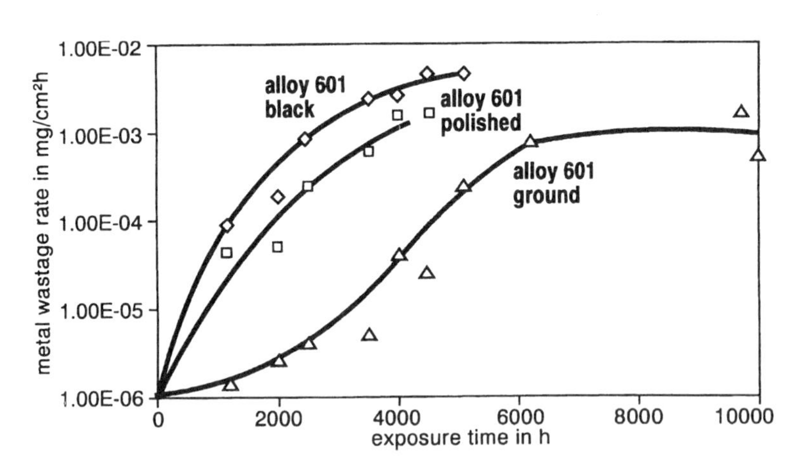

Figure 4.21: Metal wastage rate of alloy 601 after different surface treatments and subsequent exposure in strongly carburising H_2-CO-H_2O gas at 650 °C [35]

4.3.4 Nitridation

If metallic materials are exposed to nitrogen, nitrogen/hydrogen or ammonia atmospheres at high temperatures, they suffer nitridation. The process of nitridation is very similar to the process of carburisation: nitrogen penetrates into the metal and causes the formation of internal nitrides (mostly chromium nitrides). Like internal carbides, internal nitrides may cause a severe loss of ductility during exposure or even cracks along grain boundaries [27].

The amount of internal nitrides depends on the temperature, the gas composition and the alloy composition. Low alloy steels suffer internal nitridation already at temperatures as low as 300-500°C in ammonia. Stainless steel are susceptible to nitridation at temperatures above 500 °C [38,39].

Nickel-base alloys are generally less susceptible to internal nitridation because of the low nitrogen solubility [40]. Investigations of Fe-Ni-Cr and nickel-base alloys after exposure in 10%H_2/N_2 (dewpoint: -65 °C) have shown, that Fe-Ni-Cr alloys like alloy 800H and alloy DS suffer internal nitridation and ductility loss already at 1000 °C, while nickel-base alloys like alloy 600H, alloy 602 CA and alloy 601 are resistant to nitridation at this temperature [39]. Even if small amounts of chromium nitrides are formed, they do not effect significantly the mechanical properties of the materials. At 1100 °C and 1200 °C both iron-base alloys and nickel-base alloys suffer internal nitridation and ductility loss. Table 4.5 shows the Charpy-notch impact toughness after 1000 cyclic hours exposure in 10%H_2/N_2 at 1200 °C. It can be seen that Fe-Ni-Cr-alloys suffer a dramatic decrease in

Table 4.5: Impact toughness of ISO V-notch-specimens measured at room temperature after 1000 hours of exposure at 1200 °C in comparison to the impact strength in the solution annealed condition [39].

VDM Designation	Alloy	German material No.	Impact toughness prior to exposure in J/cm^2	Impact toughness after1000 hours of exposure at 1200 °C, in J/cm^2
Cronifer 2520	alloy 314	1.4841	275	10
Nicrofer 3220 H	alloy 800 H	1.4958	385	13
Nicrofer 7216 H	alloy 600 H	2.4816	345	238.5
Nicrofer 6023 H	alloy 601 H	2.4851	194	83
Nicrofer 6025 HT	alloy 602 CA	2.4633	90	78.5

room temperature ductility, while the nickel-base alloys show only small ductility losses [39]. Generally alloy 600H provides the best nitridation resistance of all nickel-base alloys in purely nitriding atmospheres. If, however, alternating oxidising/nitriding conditions are used, or if, like in tube furnaces, one furnace wall is exposed to air and the other to the nitriding gas, alloys like Nicrofer 6025HT or 6023H should be used because of their higher oxidation resistance.

4.3.5 Sulphidation

Sulphidising gases are encountered in processes in chemical industry and in coal conversion processes. Provided the sulphur concentration is high enough, metal sulphides may be formed in contact with sulphur. The growth rate of metal sulphides is several orders of magnitude higher than that of oxidation, which leads to very high corrosion rates if the process is once started [18]. Some metal sulphides, furthermore, tend to form low-melting compounds; the melting point of an eutectic Ni_2S_3-NiS compound for example is as low as 635 °C [18]. Consequently Fe-Ni-Cr alloys should be used instead of nickel-base alloys in highly sulphidising environments to avoid the formation of liquid nickel sulphides.

There is a general agreement that the resistance to sulphidation/oxidation in SO_2-containing gases increases with increasing chromium concentration [41]. Fig. 4.22 shows extrapolated maximum corrosion attack at 650 °C as a function of the SO_2-concentration of the gas. At an SO_2 - concentration of 1 and 5 % all tested nickel-iron-chromium alloys are applicable. If the SO_2-concentration is increased to 10%, however, alloy 800H (with only 20 Cr) suffers a much higher

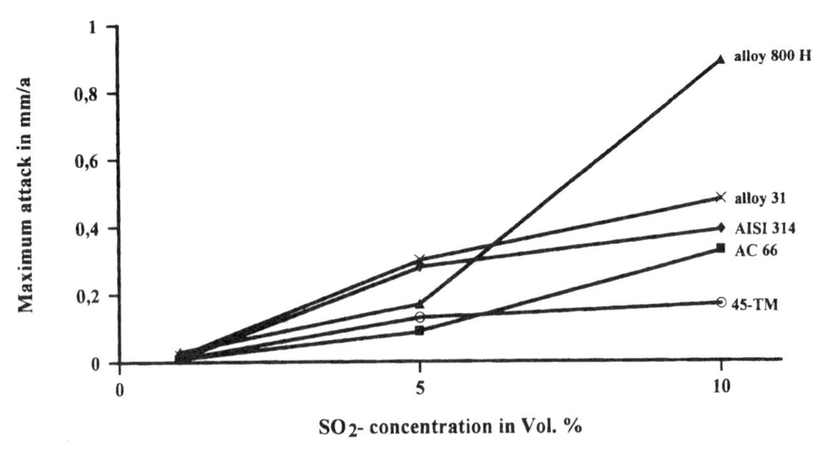

Figure 4.22: Extrapolated maximum corrosion attack of materials after exposure in SO_2/air at 650 °C as a function of the SO_2-concentration

sulphidation attack than the alloys 31, 314, AC66 and 45-TM (all with chromium concentration of 25-28 %). An increase in the temperature has a similar effect as an increase in the SO_2-concentration; at 750 and 850 °C and 1 % SO_2 the high-chromium alloys are still applicable. At very high temperatures and high SO_2 -concentrations only cast Ni-Cr alloys, alumina-forming Fe-Cr-Al alloys [42] and Fe-Ni-Cr-Al alloys (Nicrofer 6009 Al10) are resistant [43].

If materials exposed to sulphidising conditions are frequently cooled down to room temperature and suffer both high temperature sulphidation and wet corrosion by sulphuric acid (dew point corrosion), the application of Nicrofer 3033 - alloy 33 (1.4591) should also be considered. This material shows high resistance to sulphidation due to the high chromium concentration of 33 % and also to corrosion by sulphuric acids [44,45].

Reducing, sulphur-containing gases, as they are encountered in coal gasification processes, are particularly damaging to metallic materials. Extensive laboratory and field tests have shown, that Ni-Fe-Cr alloys with chromium concentrations of at least 25% withstand coal gasification environments [17,46]. Additions of silicon were found to additionally improve the resistance to sulphidation. Field test carried out in the PRENFLO-pilot plant of Krupp Koppers in Fürstenhausen, Germany, showed that alloy 45-TM with 27% chromium and 2.7 % silicon had the highest sulphidation resistance of all alloys investigated [46]. Fig. 4.23 shows the corrosion attack measured after 2000 hours of plant exposure in different locations with different temperatures.

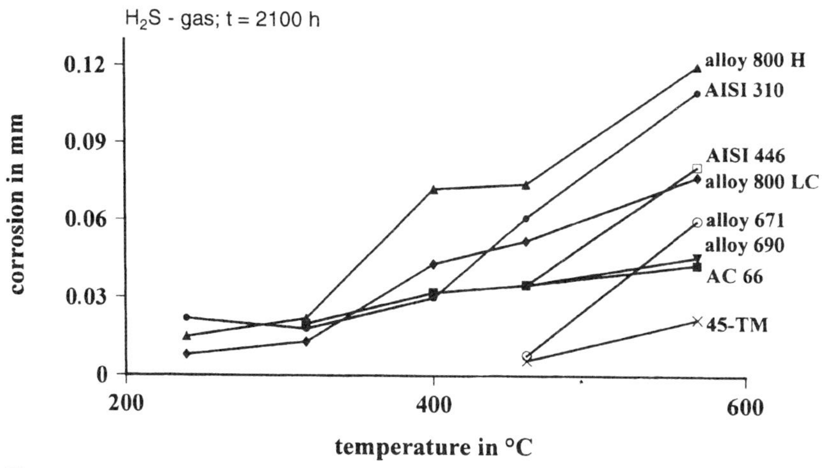

Figure 4.23: Corrosion attack of stainless steels and nickel base alloys after exposure in the coal gasification pilot plant of Krupp Koppers, Fürstenhausen as a function of temperature

4.3.6 Corrosion by halides

The exposure of metallic materials to chlorine, hydrogen chloride or other halogen-containing atmospheres can cause both the formation of metal-halogenides and the penetration of chlorine or fluorine along grain boundaries into the metal.

In oxygen-deficient HCl atmospheres the chlorides of iron, nickel or chromium may be formed. Since most metal chlorides have a high vapour pressure even at moderate temperatures (see also chapter 7), evaporation of metal chlorides may take place. Iron chlorides tend to evaporate already at temperatures as low as 167 °C ($FeCl_3$) and 536 °C ($FeCl_2$) [47]. Hence alloys with a low iron content should be used for application in HCl-containing gases. Because of the much lower vapour pressure of nickel chlorides, nickel-base materials are used wherever corrosion by halogens is expected. The highest resistance to chlorination by HCl is shown by Ni-Cr-Mo alloys. The alloys Nicrofer 6020hMo -alloy 625 (2.4856), Nicrofer 6616 -alloy C-4 (2.4610), Nicrofer 5923hMo -alloy 59 (2.4605) show low corrosion rates even at 850 °C in 10 %HCl/H_2 - atmospheres [48]. In gas atmospheres with a constantly low oxygen partial pressures even nickel 99.2 (alloy 200) or Ni-Mo alloys (Nimofer 6928-alloy B-2) might be considered.

In oxygen-containing Cl_2/O_2 - gas atmospheres or in the presence of chloride containing deposits chloride-induced oxidation is frequently observed. This process, which is a typical failure mechanism of boiler tubes in waste incineration plants has been extensively studied by Grabke and co-workers [49] and is described in chapter 7 of this book. Depending on the amount of sulphur in the gas atmosphere, alloys Nicrofer 7216H -alloy 600H.(2.4816), Nicrofer 6020hMo - alloy 625 (2.4856) and Nicrofer 45-TM -alloy 45-TM (2.4889) were found to have the highest resistance in Cl_2-containing environments. The reader is referred to chapter 7 for more detailed information.

It should be mentioned that alumina-forming alloys like Nicrofer 6025HT can be used in oxidising chlorine-containing gas atmospheres as well. Tests in a gas containing 0.25%Cl_2, 20% O_2 in Ar showed a smaller corrosion attack of alloy 602CA than of alloy 600H [50]. In oxygen-deficient HCl-atmospheres, however, alumina-forming materials should not be used, since they tend to form aluminium chloride $AlCl_3$, which becomes volatile at a temperature as low as 173 °C .

4.3.7 Corrosion in complex environments and molten salt corrosion

In real in-plant services, often more than one corrosive component is encountered and additionally wet corrosion (dewpoint corrosion), stress-induced corrosion, cloride-containing deposits, molten salts, molten metals and erosion may increase the metal wastage rates significantly. Because of the variety of possible metal-metal, gas-metal and gas-salt reactions it is impossible to predict material behaviour and corrosion rates in complex corrosive environments at high temperatures. Often laboratory tests or even field tests are necessary prior to selecting a suitable alloy. Some common corrosion mechanisms are described below:

So-called "Oil-ash corrosion" is caused by deposits of sulphates in the presence of vanadates (V_2O_5). It is generally assumed that vanadate renders the sulphate-deposit acidic and hence favours acidic dissolution of a metal oxide, e.g. Cr_2O_3 according to the reaction:

$$Cr_2O_3 + 3\ SO_2 = 2\ Cr^{3+} + 3\ SO_4^{2-}$$ [18]

Since chromia is constantly consumed during this reaction, alloys with a high chromium concentration like NiCr50 or Nicrofer 45-TM -alloy 45-TM (2.4889) should be used to prevent chromium depletion.

So-called "Hot corrosion", i.e. basic fluxing or acidic fluxing of oxide scales and metals is encountered in industrial gas turbines [51] at temperatures between 650 °C and 900 °C, especially if fuels with a low impurity are burned. Since it has been shown that molybdate MoO_3 has the same effect like vanadate and it enhances the process of acidic fluxing [51, 52, 53], the use of molybdenum-free high temperature-high strength alloys like Nicrofer 6125GT-alloy 603GT (2.4647) is recommended for industrial gas turbines and similar applications with impurities of SO_2/SO_3.

Under molten carbonates in the recovery boilers of the Pulp&Paper industry, the molybdenum containing alloys Nicrofer 3127hMo -alloy 31 (1.4562), Nicrofer 6020hMo -alloy 625 (2.4858) and the silicon-containing alloy Nicrofer 45-TM -alloy 45-TM (2.4889) have high resistance to basic fluxing [54].

Chloride-containing salts are encountered in a variety of technical processes. Typical examples are hazardous waste treatment, biomass incineration, bellows in the chemical industry and decouplers in the exhaust pipes of motor vehicles. The best resistance to corrosion by chloride-containing deposits was found in the nickel-chromium-molybdenum alloy 625. To overcome the drawbacks of alloy 625, i.e. ductility loss at intermediate temperatures and susceptibility to acidic fluxing, an advanced version of alloy 625, alloy 625Si has been developed recently. Due to its silicon content of 1 %, alloy 625Si was found to show higher resistance to chlorides than even alloy 625.

The material selection for application in molten salts and molten metals depends on the solubility of alloying elements in the melt. High chromium concentrations prevent corrosion in molten sulphates. In some molten salts alumina-forming materials like alloy 10Al and Nicrofer 6025HT -alloy 602CA (2.4633) may show lower corrosion rates than chromia-forming materials. Alloy 10Al was found to be not attacked by molten chromates at 1100 °C. In molten aluminium fluoride (aluminium production), on the other hand, the high chromium alloy Nicrofer 3033 -alloy 33 (1.4591) has shown high corrosion resistance.

4.4 Areas of application for heat-resistant and high-temperature materials

From the wide range of known uses, naturally only a brief review of the existing application areas can be given below. Plants and components are arranged according to the corrosive media present, providing this can be done simply [55] :

Oxidation

- Heat exchangers for heating steel (T = 1200°C)
- Ceramic firing furnaces and internal furnace components (T < 1200°C)
- Radiant heater tubes (T < 1100°C)
- Furnace components including furnace rolls, hangers, holders, flanges, transport rails (T < 1100°C)

Carburisation

- Furnace components for gas carburisation (T < 1100°C)
- Carbon regeneration and reactivation (T < 1000°C)
- Pyrolysis tubes in ethylene crackers (T < 1100°C)
- Heat exchangers in coal gasification and combustion plants (T < 900°C)

Sulphidation

- Heat exchangers in petroleum coal calcination plants (T > 900°C)
- Components in gas-fired furnaces
- Heat exchangers in combustion gases

Combustion gas corrosion (Salt, sulphur, halides)

- Heat exchangers in aluminium melting furnace (T < 1000°C)
- Heat exchangers in fibre glass production (T < 1200°C)
- Heat exchangers and superheater tubes in waste incineration plants (T < 900°C)
- Furnace components in waste pyrolysis plants (T < 800°C)
- Components and superheater tubes in the pulp and paper industry (T < 900°C)
- Heat exchangers and superheater tubes for coal gasification and combustion plants (T < 900°C)

Corrosion from glass melting

- Melting tanks and spinners in fibre glass production (T > 1000°C)
- Melting tanks and tools for vitrification of radioactive waste (T < 1200°C)
- Fittings for glass melting furnaces (T < 1000°C)

Nitrogen absorption

- Ammonia cracking plants (T < 500°C)
- Heat-treatment baths (sodium and potassium nitrate) for tempering of steel or heat treatment of aluminium (T < 600°C)
- Nitrocarburising baths for carbonitriding of steel surfaces (T < 850°C)
- Controlled-atmosphere or bright annealing furnaces for steel (T < 1200°C)

Halogen absorption

- Furnaces for ethylene dichloride production (T < 500°C)
- Materials for uranium separation (T = 600°C)
- Plants for vinyl chloride monomer production (T = 600 - 700°C)
- Materials for fuel cells (T = 600 - 700°C)
- Heat exchanger tubes in waste incineration plants with a high proportion of household refuse (T < 900°C)
- Heat exchanger tubes for fluidised bed combustion (T < 900°C)
- Plant components for TiO_2 production (T < 900°C)
- Salt bath furnaces for heat treatment of steels (T = 750 - 1000°C)

Unfortunately it is not possible to directly recommend materials for these fields of application, since it is not only the high-temperature corrosion behaviour which determines the selection of a material (and even this changes from one plant type to another because of different temperatures and gas compositions); the mechanical property requirements and, last but not least, the cost, of the materials or, to be more precise, the cost relative to the utilisation period also plays a decisive role.

4.5 Examples of practical industrial application

4.5.1 Petrochemical engineering

The petrochemical industry is a broad field of application for high-temperature materials, in which particularly high demands are made on the materials resistance to corrosion and high-temperature mechanical properties. An important process in this field is ethylene pyrolysis in heated tube coils. In view of the fact that the chemical reactions necessitate high temperatures, it is particularly important for the tube material to have high creep resistance. In addition to plant designs featuring horizontally arranged cast pyrolysis tubes, newer plants are being designed with freely suspended vertical reaction tubes. The growth of these tubes due to creep presents problems, making it necessary to shorten them from time to time. This involves downtime in the plant, since extensive welding is necessary. The use of Nicrofer 3220 H- alloy 800 H (1.4958), with its excellent creep resistance properties, has made it possible to shorten the downtime necessary for repairs considerably.

Figure 4.24: Pigtails and headers in reformer furnaces in Nicrofer 3220 H -alloy 800H (1.4958)

Another area of application for Nicrofer 3220 H - alloy 800 H (1.4958) is connecting tubes (pigtails), as shown in Fig. 4.24 and headers in catalytic hydrocarbon reforming. Whereas cast tubes were used for many years for headers, extruded and forged tubes are nowadays also used. Due to the greater ductility of the wrought material, these tubes can cope better with thermal stresses. One of the main requirements for pigtail materials is, in addition to high creep strength, resistance to alternating oxidising/reducing atmospheres. Here, the inevitable carburisation of the material must not be allowed to cause too severe a drop in its residual ductility. The reason is that, in case of leakage from a split tube the tube must be isolated from the rest of the system by pinching off the supply and return tubes. This deformation process must produce a gas-tight seal on the tube. It is also often necessary in the petrochemical industry to flare off undesirable reaction products, burners made of Nicrofer 3220 H - alloy 800 H (1.4958) are being used for this purpose.

Another branch of petrochemicals is the manufacture of styrene, the chemical reactions taking place in steam-heated tubular reactors. Nicrofer 3220 H - alloy 800 H (1.4958) has proved suitable both for tubing and for tube sheets.

Nicrofer 4722 Co - alloy X (2.4665) has been used for many years in nitric acid production. This process involves the reaction of ammonia to form nitrous oxides on platinum catalysts at approximately 800 - 1000°C. This means that the catalyst supports must have a high creep resistance with adequate resistance to nitrogen pick-up. Nicrofer 4722 Co - alloy X (2.4665) meets both requirements to the necessary extent.

A high degree of resistance to nitrogen pick-up is also required in the cracking of ammonia as a stage in the manufacture of heavy water. High-strength materials such as Nicrofer 6020 hMo - alloy 625 (2.4856) and Nicrofer 7016 TiNb - alloy X - 750 (2.4807) are again necessary here, due to the temperatures involved of approximately 800°C.

4.5.2 Power station construction

Strict official regulations regarding the reduction of noxious emissions, in particular of sulphur dioxide, must be adhered to in the construction of coal-fired power stations. An attractive means of solving this problem is, besides flue gas scrubbing, combustion of coal in a fluidised bed. This process has the twin advantages of a low degree of nitrous oxide formation, due to the low combustion temperature of approximately 850 - 900°C, and the low amount of sulphur dioxide emitted. By means of limestone addition, most of the sulphur dioxide is fixed as gypsum at the combustion stage. A high degree of heat transfer is achieved in this process, since the heat-transfer surfaces - generally tubes, as shown in Fig. 4.25 - extend directly into the combustion zone. This makes it possible to con-

Figure 4.25: Tubular heat exchanger in Nicrofer 3220 H -alloy 800H (1.4958) for fluidised bed combustion of coal

struct small, compact power stations, since the heat-transfer surfaces can be relatively small. In the area of the fluidised bed, the tube material may be exposed to various corrosion reactions, since the high oxygen partial pressure necessary for the formation of protective oxide layers fluctuates heavily. For this reason, both sulphidation and carburisation may occur, with the corresponding negative consequences for the material. Tests carried out on Nicrofer 3220 H - alloy 800H (1.4958) in a pilot plant indicated after several thousand hours of testing that a sulphidation/oxidation reaction initially occurred at the grain boundaries [1] . Because of the serious corrosion attack alloy Nicrofer 45-TM - alloy 45-TM (2.4889) is also considered to be a candidate alloy for this application.

Nicrofer 5520 Co - alloy 617 (2.4663) has also established itself alongside Nicrofer 4722 Co - alloy X (2.4665) and Nicrofer 7520 - alloy 75 (2.4951), the classic materials employed until now, for the production of gas turbine combustion chambers. The reason is again its excellent high-temperature mechanical properties combined with a high degree of resistance to oxidation. In gas turbine construction particularly, the use of relatively expensive materials became established at an early stage, since the high reliability thus achieved as well as the high temperatures applied led to a substantial increase in cost efficiency.

Precipitation hardened materials are used in power station construction where particularly good mechanical properties are required. Nicrofer 7520 Ti - alloy 80 A (2.4952) is used, for instance, for bolts and screws in steam turbines. Springs for the temperature range up to 650°C, required on control rods in nuclear reactors, are normally made of Nicrofer 7016 TiNb - alloy X-750 (2.4669). Spacer grids for the securing of nuclear fuel rods are also made of a precipitation hardened material, namely Nicrofer 5219 Nb - alloy 718 (2.4668).

4.5.3 Heating systems engineering and furnace industry

Heating systems engineering, particularly furnace construction, is a classic application for high-temperature materials. Today, heat-resisting and high-temperature stainless steels are more and more, wherever possible, replaced by high-strength Ni-base alloys because of economic considerations. Especially the alloys Cronifer 2012 -alloy 308 (1.4828) and Cronifer 2520 -alloy 314 (1.4841) very popular in the past are replaced more frequently because of their high tendency for σ-phase embrittlement and lack of creep strength and corrosion resistance at higher temperatures. More highly alloyed high-temperature materials are used for a large number of applications where special demands are made on strength and on resistance to furnace atmospheres. Due in particular to its greater creep resistance when compared with heat-resistant stainless steels, Nicrofer 3220 H - alloy 800 H (1.4958) is often used as a constructional material in special cases, such as furnace fans, as shown in Fig. 4.26. For high duty convection furnaces even the age-hardenable Ni-base alloy Nicrofer 5120 CoTi

Figure 4.26: Furnace fan in Nicrofer 3220H - alloy 800H (1.4958)

- alloy C-263 (2.4650) as fan and Nicrofer 6025 HT - alloy 602 CA (2.4633) as a shaft material are more frequently considered.

A material with many applications in furnace construction is the alloy Nicrofer 7216 H - alloy 600 H (2.4816). This material is especially suitable for nitrogen bearing bright-annealing units due to its relatively low tendency towards nitrogen pick-up. Nicrofer 7216 H - alloy 600 H (2.4816) is also a popular material for halogen-bearing atmospheres up to temperatures of approximately 800°C.

Nicrofer 7216 H - alloy 600 H (2.4816) as a traditional material for bright annealing muffles is more and more replaced by Nicrofer 6025 HT - alloy 602 CA (2.4633) since the nitrogen bright annealing atmospheres are replaced by hydrogen. Due to the higher creep strength and oxidation resistance of Nicrofer 6025 HT - alloy 602 CA (2.4633) this alloy has a temperature capability of up 1200°C.

Figure 4.27: Bright annealing muffle in the prefabricated condition

The Figs. 4.27 and 4.28 present the appr. 30 m long vertical bright-annealing muffle in the just assembled condition and after insertion in the bright-annealing tower of AST in Italy [56].

Another important field of application for high-temperature materials is provided by radiant heater tubes (Fig. 4.29) for the heating of annealing furnaces. This application places extremely severe demands on the materials resistance to oxidation. Cronifer 2520-alloy 310/314 (1.4841/1.4845) used for a long time for such purposes, has in recent years been replaced by the nickel-base alloy Nicrofer

Figure 4.28: Bright annealing tower of AST, Turin with a Nicrofer 6025 HT- alloy 602 CA (2.4633) bright annealing muffle. Courtesy of EBNER, Industrieofenbau, Linz

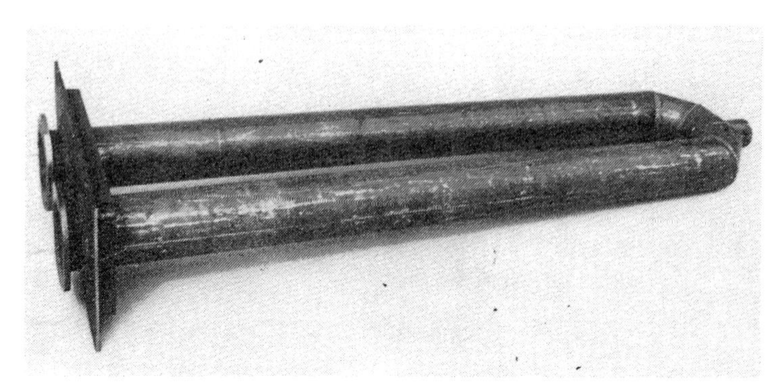

Figure 4.29: Radiant heater tube in Nicrofer 6023 H -alloy 601H (2.4851) for the indirect heating of furnaces

6023 H - alloy 601 H (2.4851). Due to its aluminium content, this material has excellent resistance to oxidation above 1000°C. For heavy loaded radiant heater tubes in nitrogen bearing atmospheres Nicrofer 4722 Co - alloy X (2.4665) has proved itself to be economic.

An excellent oxidation resistance under cyclic conditions is absolutely necessary for enamelling furnaces. The use of a high-temperature, high-strength materials allows a reduction of the load bearing cross sections thus reducing the total weight of the component, resulting in significant energy savings [57]. Fig. 4.30 shows an overhead conveyor for annealing of enamelled parts made of Nicrofer 6025 HT - alloy 602 CA (2.4633), which has already replaced the traditional alloy Nicrofer 6023 H - alloy 601 H (2.4851).

A new application for wrought high-temperature, high-strength nickel-base alloys is their use as uncooled furnace rolls in continuos roller-hearth annealing furnaces. These furnaces are used for annealing of sheets, plates and slabs. Temperatures applied are predominantly in the range of 1100 - 1200°C. Because of its excellent high-temperature strength characteristic Nicrofer 6025 HT - alloy 602 CA (2.4633) has already proved its suitability for now almost 4 years. The rolls have been operated without any cooling, that means neither barrel nor bearing cooling has been applied. No pick up on the rolls or scratches and marks on the bottom of the sheets or plates could be observed. Fig. 4.31 gives a view in the inside of the roller - hearth furnace operated with uncooled furnace rolls made of Nicrofer 6025 HT - alloy 602 CA (2.4633) [58-60].

Figure 4.30: Overhead conveyor for annealing of enameled parts

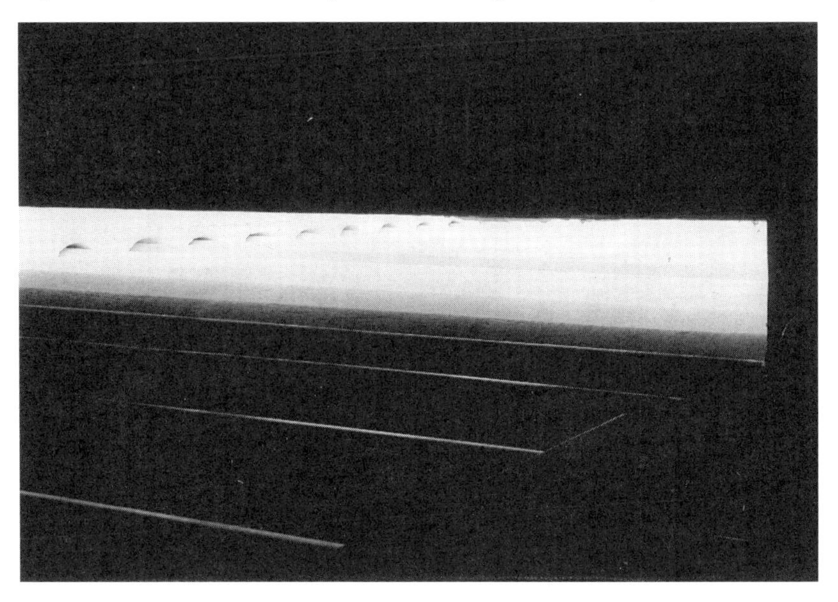

Figure 4.31: View in the Drever roller-hearth furnace with uncooled furnace rolls in Nicrofer 6025 HT -alloy 602 CA (2.4633) of Krupp VDM GmbH, Altena plant

4.5.4 Metallurgical process industry

For the direct reduction of iron ores Ni-base alloys like Nicrofer 7216 H - alloy 600 H (2.4816) and Nicrofer 6023 H - alloy 601 H (2.4851) have been used for parts subjected to severe carburisation and metal dusting. More frequently occurring shut-down of plants due to severe metal dusting and latest scientific examinations on the mechanism of metal dusting and the corrosion behaviour of commercial Ni-base alloys under metal dusting conditions, have proved Nicrofer 7216 H - alloy 600 H (2.4816) and Nicrofer 6023 H - alloy 601 H (2.4851) not to be totally resistant to metal dusting. On the other hand no metal dusting attack could be observed on Nicrofer 6025 HT - alloy 602 CA (2.4633).This is the reason why this alloy is more and more considered for structural parts in direct reduction plants. Fig. 4.32 shows prefabricated parts made of Nicrofer 6025 HT -alloy 602 CA (2.4633) as inserts for a COREX - direct reduction plant [56] .

Figure 4.32: Prefabricated components in Nicrofer 6025 HT -alloy 602 CA (2.4633) for inserts of a COREX plant. Courtesy of MCE, Linz

4.6 References Chapter 4

[1] U. Brill: Hochtemperaturwerkstoffe, in: Nickellegierungen und hochlegierte Sonderedelstähle, ed. U. Heubner, 2. Auflage, expert verlag, Ehingen, 1993, 105 - 130

[2] N. S. Stoloff: Fundamentals of Strengthening, in: The Superalloys, eds. C. T. Sims and W. C. Hagel, John Wiley & Sons, New York, 1972, 79 - 117

[3] R. F. Decker and C.T. Sims: The Metallurgy of Nickel-Base Alloys, in: The Superalloys, eds. C. T. Sims and W. C. Hagel, John Wiley & Sons, New York, 1972, 33-77

[4] U. Brill: Neue warmfeste und korrosionsbeständige Nickel-Basis-Legierung für Temperaturen bis zu 1200°C, Metall 46 (1992), 778 - 782

[5] U. Brill and D. C. Agarwal: CORROSION/93, Paper No. 226, NACE International, Houston, TX, 1993

[6] U. Brill and D. C. Agarwal: CORROSION/97, Paper No. 138, NACE International, Houston, TX, 1997

[7] U. Brill, U. Heubner, K. Drefahl und H.-J. Henrich: Zeitstandwerte von Hochtemperaturwerkstoffen, Ingenieur-Werkstoffe 3 (1991), Nr. 4, 59 - 62

[8] K. Drefahl and F. Hofmann: Austenitic High-Performance Materials for Non-Rotating Components in Stationary Gas Turbines, ASME Technical Publication 88-GT-307, 1988

[9] N.N.: Werkstoffdaten - Hochleistungswerkstoffe, Krupp VDM GmbH Druckschrift No. 53093-08, 1993

[10] Unpublished data of Krupp VDM GmbH

[11] U. Heubner, K. Drefahl and H.-J. Henrich: Creep Strength and Creep Ductility of Welded Heat Resistant Materials in a Carburising Environment, Heat Resistant Materials, Proc. of the 1st. Conf., ASM International, Materials Park, Ohio, 1991, 495 - 504

[12] K. Drefahl: Auslegung zeitstandbeanspruchter Bauteile für den Hochtemperaturbereich, demonstriert anhand praktisch ausgeführter Anlagen, VDM Report No. 14, October 1990

[13] K. Drefahl, K. H. Matucha und F. Hofmann: Determination and Valuation of Extrapolation Parameters for Creep-Stressed, Nickel-Based, High-Temperature Materials, CORROSION/88, Paper No. 376, NACE International, Houston, TX, 1988

[14] S. Leistikow: Aufbau von Oxidschichten und ihre technisiche Bedeutung, ed. A. Rahmel, DGM, Oberursel, Germany, (1983) 33.

[15] I. Barin: Thermodynamical data of pure substances, VCH Weinheim, 1993.

[16] V.L. Hill and H.S. Meyer: High temperature corrosion in energy systems, ed. M.F. Rothman, AIME, Warrendale PA, (1986) 29.

[17] W. Bakker: Mixed Oxidant Corrosion in Nonequilibrium Syngas at 540 °C, Doctoral thesis, Delft, NL, 1995.

[18] P. Kofstad: High Temperature Corrosion, Elsevier Applied Science, London, New York, 1988.

[19] A.J. Sedriks: Corrosion of Stainless Steels, J. Wiley&Sons, New York, Chichester, Brisbane, Toronto (19979) 252.

[20] H. Pfeiffer, H. Thomas: Zunderfeste Legierungen , Springer Verlag, Berlin, 1963.

[21] E. Lang: The role of active elements in the oxidation behaviour of high temperature metals and alloys, Elsevier Applied Science, London, 1989.

[22] J. Klöwer, G. Li: Materials and Corrosion 47 (1996) 545.

[23] G.C. Wood and F.H. Stott: High Temperature Corrosion,ed.R.A. Rapp, NACE, Houston, 1983, 227.

[24] J. Klöwer, U. Brill, D.C. Agarwal: CORROSION/94, Paper No.535, NACE International, Houston, TX, 1994.
[25] D.C. Agarwal, U. Brill, H.W. Kettler and G. Giersbach: Heat Resistant Materials, (Proc. Conf.), eds. K. Natesan, P. Ganesan and G. Lai, ASM, Materials Park, OH, (1995), p.63.
[26] C. Leistikow: Grundlagen und Erscheinungsformen der Hochtemperaturkorrosion, VDI-Bericht No. 235 (1975) 125.
[27] H.J. Grabke: Aufbau von Oxidschichten auf Hochtemperaturwerkstoffen und ihre technische Bedeutung, ed. A. Rahmel, DGM Frankfurt (1983) S. 89.
[28] S.K. Bose and H.J. Grabke: Z. Metallkunde 69 (1978) 8.
[29] U. Brill, U. Heubner, M. Rockel: Metall 44 (1990) 936.
[30] H.J. Grabke, A. Schnaas: Oxidation of Metals 12 (1978) 387.
[31] J.C. Nava Paz, H.J. Grabke: Oxidation of Metals 39 (1993) p. 437.
[32] H.J.Grabke, R.Krajak, J.C.Nava Paz: Corrosion Sci. 35 (1993) p. 1141.
[33] H.J.Grabke, R.Krajak, E.M. Müller-Lorenz: Materials and Corrosion 44 (1993) p. 89.
[34] H.J.Grabke, R.Krajak, E.M. Müller-Lorenz, S. Strauß: Materials and Corrosion 47 (1996) 495.
[35] J. Klöwer, H.J. Grabke, E.M. Müller Lorenz and D.C. Agarwal: CORROSION/97, Paper No.139, NACE International, Houston, TX, 1997.
[36] H.J. Grabke, E.M. Müller-Lorenz: steel research 66 (1995) 254.
[37] D. Monceau, E.M. Müller-Lorenz, H.J. Grabke: Proc. 4th Int. Symp. High Temp. Corrosion, Les Embiez, France, May 1996, 18-1.
[38] U. Brill: Hochtemperaturwerkstoffe der VDM Nickeltechnologie und ihr Einsatz im Ofenbau, VDM Report No. 15, (1990).
[39] J. Klöwer, U. Brill, M.Rockel: Proc. 4th Int. Symp. High Temp. Corrosion, Les Embiez, France, May 1996, 1-1.
[40] I.C. Humbert und I.E. Elliott: Trans. ASME 218 (1960) 1076.
[41] W.Z. Friend: Corrosion of nickel and nickel base alloys , John Wiley &Sons (1980),178.
[42] J. Klöwer: Werkstoffe und Korrosion 47 (1996) 685.
[43] J. Klöwer, G. Sauthoff and D. Letzig: CORROSION/96, Paper No.144, , NACE International, Houston, TX, 1996.
[44] M. Köhler, U. Heubner, K.W. Eichenhofer and M. Renner: Corrosion 1995, Paper No.338, NACE Int., Houston, TX, 1995.
[45] M. Köhler, U. Heubner, K.W. Eichenhofer and M. Renner: Corrosion 1996, Paper No.428, NACE Int., Houston, TX, 1996.
[46] N. Ullrich, W. Schellberg and W.T.Bakker: 12th EPRI Conf. on Gasification Power Plants, October 1993, San Francisco, CA.
[47] P. Elliot et al.: High Temperature Corrosion in Energy Systems, ed. M.F.Rothman, AIME, Warrendale PA, (1985), p. 437.
[48] U. Brill, J. Klöwer, D.C. Agarwal: Corrosion/96, Paper No.439, NACE International, Houston, TX, 1996.
[49] E. Reese and H.-J.Grabke: Werkstoffe und Korrosion 43 (1992), 547.
[50] J. Klöwer, U. Brill: to be published
[51] R.A. Rapp: Mat. Sci. and Eng. 87 (1987), p. 319.
[52] J. Klöwer: Heat Resistant Materials, eds. K. Natesan, P. Ganesan and G. Lai, (Proc. Conf.), ASM, Materials Park, OH, (1995), p. 245.
[53] J. Klöwer: CORROSION /96 Paper No. 173, NACE International, Houston, TX, 1996.
[54] J. Klöwer and F. White: Proc. Conf. Int. Symp. on Corrosion in the Pulp and Paper Industry Stockholm, Sweden, 14.-16. May 1995, NACE, TAPPI (1995), p. 179.

[55] U. Brill: Krupp VDM AG's high temperature alloys and their use in furnace construction, VDM Report No. 15, 1991
[56] U. Brill und M. Rockel: Auslegung hochtemperaturbeanspruchter Bauteile, VDI-Z 138 (1996), Nr.11/12, S.28-30
[57] U. Brill, D. C. Agarwal and M. Metzler: Practical Experience with the New Alloy 602 CA (NiCr25FeAlY)-Applications in Heat Treat Industry, CORROSION 93, Paper No. 235, NACE International, Houston, TX, 1993
[58] U. Brill and D. C. Agarwal: Alloy 602 CA - A New Alloy for the Furnace Industry, Proc. of the 2nd Conf. on Heat Resistant Materials, ASM, Materials Park, Ohio, 1995, 153 - 161
[59] U. Brill, D. C. Agarwal, H.-W. Kettler and G. Giersbach: Innovations in Alloy Metallurgy for Furnace Rolls and Other High Temperature Applications, Proc. of the 2nd Conf. on Heat Resistant Materials, ASM, Materials Park, Ohio, 1995, 63 - 66
[60] U. Brill: Praktische Erfahrungen mit dem neuen Werkstoff Nicrofer 6025 HT-alloy 602 CA- (2.4663) im Ofen- und Wärmebehandlungsanlagenbau, Zeitschrift Stahl, Heft 6, 1995, 37 - 40

5 Use and working of nickel alloys in chemical plant construction and in environmental technology

R. Koecher

5.1 Introduction

Nickel and nickel alloys have received extensive and varied use in chemical plant construction and environmental technology because of their good mechanical properties and their good resistance in acidic and basic aggressive media.

Composite materials with nickel alloy cladding offer numerous material and property combinations to equipment and plant designers. Clad components also offer technical and economic advantages for numerous duty conditions compared to solid designs. In environmental technology (flue gas desulphurisation), nickel alloys have been successfully used to a considerable extent as well as special stainless steels. Absorber plants are constructed in solid designs as well as in clad plate. Lining with thin sheets (wallpaper technique) is the preferred method in refurbishment work. The descriptions below give some general and specific fabrication details that enable a significant improvement in the quality of fabrication to be achieved.

5.2 Nickel alloys in chemical plants

The advances in welding have substantially improved fabrication methods and the fabrication quality of apparatus made from nickel-base materials.

Shown in Figures 5.1 and 5.2 are some components of relatively simple design. Setting aside the necessity for ensuring that the hot forming, heat treatment and pretreatment techniques are suitable for welding the semi-finished parts, this equipment does not nowadays make any particularly great demands on the equipment fabricator. Nickel-base alloys of the type NiCr15Fe/Alloy 600(2.4816) have good weldability. Optimum welding conditions must be used and extensive nondestructive testing (surface crack detection by means of dye penetrant test, radiographic examination) performed where high weld quality is required. Figure 5.3 shows an example.

When working nickel alloys of the type NiCr23Fe/Alloy 601 (2.4851), a material with outstanding oxidation resistance and good resistance in carburizing conditions in oxidizing and sulphur containing atmospheres, heavy demands in terms of welding are made on the fabricator because of the Al content of 1.1 to 1.6 %. An example is shown in Fig 5.4.

Figure 5.1: Reaction furnace in Alloy 201 (LC-Ni99 Material No. 2.4068), dimensions: dia. = 2000 mm, length = 7700 mm, thickness = 9 mm for the cylindrical shell, thickness = 13 mm for dished ends; photograph by Carl Canzler GmbH & Co., Düren

Figure 5.2:
Column sections in Alloy 400 (NiCu30Fe Material No. 2.4360), dimensions: dia. = 2000 mm, height = 5940 mm, thickness = 6 mm, photograph by Carl Canzler GmbH & Co., Düren

Figure 5.3: Regeneration tower in Alloy 600 (NiCr15Fe Material No. 2.4816), dimensions: dia. = 2000/4100/1535 mm, height = 13360 mm, thickness = 5 and 6 mm; photograph by Carl Canzler GmbH & Co., Düren

Figure 5.4: Melting pot in Alloy 601 (NiCr23Fe Material No. 2.4851), dimensions: dia. = 1860 mm, thickness = 10 mm for cylindrical shell, thickness = 12 mm for diffuser bottom; photograph by Carl Canzler GmbH & Co., Düren

Figure 5.5: Heat-exchanger stack in Alloy 825 (NiCr21Fe Material No. 2.4858), stack dimensions: dia. = 980 mm, thickness = 4 mm; built in spiral heat exchangers, Rosenblad system: area a= 35 m^2/20m^2/8.5m^2, design: p = 0,4 bar, t = 850 °C; photograph by Carl Canzler GmbH & Co., Düren

The heat-exchanger stack in NiCr21Mo/Alloy 825 (2.4858) shown in Fig. 5.5 is employed for the regeneration of catalysts in a petrochemical plant.The waste gases leaving the combustion chamber at a temperature of approximately 850 °C are used to heat steam or nitrogen to about 500 °C in the first stage and then to heat combustion air to the combustion chamber and to the burners to the highest possible temperature in the second and third stages. The spiral heat exchangers in the stack operate in a cross flow system. In the process described the material is severely attacked both thermally as well as by corrosion (H_2S, SO_2, chloride-bearing impurities).

The manufacture of the equipment was carefully optimized from the viewpoint of material selection, welding process, filler metal and design. In the course of this optimization special test pieces were tested in operationally similar conditions in cooperation with the material supplier.

Nickel and nickel alloys are also frequently used as liner materials for chemical plant equipment. Where components are to be used at elevated temperatures, consideration must be given to the coefficients of thermal expansion of the liner material and of the pressure-bearing external shell. The customary nickel alloys in chemical plants have a coefficient of thermal expansion 20-36% higher than that of structural steels, resulting in compressive stresses in the liner material at high temperatures. Mechanical overloading resulting from tensile stresses is thus unlikely to occur. Liners cannot be used in plant subjected to cycling pressures and temperatures or in vacuum equipment. If corrosion attack takes place simultaneously, these cycling stresses can result in cracking, particularly in weld regions, as a result of fatigue induced crevice corrosion. Cycling internal pressure and temperature should therefore be avoided in lined plant.

Operational experience has shown that the nickel alloy NiMo16Cr16Ti/Alloy C-4 (2.4610), for example, is very suitable as a liner material for carbon steel chemical reactors, even under high temperatures and pressures (p = 160 bar, t = 200 °C).

The highly corrosion-resistant nickel alloys NiMo16Cr16Ti/Alloy C-4 (2.4610), NiMo16Cr15W/Alloy C-276 (2.4819), NiCr23Mo16Al/Alloy 59 (2.4605), NiMo28/ Alloy B-2 (2.4617) and NiCr22Mo9Nb/Alloy 625 (2.4896) are being used to an increasing extent in chemical plant.

When cold-forming nickel alloys, allowance must be made for the tendency of these materials to work-harden. This is of particular importance for components which, due to operational stresses, need to be ductile to ensure adequate reliability in operation. The strength and toughness behaviour of sheet material made of NiCr15Fe/Alloy 600 (2.4816) in cold-formed and subsequently heat-treated condition have been investigated, the results being shown in Table 5.1.

Table 5.1: Strength properties of Alloy 600 (NiCr15Fe Material No. 2.4816) as 6.5 mm thick plate at room temperature, cold-formed by bending through approx. 12 %, subsequently annealed

Position of specimen	Condition	Rp 0,2 (N/mm^2)	Rm (N/mm^2)	A5 (%)	Hardness HV1
Flat plate	Delivery condition	417	835	37	210 – 229
Curved flange	12 % cold-formed	626	830	10	248 – 313
Flat plate	10 min. 940 – 950 °C	376	682	37.5	
Curved flange	10 min. 940 – 950 °C	347	747	41	224 – 251

It can be seen that subsequent heat treatment is essential for cold-formed, mechanically stressed components, particularly for components with toughness requirements (e.g. expansion joints) to ensure adequate operational reliability and to simultaneously meet the requirements of the AD regulations.

5.3 Composite materials in chemical plant construction

Composite materials have found a wide range of applications in equipment and plant construction. For numerous applications the clad design offers economic and technical advantages compared to solid designs. Therefore clad plates are very important in equipment fabrication [1].

5.3.1 Material combinations for composite materials by method of manufacture

5.3.1.1 Roll-bond cladding

In roll-bond cladding a clad packet (sandwich) is manufactured from plates of the base and cladding materials and then rolled at approx 1250-750 °C depending on the material combination. The bond between the base and cladding materials is formed under the influence of pressure and temperature by diffusion and solid solution formation. After rolling the clad plate is usually heat-treated as a packet. Roll-bond cladding is only suitable for material combinations in which the thermal effects do not give rise to any brittle intermetallic phases. Intermediate layers can be used that do not form intermetallic brittle phases in the rolling temperature range in question.

In the roll-bond cladding procedure recently developed by Voest-Alpine Stahl Linz Ges. m.b.H. [2] the clad plate is rapidly cooled from the rolling temperature range. This effectively suppresses the formation of resistance-reducing structural precipitates in precipitation-susceptible cladding materials, e.g. the austenitic steel Cronifer 1925 hMo (1.4529), so that the cladding material has an optimum corrosion resistance after roll-bond cladding [3]. Composite materials with highly corrosion resistant claddings such as alloy C-276 (2.4819) also do not require subsequent heat-treatment after rolling with this manufacturing technique.

With certain material combinations intermediate layers of nickel prevent or reduce undesired diffusion (e.g. C diffusion) [4].

Clad plates with these cladding materials that have to be subjected to special heat treatments, e.g. solution heat treatment, for reasons of corrosion resistance require special base materials [2]; see 5.3.3.2.

Using roll-bond cladding it is possible to produce very large formats (2.8 x 10 m) with unit weights up to 20 tonnes. Table 5.2 summarizes material combinations of roll-bond clad steel plates with nickel alloy cladding.

Table 5.2: Rollbond-clad composite materials with nickel alloy claddings; tested material combinations

Base material	**Clad material**		
	Material No.	Symbol	Alloy
Structural steels to DIN 17100	2.4066	Ni 99.2	200
	2.4068	LC-Ni 99	201
Boiler steels to DIN 17155 as well as VdTÜV Werkstoffblatt 427/1 resp. 440/1 (extracts)	2.4360	NiCu30Fe	400
	2.4617	NiMo28	B-2
	2.4816	NiCr15Fe8	600
Fine grained structural steels to DIN 17102	1.4876	X10NiCrAlTi3220	800
	2.4858	NiCr21Mo	825
Tube steels	2.4856	NiCr22Mo9Nb	625
Shipbuilding steels	2.4610	NiMo16Cr16Ti	C-4
	2.4819	NiMo16Cr15W	C-276
	2.4605	NiCr23Mo16Al	59

5.3.1.2 Explosive cladding

The process of explosive cladding is basically characterized by a cladding sheet coated with explosive (the “flying plate”) being located over the base material at a constant spacing (acceleration path), and “fired” onto the base plate after detonating the explosive at a high velocity of approximately 100 to 1000 m/s.

The spacing depends on the material combination and cladding dimensions and varies from 0.1 to 60 mm. In the explosive process pressures of 10 to 100 kbar arise at the collision point, causing the metal surfaces to deform plastically and the top layers of atoms to be ejected. The metallurgically clean surfaces approach each other to within the lattice spacing and a metallic bond forms between the metal partners. In the interface area there is a very narrow (approx 1 to 5 μm wide) structural zone of high strength consisting of components of the two metal partners [5].

Explosive cladding is carried out in the atmosphere and in vacuum chambers at a vacuum of 40 mbar. The latter process has the advantage that the energy expended to overcome the air resistance in the wedge-shaped explosion gap is significantly lower and consequently reduces the amount of explosive required by up to 40 % compared with the conventional explosive process. Further advantages are lower shock-wave stresses, a reduction in interface area defects, better cladding performance of thin large-format cladding sheets and a fabrication procedure which is not dependent on the weather [6].

The quality of the cladding, characterized by the form of the interface area (flat, wavy), shear strength and bond strength, is dependent on the cladding parameters such as collision point velocity, impact velocity and collision angle. Using explosive cladding, basically all the common clad materials used in chemical plant construction can be produced, i.e. also those in which intermetallic phases form under the influence of heat, e.g. steel/zirconium and steel/tantalum.

With certain material combinations, again undesired diffusion processes can be avoided or reduced by the use of suitable intermediate layers. An example is the material combination known as boiler plate H II/NiMo28 (alloy B-2). Using a C-diffusion-retarding intermediate layer of the austenitic steel 1.4541, crack formation in the cladding during heat treatment and hot working is prevented [7].

Special base materials are necessary, as in roll-bond cladding, for explosively clad sheets with cladding materials that have to be subjected to special heat treatments for reasons of corrosion resistance.

The sizes of the explosively clad sheets are restricted by the limited measurements of the clad materials and by the limitation of the explosion chambers and the quantity of explosive. Table 5.3 gives data on material combinations of explosively clad composite materials with nickel alloy claddings.

Table 5.3: Explosively clad composite materials with nickel alloy claddings; tested material combinations

Base material	Cladding material	Comment
Structural steels to DIN 17100 Fine grained structural steels to DIN 17102 Boiler plate to DIN 17155	Ni99.2 - alloy 200 LC-Ni99 - alloy 201 NiCu30Fe - alloy 400 NiMo28 - alloy B-2* NiCr15Fe8 - alloy 600 NiCr21Mo - alloy 825 NiCr22Mo9Nb - alloy 625 NiMo16Cr16Ti - alloy C-4 NiMo16Cr15W - alloy C-276 NiCr23Mo16Al - alloy 59	* with intermediate layer of 1.4541
	Special claddings	
Stainless austenitic steels to DIN 17440	LC-Ni99 - alloy 201 NiMo28 - alloy B-2 NiCr21Mo - alloy 825 NiMo16Cr16Ti - alloy C-4	
Austenitic special stainless steels X10NiCrAlTi3220 - alloy 800 H	LC-Ni99 - alloy 201 NiMo16Cr16Ti - alloy C-4	
Nickel-base alloy NiCr15Fe8 - alloy 600	LC-Ni99 - alloy 201	
E-Cu	NiCr19NbMo - alloy 718	clad both sides

5.3.1.3 *Rolling after explosive cladding*

In this process explosively clad slabs are hot rolled into large-format clad plates. When cladding with stainless steels, the thermal effect intensifies the diffusion in the interface area, which has a favourable effect on the bond strength of the cladding (8). This process is also used for the manufacture of sheets clad with nickel, NiCu30Fe and NiCr15Fe [8].

5.3.1.4 *Cladding by deposit welding*

Cladding by deposit welding is described in detail in Chapter 3. It should be mentioned that this process can be manually carried out for small-scale operations, e.g. seal faces on nozzles, whereas mechanized high performance processes such as electroslag welding, GTAW hot wire, or plasma hot-wire welding are available for deposit welding components with large surfaces.

5.3.2 Quality control

The AD specification W8 (07.87) controls the materials, specifications, heat treatment, testing and quality verifications of clad plate used for the construction of pressure vessels. The following tests are carried out on clad plate:

- Ultrasonic testing, bend test with side bend specimen and shear test for quality verification in the interface area.
- Tensile test on unclad specimens to test the mechanical properties of the base metal.

NOTE: In composite materials in which the clad material is included in the stress calculations specimens with cladding are tested.

- Notched-bar impact tests on the base metal with specimens taken close to the cladding to verify the toughness properties in the base metal.
- Structural investigation and hardness testing to test the structural state and the hardness as well as the form of the interface area.
- Wall thickness measurements on materials clad by roll-bond cladding.

Further tests may be arranged, including surface crack testing by means of dye penetrant test and testing of the corrosion resistance.

Moreover, it is necessary from case to case to simulate the working operation and final state of working on suitable test pieces, before commencing fabrication, e.g. temperature control in hot forming of heads, final heat treatment and welding. The results of the simulation tests enable knowledge to be obtained about the effect of further manufacturing (e.g. hot forming, heat treatment and welding) on the material properties of the components in the final state with sufficient reliability.

5.3.3 Properties and working of composite materials with particular reference to plate clad with nickel and nickel alloys

5.3.3.1 Properties

In roll-bond clad plate, the base and cladding materials are usually in a recrystallized state. The interface area is flat. Localized work hardening (hardness peaks) in the base and cladding material do not occur.

Explosively clad composite materials work-harden in both the base and the cladding material because of the explosion shock loading. This is characterized by a decrease in the elongation and notch impact toughness in the base metal, sometimes to values below the minimum requirement.

This work-hardening of the cladding material is revealed by pronounced hardness peaks in the interface area, particularly with high-molybdenum nickel-base alloys. In these cladding materials hardness peaks of HV1 = 500 - 550 are observed (Fig. 5.6). The work-hardening is reduced by an appropriate heat treatment, which has to be compatible with both the base and cladding materials. This largely restores the properties to the original state [9]. The structure of the previously work-hardened zones exhibits fine-grained recrystallization in the heat-treated state. Resistance to intercrystalline corrosion and shear strength are not impaired by the heat treatment. It should be pointed out that steel sheets clad with NiMo28 have for some time only been manufactured with a diffusion-retarding intermediate layer (nickel or 1.4541), see Section 5.3.1.2.

Table 5.4: Mechanical properties of an explosive-clad material in Alloy 600/201 (NiCr15Fe/LC-Ni99), wall thickness 15 + 3 mm, strength and toughness properties of base material Alloy 600 (NiCr15Fe8) in initial condition, clad condition and after heat treatment 980 °C/30 min/air cooling. Round tensile specimens: B 8 x 40, DIN 50125/DVM specimens, specimen position q, notch perpendicular to plate surface. Specimen taken from directly below cladding layer.

	Rp 0,2 (N/mm^2)	Rp 1,0 (N/mm^2)	Rm (N/mm^2)	A5 (%)	Impact energy (J/cm^2)
Condition, minimum requirement	Minimum requirements as per VdTÜV Material Data Sheet 305/1.79 and/or VDM Material Data Sheet				
	180	–	550–750	30	150
before cladding	378	424	734	34	186
after cladding	551	580	756	28	156
after cladding and annealing	372	414	720	35	196

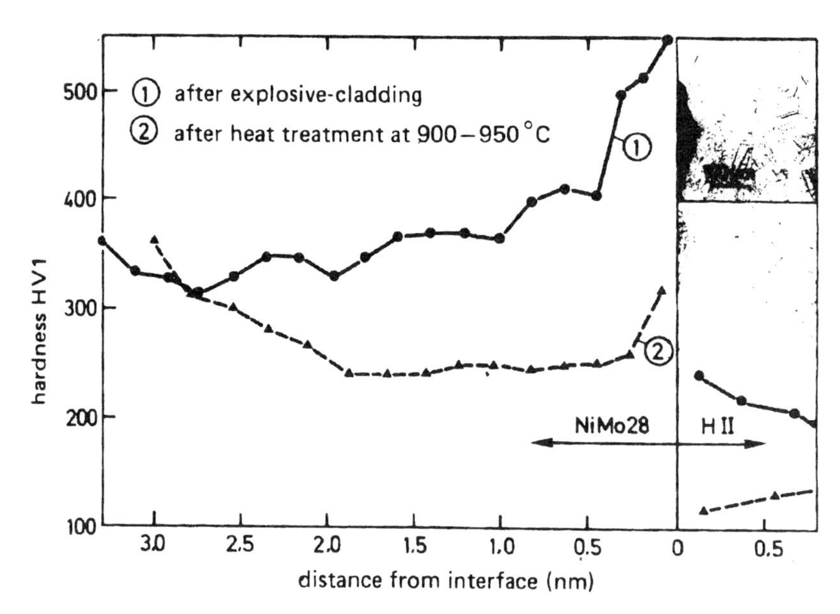

Figure 5.6: Explosive-clad composite material comprising boiler plate H II Alloy B-2 (NiMo28 Material No. 2.4617), hardness curve in explosive-clad and subsequently heat-treated condition, and structure of Alloy B-2 (NiMo28) at interface

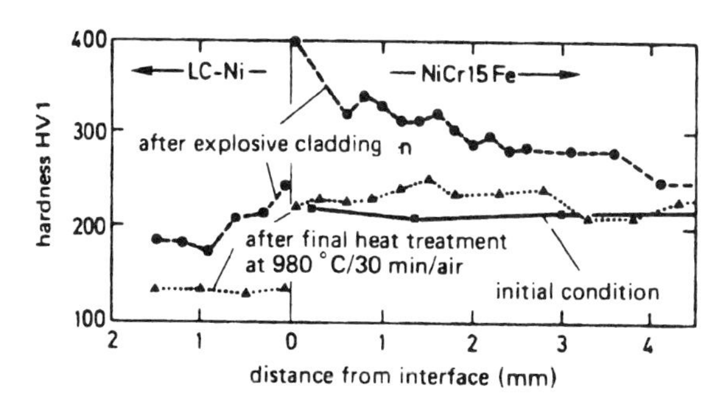

Figure 5.7: Explosive-clad composite material comprising Alloy 600 (NiCr15Fe8 Material No. 2.4816)/Alloy 201 (LC-Ni99 Material No. 2.4068), wall thickness 15 + 3 mm, hardness curve in the area of the interface for explosive-clad condition and after subsequent heat treatment at 950 °C/30 min/air

Tests were carried out on an explosively clad plate made of the base material NiCr15Fe (2.4816) clad with LC-Ni 99 (2.4068) in the thicknesses 15 + 3 mm in order to determine the toughness loss in the base material. Figure 5.7 indicates a clear increase in hardness in the interface area. As Table 5.4 shows, the results of tensile testing indicate a clear rise in the yield strength and a decrease in elongation at fracture to a value slightly below the minimum requirement. It can also be seen that a reduction in notch impact toughness occurs, although this does not fall below the minimum requirement. Shear strength tests showed a reduction from 334 to 283 N/mm^2 in the annealed state, a drop of approximately 15% (minimum required: 140 N/mm^2). Side bend tests produced bending angles of 180 degrees without cracking. The condition of the structure is shown in Figure 5.8. A clear grain structure elongation can be discerned in the interface area in the non-heat-treated condition. The base material NiCr15Fe exhibits, in heat-treated condition, fine-grained recrystallization in the previously work-hardened area. As anticipated, pronounced grain growth has occurred in the nickel cladding. The interface area has a marked wavelike profile, which is not an optimum condition. Summarizing, it can be stated that in this case no critical impairment of toughness characteristics is caused to the base material by explosive cladding. These results indicate the justification of dispensing with subsequent heat treatment of explosively clad plate, since heat treatment in the temperature range around 980 °C can be regarded as critical for the nickel cladding due to the danger of embrittlement resulting from intergranular oxidation and sulphur absorption.

The effects of soft annealing on the mechanical properties of the base material and on the IC resistance of the cladding material were investigated for an explosively clad composite material consisting of X10NiCrAlTi3220/Alloy 800 (1.4876) base material clad with NiMo16Cr16Ti/Alloy C-4 (2.4610).

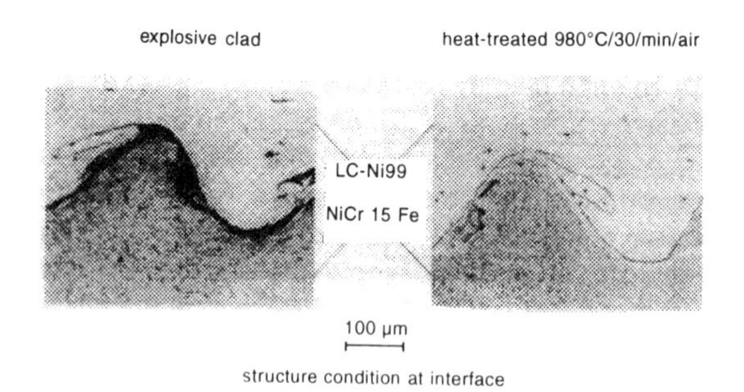

Figure 5.8: Explosive-clad composite material comprising Alloy 600 (NiCr15Fe Material No. 2.4816)/Alloy 201 (LC-Ni99 Material No. 2.4068), structure condition at interface

As was the case with tests on other explosively clad composite materials with claddings of NiMo alloys, clear work-hardening caused by explosion loading occurred in the interface area, which is removed almost completely by means of heat treatment in the temperature range 920-980 °C. The effect of heat treatment in the soft annealing range (920-980 °C) on the mechanical properties of the base material is shown in Figure 5.9. The test results indicate that the toughness properties (A5, ak) in the explosively clad state meet the requirements of the relevant codes, and a significant reduction of shock-wave hardening can be achieved by means of heat treatment at 920 °C. The resistance of the cladding material of the heat-treated composite to intercrystalline corrosion was tested in accordance with Stahl-Eisen-Prüfblatt 1877 (June 1977), Verfahren II.
Result: resistant to intercrystalline corrosion.
The weldability of the work-hardened base material was not examined.

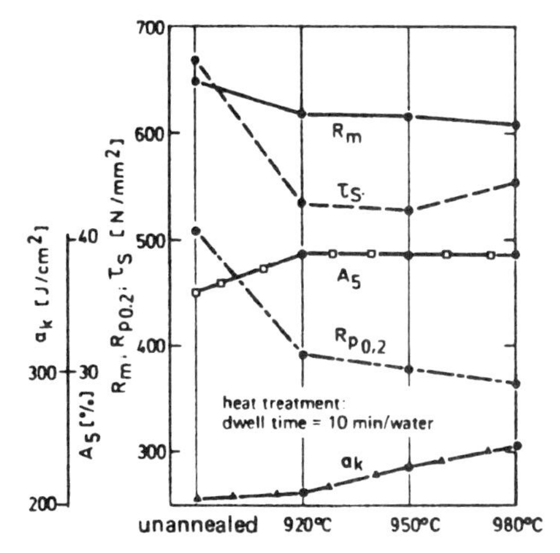

Figure 5.9: Explosive-clad composite material comprising Alloy 800 (X10NiCrAlTi3220 Material No. 1.4876)/Alloy C-4 (NiMo16Cr16Ti Material No. 2.4610), wall thickness 10 x 3.2mm. Material properties of the Alloy 800 base material (X10NiCrAlTi3220) at room temperature related to heat treatment (soft annealing). The base material's initial condition was: soft annealing, grain size ASTM 6–7

The influence of the as-delivered condition of the base material (soft annealed or solution heat-treated) on the condition of the interface area was examined for explosively clad plate with the previously mentioned material combination, both for the explosively clad and for the subsequently heat-treated conditions. Figures 5.10 and 5.11 show the structure of the interface area in the region of the base material, and the hardness curve is shown in Figures 5.12 and 5.13. Hardness testing indicated that the extent of the structure region showing definite work-hardening in the base material tends to be related to the heat-treated con-

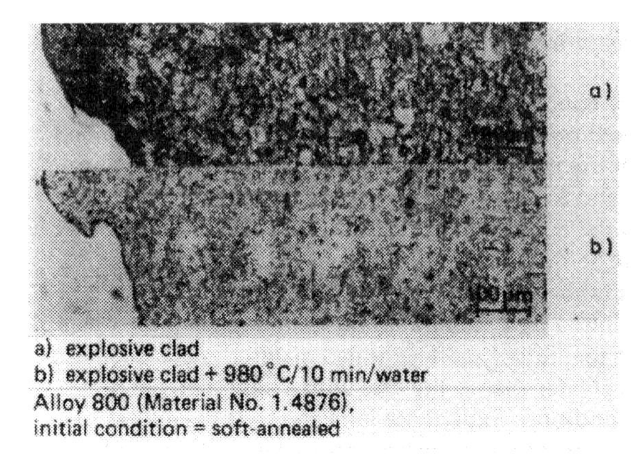

Figure 5.10: Explosive-clad composite material comprising Alloy 800 (X10NiCrAlTi3220 Material No. 1.4876)/Alloy C-4 (NiMo16Cr16Ti Material No. 2.4610). Structure of base material, soft annealed for initial condition

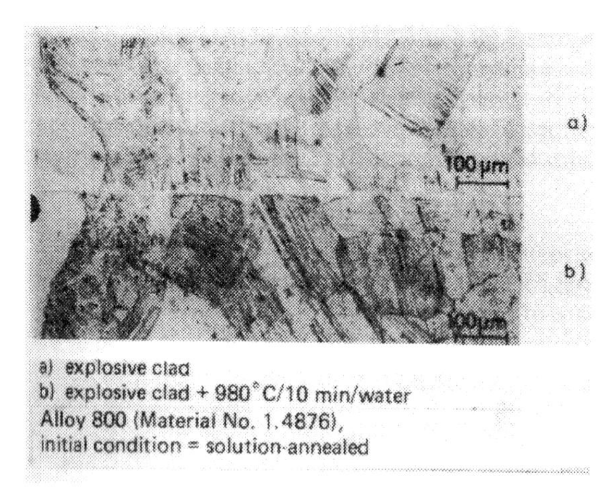

Figure 5.11: Explosive-clad composite material comprising Alloy 800 (X10NiCrAlTi3220 Material No. 1.4876)/Alloy C-4 (NiMo16Cr16Ti Material No. 2.4610). Structure of base material, solution annealed for initial condition

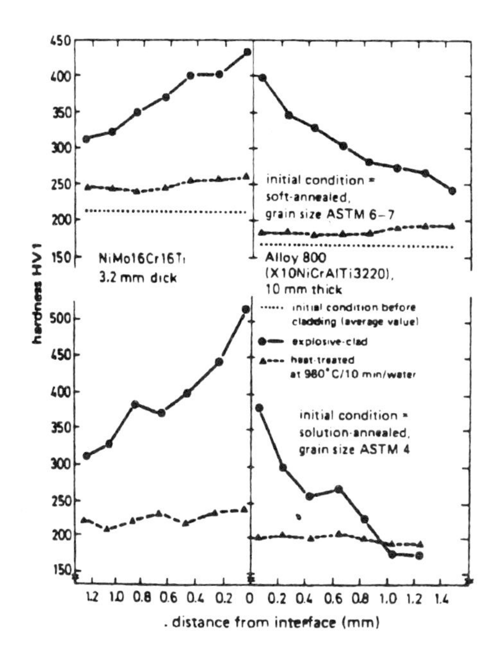

Figure 5.12 and 5.13: Explosive-clad composite material comprising Alloy 800 (X10NiCrAlTi3220 Material No. 1.4876)/Alloy C-4 (NiMo16Cr16Ti Material No. 2.4610). Hardness curve in the soft-annealed and solution-annealed initial conditions of Alloy 800 base material (X10NiCrAlTi3220)

dition. The work-hardening of the originally soft-annealed base material is characterized by the hardened zone being greater than is the case with a base material in originally solution heat-treated condition. Experience indicates that the extent of the fine-grained recrystallized structure in the heat-treated condition can be used as a criterion for assessing a critical cold work-hardening caused by explosive cladding. As roughly indicated by the results of the hardness tests, the fine-grained recrystallized area of the structure is very narrow (approximately 150-200 µm) in the case of the solution heat-treated original condition, and significantly larger (approximately 700-800 µm) in the case of the orignally soft-annealed condition. These observations permit the deduction that the base material X10NiCrAlTi3220 shows in coarse-grained, solution heat-treated condition, a reduced tendency towards work-hardening as a result of explosive cladding with nickel alloys than it does in fine-grained, soft-annealed condition.

The cladding material suffers pronounced work-hardening in the interface area. This is reduced by heat treatment in the soft annealing temperature range (920-980 °C). Fine-grained recrystallization of the previously work-hardened structural zone is again found in this case. The mechanical properties have to satisfy the requirements of AD-Merkblatt W 8 (07.87) for use in pressure vessel construction. This also has to be guaranteed in the worked state. In the final worked state the cladding should have comparable corrosion resistance to the material in the non-clad condition. The corrosion resistance is not affected by explosive cladding; this has been confirmed by numerous experiments [5].

5.3.3.2 Fabrication

In process plant construction, the most important operations are cold and hot forming, heat treatment and welding. Special claddings (see Table 5.3) are excluded from the following descriptions. Special fabrication specifications have to be produced for each case.

Cold working is preferentially used for the manufacture of cylindrical shells and dished ends or segments for dished ends with thin and meduim wall thicknesses. It is done by roll forming, pressing, flanging and rolling. If heat treatment is necessary, for instance normalization or stress relieving, the properties of the clad material have to be taken into account. According to the AD specifications; a final normalization treatment is necessary after cold working with a degree of deformation > 5 %.

Hot working is used mainly for the production of flanged and pressed components.

Standard claddings are hot-worked at approx 950-1050 °C. Standard clad materials with nickel alloy claddings include composite materials with base metals of common structural steels (boiler plate, fine-grained structural steel) and clad materials of nickel, NiCu30Fe and NiCr15Fe. The last hot-working procedure should finish in the region of the normal annealing temperature of approx. 900 °C ("normalization pressing"). Subsequent annealing is then unnecessary. Clad sheets with precipitation-susceptible clad materials are hot-worked at approx. 1100 °C to achieve optimum corrosion resistance. The finished components are finally heat treated in the region of the solution heat-treatment temperature for the cladding (1050-1150 °C) and quenched in water. The common base metals (boiler plate, fine-grained structural steel) are not suitable or only suitable to a limited degree for this high-temperature annealing because of the expected coarse grain formation and loss of toughness. Special steels therefore have to be used that do not have a tendency to form coarse grains during the high-temperature annealing mentioned above and do not undergo any reduction in toughness properties. The low carbon Mn- and Ni-alloyed steel type 12MnNi63 has proved to be well suited [10]. Voest-Alpine Stahl Linz Ges. m.b.H has developed special base metals for this purpose.

Stress relieving is carried out at 540-620 °C and in special cases on cold worked components as an alternative to normalization treatment as well as on welded pressure vessels with wall thicknesses ≥30 mm and ≥38 mm (AD specification HP7/2). This heat treatment can have a negative influence on the material and properties of clad components with precipitation susceptible cladding materials, including reduction of the corrosion resistance. For this reason it is recommended that the manufacturer of the semi-finished product (e.g. sheet material) is consulted in every case.

Composite materials with nickel alloy claddings are welded following the guidelines in DIN 8553. Details on edge preparation and welding are given in Section 3.2.1. For welds that can be accessed from both sides, the base metal is usually welded first and then the clad side after grinding out the weld root. Generally, the first layer on the clad side is applied with a higher-alloyed filler metal as a buffer layer. Filler metals that are matched to the properties of the base and cladding materials are used. In the case of clad plate with claddings of nickel, alloy C-4, alloy 59 and alloy 400, the filler metal for the buffer layer is also used for the subsequent and capping passes. Basically, care should be taken to ensure that when welding the cladding, dilution by the base metal is kept as low as possible to achieve the maximum degree of corrosion resistance. As an example, for Ni-clad plates the iron content of the capping pass of the cladding material weld is limited to $\leq$ 0.4 %. The weld required metal purity can be achieved by multi-pass stringer bead welding.

Filler metals for nickel and nickel alloys are standardized in DIN 1736 (08.85) Recommendations for the selection of filler metals for welding of composite materials with nickel and nickel alloy cladding are given in Table 5.5.

For the manufacture of mechanically highly stressed welded joints on the cladding side, e.g. on mechanically heavily loaded internal components, welding of the components made of nickel or nickel alloys is preferably carried out directly with the base metal in order to guarantee a strong joint with the usually thick-walled structural steel vessel shell. For this purpose the cladding is machined off and a deposit welded cladding is applied using suitable nickel alloy filler metals prior to welding the internal components. Dilution with the base metal should be as low as possible to retain the corrosion resistance. With roll-bond clad plate from Voest-Alpine Stahl Linz Ges. m.b.H, removal by machining of the roll-bonded cladding followed by weld-cladding as a preparation for fillet welding is only necessary when local bonding defects of the cladding have been found by US testing [2].

Practical examples of the welding of nickel-clad plate are shown in Fig. 5.14. The mechanical properties of the welded joints satisfy the requirements of the AD guidelines, while the structure and alloy condition of the cladding-side welds meet the requirements for preventing corrosion attack of LC-Ni 99.

Table 5.5: Welding of clad sheets with nickel alloy claddings. Recommended welding filler metals [2]

Cladding		Filler metal type					
DIN Mat. No.	Alloy Type	muiti pass (GTAW/GMAW)				single pass	
		Buffer pass		Following passes		RES-(SA)-strip	
		DIN	AWS	DIN	AWS	DIN	AWS
2.4066	Ni200	SG-NiTi4	ERNi-1	SG-NiTi4	ERNi-1	UP-NiTi4	ERNi-1
2.4068	Ni201						
2.4360	alloy 400	SG-NiCu30MnTi	ERNiCu-7	SG-NiCu30MnTi	ERNiCu-7	SG-NiCu30MnTi	ERNiCu-7
1.4876	alloy 800			SG-NiCr21Mo9Nb	ERNiCrMo-3	UP-NiCr20Nb	ERNiCrMo-3
2.4858	alloy 825	SG-NiCr21Mo9Nb	ERNiCrMo-3				
2.4816	alloy 600			a) ERNiCrMo-3 b) SG-NiCr20Nb	a) ERNiCrMo-3 b) ERNiCr-3	UP-NiCr20Nb	ERNiCr-3
2.4856	alloy 625	a) SG-NiCr21Mo9Nb b) SG-NiMo16Cr16Ti	a) ERNiCrMo-3 b) ERNiCrMo-7	SG-NiMo16Cr16Ti	ERNiCrMo-7	UP-NiMo16Cr16Ti	ERNiCrMo-7
2.4605	alloy 59	SG-NiCr23Mo16Al	---	SG-NiCr23Mo16Al	---	UP-NiCr23Mo16Al	---
1.4610	alloy C-4	SG-NiMo16Cr16Ti	ERNiCrMo-7	SG-NiMo16Cr16Ti	ERNiCrMo-7	UP-NiMo16Cr16Ti	ERNiCrMo-7
2.4617	alloy B-2			SG-NiMo27	ERNiMo-1	---	---

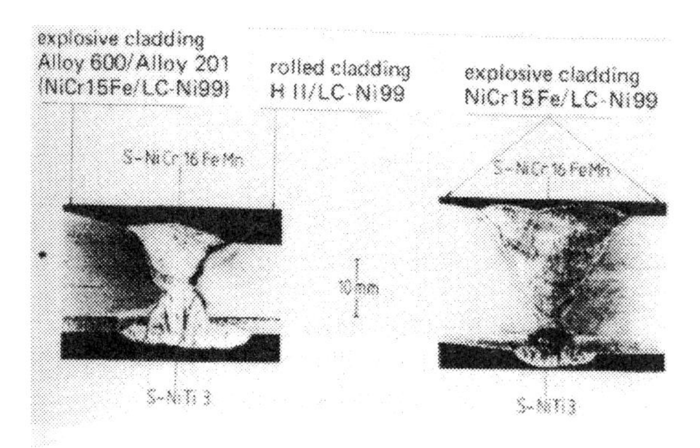

Figure 5.14: Welding of explosive-clad plate with Alloy 201 cladding material (LC-Ni99 Material No. 2.4068)

5.4 Use of composite materials with nickel alloy cladding in chemical plant

5.4.1 Vessels made of clad plates

The example shown in Fig 5.15 is a vessel of nickel clad steel plate in which the iron content of the capping pass of the cladding weld was limited to 0.3 % to meet the increased requirements for corrosion resistance.

In such cases special welding conditions (e.g. stringer bead welding and limitation of specific heat input) should be employed during the welding of the nickel cladding to ensure that dilution by the base metal is kept as low as possible. With the agitator vessel shown in Fig. 5.16 adequate resistance to intercrystalline corrosion was verified in accordance with the Stahl-Eisen-Prüfblatt 1877 Verfahren III for semi-finished products of clad plate after heat treatment, hot working and welding. Fig. 5.17 illustrates the application of explosively clad plate made of heat-resistant austenitic steel X10NiCrAlTi3220/Alloy 800 (1.4876) clad with the nickel-base alloy NiMo16Cr16Ti/Alloy C-4 (2.4610). Longitudinally welded tubes made of explosively clad plate of the material combination mentioned were used for the manufacture of the vessel nozzles. The GTAW longitudinal welding was carried out from one side with the filler metal SG-NiMo16Cr16Ti for the root pass on the cladding side and with the filler metal SG-NiCr20Nb to build up the weld in the base metal.

The clad design was selected on account of the critical operating conditions (cycling pressure and temperature loading). Lined designs are not suitable for these operating conditions. For the shell plates the base metal was explosively

Figure 5.15:
Reactor in roll-clad composite material WSTE 355/Alloy 201 (LC-Ni99 Material No. 2.4068), dimensions: dia. = 2000 mm, height = 2900 mm, thickness = 32 + 3 mm for cylindrical shell, thickness = 40 + 3 for dished ends. Design data: pressure = 30 bar/ temperature = 350 °C. Particular requirements: Fe content in cap pass of Ni cladding weald < 0.3 %.
Photograph by Carl Canzler GmbH & Co., Düren

Figure 5.16: Agitator vessel in explosive-clad composite material boiler-plate H II/Alloy B-2 (NiMo28 Material No. 2.4617) with external shell of boiler-plate H II. Dimensions of clad vessel: dia. = 1792 mm, height = 2254 mm. Design data: internal chamber pressure = 10/20.7 10/ 20.7 bar, temperature = -20/+200 °C, shell chamber p = 6.9 bar/t = -20/+200 °C. Photograph by Carl Canzler GmbH & Co., Düren

Figure 5.17: Components of a reaction vessel in explosive-clad composite material Alloy 800 (X10NiCrAlTi3220 Material No. 1.4876)/Alloy C-4 (NiMo16Cr16Ti Material No. 2.4610). Dimensions: dia. = 1400 mm, height = 5600 mm, thickness = 16 + 3 mm for cylindrical shell, thickness = 10 + 3 for henispherical ends

clad in solution heat treated condition. The clad semi-finished products were then heat-treated (soft annealed) at 960 °C. The properties of the clad semi-finished products meet the requirements of AD-Merkblatt W8 and VdTÜV-Werkstoffblatt 434. Adequate resistance to intercrystalline corrosion was verified for the clad material according to Stahl-Eisen-Prüfblatt 1877, Verfahren II. The base material of the clad plate for the hemispherical ends was explosively clad in soft annealed condition with NiMo16Cr16Ti/Alloy C-4 (2.4610). The ends were manufactured using multi-stage hot forming in the temperature range 1100-850°C followed by solution heat treatment at 1150 °C.

The mechanical properties of the finished component were tested according to AD guidelines. The results complied completely with the specification requirements. The intercrystalline corrosion resistance of the cladding material was tested according to Stahl-Eisen-Prüfblatt 1877, Verfahren II. The cladding material was also proved to be resistant to grain disintegration. For the pressure vessel shown in Fig. 5.18, constructed of roll bond-clad sheets of the material combination 12MnNi63/Nicrofer 6020 hMo - alloy 625 (2.4856), the dished ends were manufactured by hot-forming, annealed at about 1050°C and finally quenched in water. The cladding material proved to be free of precipitation, while the mechanical properties of the composite material completely fulfilled the requirements of the guidelines.

Figure 5.18: Autoclave, manufactured from roll-clad steel plate, 24 mm 12MnNi63 steel clad with 3 mm Nicrofer 6020 hMo - alloy 625 (2.4856). Dimensions 250 mm diameter, 5860 mm height, working at 19.6 bar at 214 °C. Photograph by BEA Canzler GmbH & Co., Düren

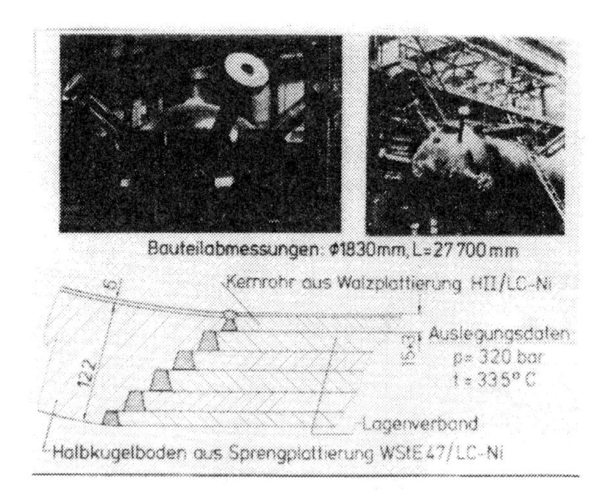

Figure 5.19: Oxidation Reactor of multi-layer design, with core tube in roll-clad composite material boiler-plate II/Alloy 201 (LC-Ni99). Krupp system multilayer design. Photograph by Carl Canzler GmbH & Co., Düren

Fig. 5.19 shows an example of the use of nickel-clad steel plate as a liner material for thick-walled pressure vessels with high working pressures and temperatures. In this design the nickel material subjected to heavy corrosion attack is relieved of mechanical loading as much as possible, which has a favourable influence on the corrosion resistance behaviour. Liners made of solid nickel sheet proved to be unsuitable for the service conditions present in this case because of their susceptibility to stress corrosion cracking in the weld area.

5.4.2 Shell-and-tube heat exchangers

5.4.2.1 Joining tubes to tube sheets Mechanical and hydraulic tube expansion

In tube expansion, where the connection's leak-tightness is an important requirement, the tube material, due to the high expanding forces necessary, is generally subjected to pronounced work-hardening and reduction of wall thickness. Work-hardening and reduction of the tube-wall thickness have not been observed in hydraulic tube expansion. In hydraulic expansion, the tube is initially deformed elastically, then plastically by the applied hydraulic pressure, and is pressed against the hole wall [11-13]. Figure 5.20 shows the basic arrangement for the application of this method. This technique is particularly suitable for the installation of tubes with thin walls or critical expansion behaviour.

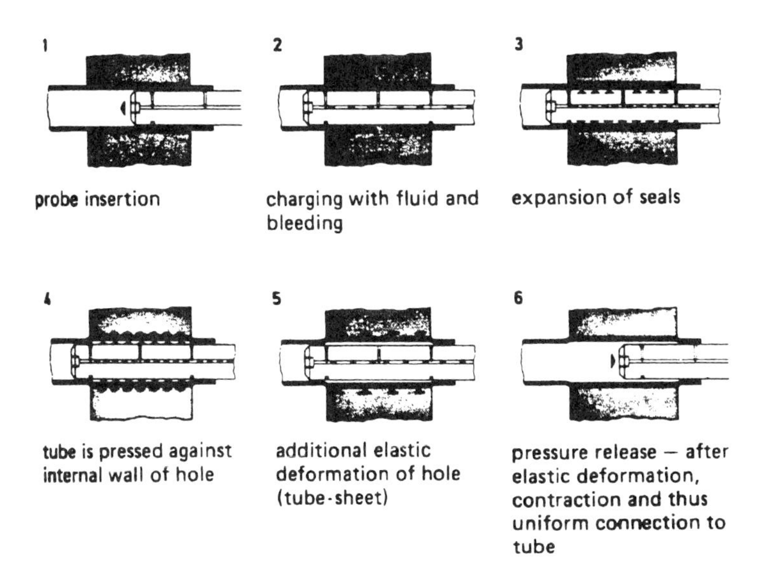

Figure 5.20: Hydraulic tube-to-tube-sheet joints, principles of the process

Important criteria for the quality of mechanical tube fixing are joint expansion, wall thickness reduction and cold work-hardening of the expanded tube [14]. Expansion tests (rolling, hydraulic expansion) were carried out on nickel tubes. Hardness measurements demonstrated clearly, as shown in Fig. 5.21, that work-hardening of the tube wall occurred in mechanical tube expansion, particularly in the area of the tube inner wall surface. In contrast, as can also clearly be seen from Fig. 5.21, no significant work-hardening of the tube wall occurred with hydraulic expansion. The work-hardening characteristics were also found in other tests. Microstructural investigation of the mechanically expanded tube revealed a clearly defined deformation structure, whilst the hydraulically expanded tube material exhibited no indication of cold work-hardening. The corresponding tests were deliberately carried out under unfavourable conditions, such as would not arise in practice, with regard to the wide expansion gaps selected (see caption to Fig. 5.21), in order to obtain reliable data regarding work-hardening behaviour.

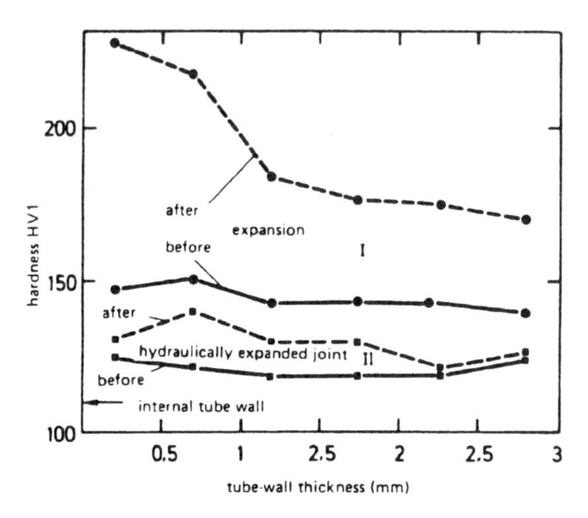

Figure 5.21: Connection of Ni tubes to tube-sheets of boiler-plate H II. Materials and dimensions: tube-sheet in H II, thickness = 90 mm; seamless tubes in Alloy 201 (LC-Ni99), dia. 57 x 3 mm; joint gap: first joint in 4 expanding stages, second joint: hydraulically expanded at p = 558 bar; tightness of joint: leakage 10-5 mbar x l/sec, verified using halogen leak test (Frigen); thickness reduction of tube wall: first joint: 7.9 %, second joint: 0.65 %

Mechanized welding of tubes to tubesheets

Mechanized tube welding is becoming increasingly popular for welding of tubes into tubesheets. The mechanized GTAW process is preferred for non-ferrous metals and austenitic steels. Extensive studies have been carried out of welding of nickel tubes into tubesheets made of nickel and nickel-clad steels.

The essential criteria for assessment of weld quality are:

- joint cross-section
- satisfactory fusion of tube and tubesheet material
- sufficient resistance of the weld to tear-out
- homogeneous weld structure, i.e. welds as free as possible of inclusions and porosity
- leak-tightness

The examination of test pieces for the determination of optimum welding parameters for production welding generally includes the following tests:

tightness testing
surface crack testing by the dye-penetrant method
radiographic examination
metallographic and hardness testing and
tear-out testing.

Depending on operational requirements, tube-to-tubesheet welds are tested on the finished heat exchanger by tightness testing (halogen or helium leak test), surface crack testing by means of dye penetrant testing and radiographic examination [15, 16].

The results of various welding tests carried out on nickel tubes are described below.

Fig. 5.22 shows the structure of a mechanized tube-to-tubesheet welded joint in LC-Ni99, carried out in two passes with SG-NiTi4 as the filler metal. The joint cross-section was adequate and the test described earlier also gave extremely satisfactory results. Fig. 5.23 shows the structure of a nickel tube-to-tubesheet weld in a nickel-clad tubesheet. The parameters selected for such welds should ensure that no fusion of the base material of the clad tubesheet occurs. Extensive tests were carried out in order to determine the optimum conditions for the welding of tubes of NiMo28/Alloy B-2 (2.4617) into tubesheets of explosively clad steel plate with NiMo28 cladding material. Mechanized GTAW welding in two passes, in this instance with SG-NiMo27 as the filler metal, again proved to be a reliable method for achieving a high and reproducible weld quality in tube-to-tubesheet welding. The resulting microstructure is illustrated in Fig. 5.24. Due particularly to the good resistance of the weld metal to highly aggressive media, high- nickel alloys and nickel-base materials have achieved considerable signifi-

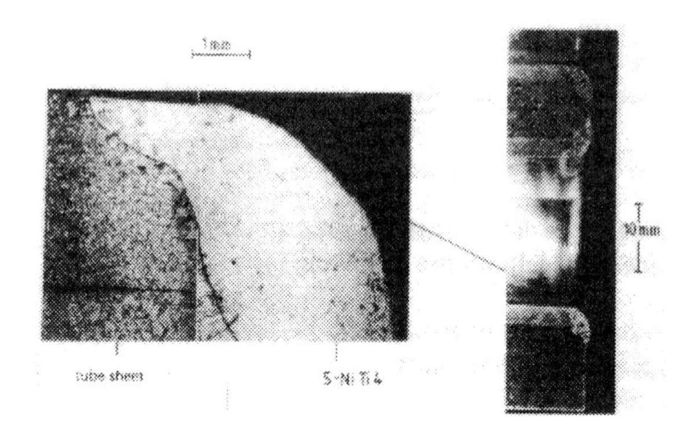

Figure 5.22: Structure of a mechanical GTAW tube-to-tube-sheet weld. Material: tubes and tube-sheet in Alloy 201 (LC-Ni99 Material No. 24068)

cance as filler metals for welding nickel-base materials to austenitic special steels, and for welding austenitic steels to one another. Table 5.6 provides a summary of tube-to-tubesheet welding tests in which good weld quality was again achieved using mechanized GTAW welding.

Table 5.6: Mechanical GTAW tube-to-tubesheet welding, proven material combinations

Component	Material	Alloy composition (indicative figures in wt. %)					
		C	Ni	Cr	Mo	Cu	Other
Tube sheet S = 30 mm	X2NiCrMoCu 25205, M.-No. 1.4539	0.015	25	21	4.8	1.5	0.07 N
Tubes 21.3x1.65 mm	NiMo16Cr16Ti M.-No. 2.4610	0.006	66	16	16	–	0.3 Ti
	X2NiCrMoCuN 25206 M.-No. 1.4529	0.015	25	21	6.2	0.8	0.14 N
Filler material for connection of 1.4539 to NiMo16Cr16Ti	S-NiCr20Mo15 M.-No. 2.4839 (wire electrode 0.8 mm dia.)	0.01	65	20	15	–	max. 1.5 Fe
Filler material for connection of 1.4539 to M.-No. 1.4529	S-NiCr21Mo9Nb M.-No. 2.4831 (wire electrode 0.8 mm dia.)	0.02	63	22	9	–	max. 3 Fe 3.4 Nb

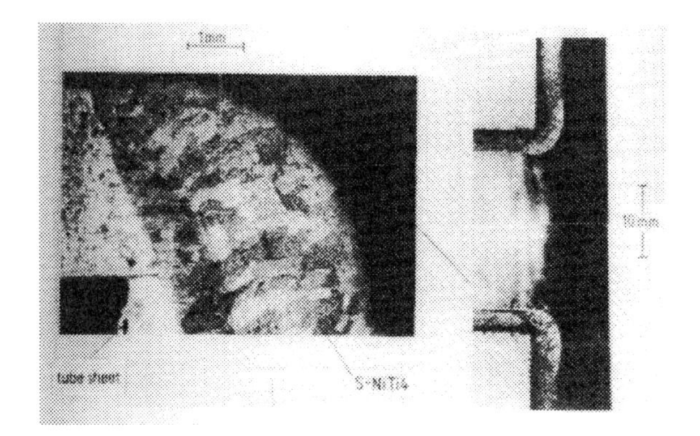

Figure 5.23: Structure of a mechanical GTAW tube-to-tubesheet weld. Materials: tubes in Alloy 201 (LC-Ni99 Material No. 24068), tube-sheet in explosive-clad, composite-material boiler-plate 15Mo3/Alloy 201 (LC-Ni99)

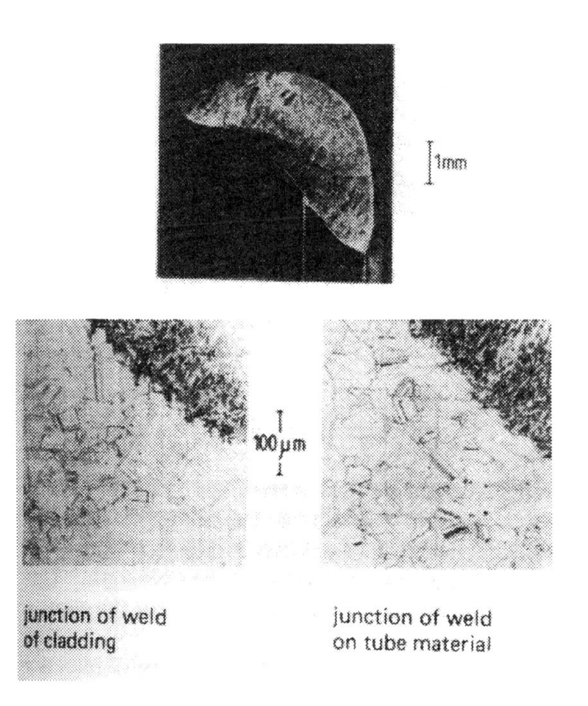

Figure 5.24: Structure of a mechanical GTAW tube-to-tubesheet weld. Materials: welded tube in Alloy B-2 (NiMo28 Material No. 24617), tube-sheet in explosive-clad, composite-material boiler-plate H II//Alloy B-2 (NiMo28), joint cross-section of tube-to-tube-sheet weld: a = 1.3 x tube-wall thickness

To summarize, mechanized tube-to-tubesheet welding produces the best possible weld quality, particularly in the case of nickel and a number of nickel alloys. Furthermore, this method provides better assurance of a reproducible high quality weld structure than does manual welding.

Explosive welding of tubes to tubesheets.

This method is characterized by the fact that the tube end is explosively expanded under special conditions so that the tube wall is accelerated against the wall of the hole to form a solid weld between the tube wall and the hole wall in a similar manner to the explosive cladding of plate [17, 18, 19]. This process has so far only been used to a limited extent for the welding of tubes into tubesheets.

5.4.2.2 Practical examples

In the shell-and-tube heat exchanger shown in Fig. 5.25 the nickel tubes have been bonded by mechanized GTAW welding into the nickel-clad tubesheet. An example of the use of high-nickel filler metals for welding of austenitic special steels to achieve a high corrosion resistance in the weld metal is shown in Fig.

Figure 5.25: Tubular heat exchanger, 80 m^2, in nickel, Materials and dimensions: shell in boiler-plate 15Mo3, dia. 1400 x 12 mm, ends and tubesheets in explosive-clad composite-material boiler-plate 15Mo3/Alloy 201 (LC-Ni99 Material No. 2.4068), wall thickness 20 + 3 and 78 + 4 mm, seamless Ni tubes dia. 57 x 3 mm, tube-to-tubesheet welding: mechanical GTAW, with filler material S-NiTi4. Design data: shell chamber 10/–1 bar, temperature = 350 °C, tube chamber 16/–1 bar, temperature = 400 °C. Photograph by Carl Canzler GmbH & Co., Düren

Figure 5.26: Tubular heat exchanger, 260 m², in X2CrNi2520 (Material No. 1.4335), materials and dimensions: shell in X10CrNiTi189 (Material No. 1.4541) and X2CrNi2520 (Material No. 1.4335), ends in X2CrNi2520, tubesheet and seamless tubes in X2CrNi2520, wall thickness 30 mm and dia. 25 x 2 mm; tube-to-tubesheet welds: mechanical GTAW, with filler material S-NiCr29Mo (Material No. 2.4656): design data: shell chamber 18 bar, temperature = 220 °C, tube chamber 4 bar, temperature = 220 °C. Photograph by Carl Canzler GmbH & Co., Düren

5.26. Tubes in X2CrNi2520/Alloy 310 (1.4335) are welded into tubesheets of the same metal using mechanized GTAW welding with SG-NiCr29Mo (2.4656) as the filler material. The joint cross-section was measured as $a \geq$ tube wall thickness. Non-destructive testing on the welds of the heat exchanger revealed no defects. Resistance to intercrystalline corrosion was proved for the welded joints.

In the case of the shell-and-tube heat exchanger made of LC-Ni (Fig. 5.27) the tubes were hydraulically expanded into the LC-Ni tubesheet under controlled conditions followed by mechanized GTAW welding. The heat exchanger is subjected to critical NaOH conditions and has been in operation for several years without difficulties. The fabrication procedure described has proved to be completely satisfactory.

Figure 5.27: Tube heat exchanger in LC-Ni 99 with 2075 m^2 heat transfer surface. Shell consists of boiler steel 17Mn4 with a compensator in NiCr15Fe8/alloy 600 (2.4816). Tubesheets in solid LC-Ni99. Dimensions: Diameter 2790 mm, length 12500 mm. Tubes are hydraulically expanded into the tubesheet followed by automatic GTAW welding. Photograph by GEA Canzler GmbH & Co., Düren.

5.5 Nickel alloys in pollution control systems (flue gas desulphurisation)

5.5.1 Basic considerations for design and material selection

Absorbers are large vessels similar in design to flat bottomed tanks with a diameter of up to 15 m and a height of up to 41 m.

They are manufactured as self-supporting components with relatively thin shell and base plates arranged in a steel supporting frame to which they are externally welded [20]. Previously some structures were manufactured entirely from structural steel and internally coated with an organic coating for corrosion protection. Severe corrosion attack was discovered after commissioning, particularly in the lower part of the absorber, which necessitated refurbishment [21] (see Section 5.5.2.3). In contrast, the all-metal concept described below has proved very successful.

The material selection for the corrosion-resistant all metal design is based on the corrosion conditions of the individual absorber components. Critical attack conditions are found in the lower part of the absorber (scrubber, clean gas ducts) and in some cases also in the flue gas chimney, it is possible for the temperature there to fall below the dew point [22].

In numerous plants, the highly corrosion resistant nickel base alloys NiCr22Mo9Mb/Alloy 625 (2.4856) and NiMo16Cr15W/Alloy C-276 (2.4819) were previously preferred and used with very good results. In a few cases the nickel-base alloys NiMo16Cr16Ti/Alloy C-4 (2.4610) and NiCr22Mo7Cu/Alloy G-3 (2.4619) were also used [23]. Special stainless steels were used for the less critically stressed components and absorber regions [22].

5.5.2 Designs of absorbers

5.5.2.1 Solid design

Hot rolled plates and strip of alloy 625 and special stainless steels in wall thicknesses from approx. 20 mm down to 5 mm were used [20]. The following welding techniques were utilized.

For butt welds:
- manual GTAW
- pulsed manual GTAW
- manual GTAW, simultaneous welding on both sides

For build-up and capping passes on butt welds
- SMAW

For fillet welds in the support structure
- GTAW to a limited extent.

For the welds in the lower part of the absorber in contact with the product, the filler metals SG-NiCr21Mo9Nb (2.4831) and SG-NiCr20Mo15 (2.4839) were used. The side in contact with the product was without exception GTAW welded. The action of heat was severely limited by controlling the heat input per unit of length to 12 KJ/cm and by water cooling on the opposite side to achieve optimum corrosion resistance [20]. The vertical-up and horizontal welding positions were used during on-site fabrication. The absorber tower was internally spray pickled over the entire area and passivated before fitting the plastic components (20).

5.5.2.2 Design with clad plates

If wall thicknesses $\geq$ 8 mm approx. in the suitable nickel-base alloys 625 and C-276 are necessary for stability reasons, the clad design offers economic advantages in terms of the cost of the semi-finished products (i.e. sheet material) [24].

Special base materials are not necessary in this case since there is no further working by hot forming and consequently subsequent heat treatment (solution annealing) is not necessary. Therefore the common structural steels shown in DIN 17100 (e.g. St 37-2) can be used as the base materials (see Section 5.3.3.2). A further advantage is that the welding of the vessel wall with the support structure is much simpler and can be carried out more efficiently than with the solid design.

Explosively clad steel sheet clad with alloy C-276 and alloy 625 as shown in Table 5.7 is used for the manufacture of absorbers. Large format roll-bond clad plates in the material combinations mentioned are now also available (see Section 5.3.1.1.).

The welding of the clad plates is mainly carried out as described in Section 5.3.3.2. The capping passes of the clad-side welds should be made using the GTAW process to produce the smoothest possible surface. Taking the thin cladding into consideration (1.6 mm), the minimum possible heat input should be used for the welds in the cladding material because of the risk of iron dilution of the weld. Particular attention should be paid to this when welding in the vertical-up and horizontal welding positions (see Section 5.5.3.).

5.5.2.3 Liners

This process, which is described in more detail in Section 3.2.2, is used preferentially for refurbishment work. "Wallpapering" involves the use of a thin-walled loose liner consisting of large-format corrosion-resistant sheets which are applied section by section to vessels of structural steel or to structural components. Data on liner sheets made of highly corrosion-resistant nickel-base alloys are given in Table 5.7.

The liner is fixed to the vessel wall using both backing strips of corrosion-resistant material as well as the overlapping method. To improve the stability of the liner system punctiform fixings in the form of plug welds were also applied using GTAW or pulsed GTAW with programmable process control [22, 25]. As can be seen in Table 5.7, extremely large-format sheets are available for thin-walled liners, the use of which gives significant savings in welding costs compared to the use of sheets with conventional dimensions [22]. An example of the application of wallpapering for refurbishing absorber towers with thin-walled large-format liner sheets of highly corrosion-resistant nickel-base alloys is shown in Fig. 5.28.

Table 5.7: Use of highly corrosion-resistant nickel-base alloys in flue-gas desulphurisation plants.

Solid range							
Material			Mill product cold rolled sheet		Use in FGD field	Source	Comments
Designation	Mat. No.	Alloy type	Wall thickness mm	Dimension mm			
NiMo16Cr15W	2.4819	alloy C-276	1.6	2500 x 8000	Absorber	(25)	
				2324 x 6121	refurbishment	(22)	
NiMo16Cr16Ti	2.4610	alloy C-4	1.6	2500 x 8000	using the		
NiCr22Mo9Nb	2.4856	alloy 625	1.6	2500 x 8000	„wallpaper“ technique	*2)	

Clad plate									
Cladding material			Base	Cladding	Wall	Dimension	Use in FGD	Source	Comments
Designation	Mat. No.	Alloy type	material	process	thickness	mm	field		
NiMo16Cr15W	2.4819	alloy C-276	St 37-2	Explosively	15 + 1.6	2110 x 4700		(6)	
			H II	clad	10 + 1.9	2000 x 4370		*1)	
NiCr22Mo9Nb	2.4856	alloy 625	St 37-2		8 + 2	2170 x 5600		*1)	
					12 + 1.6	2430 x 6000	New	(6)	Finished
					22 + 1.6	2050 x 5600	production	(6)	format
					15 + 1.6				
					12 + 1.6	2430 x 6000		*1)	
NiCr22Mo9Nb	2.4856	alloy 625	St 37-2	Rollbond clad	6 + 2	approx. 6000 x 3000	Refurbishment of clean gas duct lining	(21) *2)	---
						14000 x 3800	---	*2)	max possible *2)

*1) Dynamit Nobel AG, Werk Würgendorf, Burbach-Würgendorf *2) Krupp VDM GmbH, Werdohl

Figure 5.28a: Example of sheet lining using the „wallpaper" concept: Installation of a retrofit metallic liner in a lignite-fired power station to repair failed organic coating. Sheet fixing rig and working platform

A special feature of the liner technique is the use of roll-bond clad sheets in the material combination St 37 - alloy 625 (2.4856). A rectangular clean gas duct was lined with these plates in the wall thickness 6 + 2 mm during refurbishment work [21].

Figure 5.28b: Example of sheet lining using the „wallpaper“ concept: Execution of circumferential welds using GTAW weld heads on guide rails

5.5.3 Fabrication quality requirements

In order to achieve high resistance to pitting, the filler metal must be sufficiently highly alloyed and the weld surface must be as smooth as possible and in particular free of porosity and slag inclusions [26]. Also, unacceptable iron dilutions should be avoided when welding clad plates and other composite systems. It is therefore necessary to carry out surface crack testing and control of the iron dilution during fabrication using suitable testing equipment [26]. The pitting resistance of the capping passes of welded joints should also be monitored using procedure tests. The GTAW process has given good results in achieving high resistance to pitting and crevice corrosion. Weld spatter and rough weld surfaces must be avoided in order to counteract the formation of solid deposits which promote the formation of crevice corrosion.

5.6 Summary

Nickel and nickel-base alloys are widely used in the construction of chemical process equipment. Correct working during component manufacture is a prerequisite for the full utilization of the good properties of these materials from the point of view of corrosion resistance and mechanical properties. Appropriate quality control during the manufacturing must be used to achieve a high quality. The use of composite materials with highly corrosion-resistant cladding layers of nickel-containing materials offers both economic and technical advantages in many applications when compared to solid designs.

The further development of the roll-bond cladding process has enabled the material combinations with nickel-base alloys for clad plate to be considerably extended. Modern manufacturing procedures contribute significantly to improvements in the quality and reliability of fabrication.

In recent years the application of highly corrosion-resistant nickel alloys has significantly increased through the use of these materials in pollution control systems. Their use in the construction of chemical plants is also growing.

5.7 References Chapter 5

[1] R. Köcher: Wirtschaftlicher Einsatz von korrosionsbeständigen Werkstoffen im Apparatebau, Chem.-Ing.-Tech. 59 (1987) 564-571

[2] Voest-Alpine Stahl Linz Ges.m.b.H.: brochure "Plattierte Bleche", Nr. TR 42/20012-90 M - 5000 d

[3] R. Pleschko, R. Schimböck, E.M. Horn, P. Mattern, M. Renner, W. Heimann: Walzplattierungen mit dem hochkorrosionsbestöndigen austenitischen Stahl X1NiCrMoCuN25206, W.-Nr. 1.4529, Werkstoffe und Korrosion 41 (1990) 563-570

[4] E.M. Horn, J. Korkhaus, P. Mattern: Plattieren - eine kostengünstige Alternative für den Einsatz metallischer Werkstoffe in der Chemietechnik, Chem.-Ing.-Tech. 63 (1991) 977-985

[5] U. Richter: Sprengplattierungen im Chemie-Apparatebau, Chemie-Technik, 12 (1983) Nr. 4, 63-71

[6] Information GVM Gesellschaft für Verbundmetalle mbH & Co. KG, Bocholt

[7] German printed patent DE 300 8238 C 2 "Sprengplattierter Schichtverbundwerkstoff"

[8] H. Pircher, R. Pennenkamp: Plattiertes Grobblech in der Energietechnik - Eigenschaften und Anwendungen, Stahl und Eisen 102 (1982) No. 12

[9] R. Köcher: Nichteisenmetalle im Apparate- und Anlagenbau, Chem.-Ing.-Tech. 55 (1983) 752-762

[10] H. Pircher, G. Sussek: Plattiertes Grobblech mit Auflagen aus nichtrostenden Stählen, Thyssen Techn. Reports 2 (1983) 124 - 131

[11] H. Krips, M. Podhorsky: Hydraulisches Aufweiten - ein neues Verfahren zur Befestigung von Rohren, VGB Kraftwerkstechnik 56 (1976) 456-464

[12] H. Krips, M. Podhorsky: Hydraulisches Aufweiten von Rohren, VGB Kraftwerkstechnik 59 (1979) 81-87

[13] H. Krips, M. Podhorsky: Hydraulisches Aufweiten von Rohren für Wärmetauscher mit Rohrplatten, Jahrbuch der Dampferzeugungstechnik, 4th Edition 1980/81

[14] Hydraulische Rohrbefestigung in Wärmetauschern, Dampferzeugern und Kondensatoren, company brochure of TU-EX-CO, Wuppertal (1982)

[15] K. Gerischer, K.O. Cavalar, K.O. Hirschhäuser: Durchstrahlungsprüfung der Schweißverbindungen Rohr/Rohrboden in Wärmetauschern, Materialprüfung 19 (1977) 439-441

[16] K. Gerischer u. K.O. Cavalar: Prüftechnik und Filmbeurteilung bei der Durchstrahlungsprüfung von Rohr-Rohrboden-Schweißverbindungen in Wärmetauschern, Schweißen und Schneiden 34 (1982), No. 1, 25-28
[17] S. Yokell: Heat-exchanger tube-to-tubesheet connections, Chemical Engineering (1982) No. 2, 78-94
[18] R. Prümmer, F. Jähn: Fortschritte beim explosiven Einschweißen von Rohren in Rohrböden, Bänder, Bleche, Rohre 10 (1976) 415-418
[19] H. Hampel u. H. Luft: Explosives Einschweißen und Aufweiten von Rohren in Rohrböden, lecture given in Düsseldorf on 1st December 1983 at the meeting "Explosivplattieren - ein modernes Verfahren zur Herstellung von Hochleistungsverbundsystemen", VDI-Gesellschaft Werkstofftechnik
[20] J. Bäumen: Edelstahl und Nickelbasis-Legierungen in Rauchgasentschwefelungsanlagen, lecture at Technische Akademie Esslingen on 11th April 1989
[21] J. Chwieralski, H. Dembeck, P. Drefke, H. Hess, G. Wellnitz: Erfahrungen bei der Sanierung von Reingaskanälen in REA-Anlagen von Braunkohlekraftwerken mit metallischen Auskleidungen, VGB-Kraftwerkstechnik 71 (1991), Heft 5
[22] R. Kirchheiner, W. Römer: Korrosionsbeständige VDM-Werkstoffe für Rauchgasentschwefelungsanlagen, VDM report No. 18, Oktober 1991, available at Krupp VDM GmbH, Werdohl
[23] Krupp VDM GmbH, References Umwelttechnik, Rauchgasentschwefelungsanlagen REA, Januar 1992, available at Krupp VDM GmbH, Werdohl
[24] Information by Krupp VDM GmbH, Werdohl (February 1992)
[25] Th. Hoffmann, D.C. Agarwal: "Wallpaper" installation guidelines and other fabrication procedures for FGD maintenance, repair and new construction with VDM high-performance nickel alloys, VDM report No. 17, June 1991, available at Krupp VDM GmbH, Werdohl
[26] U. Heubner, W. Heimann, R. Kirchheiner: Korrosionsbeständige metallische Werkstoffe für Rauchgasentschwefelungsanlagen, Werkstoffe und Korrosion 38 (1987) 746-756

6 Selected examples of nickel-alloy applications in chemical plants

F. White

6.1 Introduction

The successful commercial development of chemical processes depends on a number of economic factors. These include, among many others, the type of energy and raw materials available, personnel, transportation, investment costs, environmental and safety regulations, the market situation, etc. Temperatures and pressures are defined with the aim of achieving the highest possible economic yield. It is thus frequently necessary to use materials with particular mechanical, physical and corrosion-resistance properties. In many cases, it has only been possible to scale up processes from the laboratory to a commercially feasible level after suitable materials have become available.

In what follows, examples of organic and inorganic chemical processes are used to illustrate the close relationship between the aim to ensure the greatest possible degree of completion of a reaction in a certain direction and the resulting need to use materials specifically suited to the application. Where it would appear to be of benefit to users and manufacturers, the characteristic properties of important materials are also dealt with, and their typical fabrication characteristics explained.

6.2 Materials for the production of caustic soda

Caustic soda (NaOH) is one of the most important basic chemicals produced. It finds wide use in many industries: chemical, metallurgical, paper, food, detergents, etc. It is produced by means of electrolysis of a sodium chloride solution. Three principal types of electrolytic cells are in use: the mercury cell, the diaphragm cell and the more recently developed membrane cell. The mercury cell yields a concentrated (50 %) caustic soda solution of high purity, but poses pollution risks. The diaphragm process has the disadvantage of producing a dilute caustic solution (20 %) with a high percentage of sodium chloride as well as contaminants such as chlorates. The membrane process, on the other hand, produces a caustic solution of high purity and of concentration up to 35%. The membrane process therefore is progressively replacing the mercury cell and the diaphragm cell.

Caustic soda, particularly at high concentrations and temperatures, is very corrosive to many metals. It has long been known, however, that nickel is especially resistant to corrosion by caustic soda at all concentrations and temperatures - Fig. 6.1 [1]. Furthermore, nickel is not susceptible to caustic-induced stress-corrosion cracking.

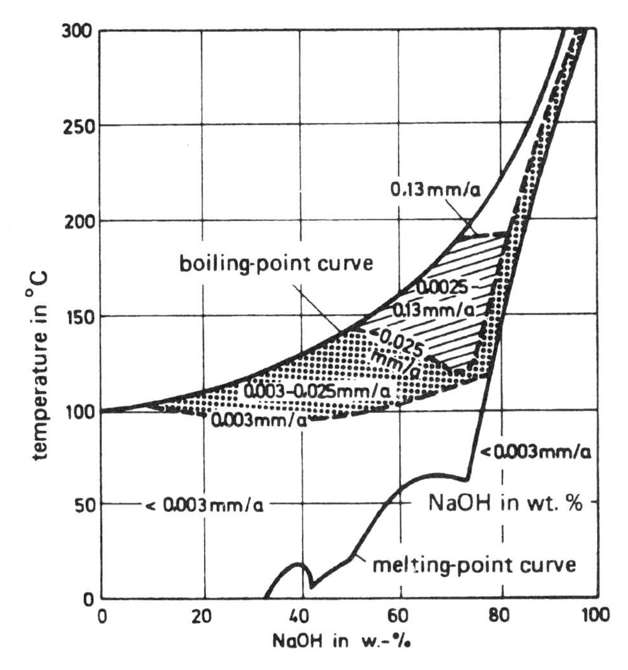

Figure 6.1: Iso-corrosion diagram for LC-Nickel and Nickel 99.2 (2.4066), according to [1]

Commercially pure nickel is used therefore at all stages of caustic soda production. Its use begins with the electrolysis cell of the membrane process, whose cathodes are of nickel sheets. The liquor is concentrated in a evaporative concentration plant, traditionally constructed in nickel. This plant may be of multistage or falling-film type.
Figure 6.2 [2] shows a flow-chart of a typical multi-stage system, of the diaphragm-cell process. Other important designs of multi-stage evaporation plant include Swenson-Whiting Fermont, Escher-Wyss and Hooker-Zaremba - Figures 6.3 and 6.4.

In many plants nickel has proven to be an ideal material for this purpose, being used for heat-exchanger tubes, as sheet and plate and as cladding for evapora-

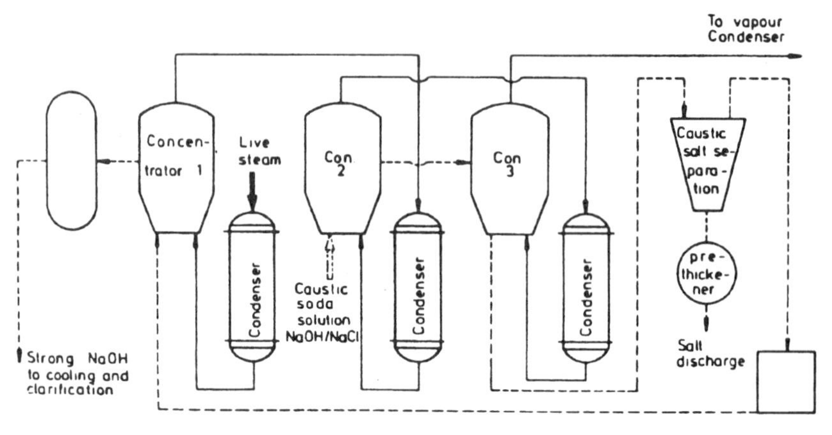

Figure 6.2: Escher-Wyss evaporative concentration process for the manufacture of caustic soda [2]

Figure 6.3: Evaporation plant at Rhône-Poulenc Pont-de-Claix (Escher-Wyss process)

tors, and for piping. In some cases operating conditions in the exchangers are particularly severe, necessitating frequent changes of the tubes. Here, a significant improvement in tube life has been obtained by using a special "caustic soda grade" of nickel tubing developed by Krupp VDM and DMV.

Falling film evaporators are used to produce highly concentrated or anhydrous caustic soda. One established process is the Bertrams process, whereby caus-

Figure 6.4: Caustic evaporators at Canadian Industries Ltd. at Bécancour (Zaremba process)

tic soda is evaporated in a falling film evaporator of LC-Nickel 99.2 - alloy 201 (2.4068), using molten salt at approximately 400°C as heating medium. LC Nickel 99.2 - alloy 201 (2.4068) is used because at temperatures above 315°C the higher carbon-containing Nickel 99.2 - alloy 200 (2.4066) is subject to embrittlement by graphite precipitation at the grain boundaries.

Whilst nickel is by far the most widely used material for caustic soda evaporation, other materials may also be used; for example, in the presence of certain oxidizing impurities such as chlorates or of sulphur compounds, the chromium-containing material Nicrofer 7216 - alloy 600 (2.4816). Here it is important to ensure that operating conditions are not likely to provoke stress-corrosion-cracking of such alloys [1]. Figure 6.5 compares weight losses of various alloys in deaerated 50% NaOH at 316°C [3].

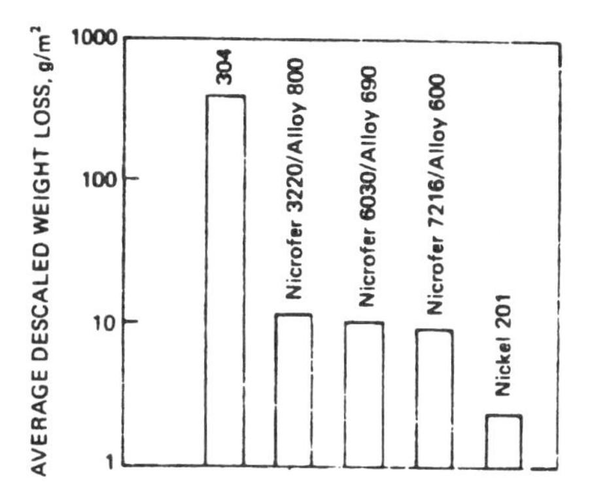

Figure 6.5:
Weight loss of various materials after two weeks exposure to deaerated 50% NaoH at 316°C according to Sedriks, Schultz and Cordovi [3]

6.3 Materials for nitric acid production

Nitric acid is nowadays produced almost exclusively by the oxidation of ammonia. There are three processing stages:

a) combustion of ammonia $4\ NH_3 + 5\ O_2 \rightarrow 4\ NO + 6\ H_2O$

b) oxidation of nitrogen monoxide $2\ NO + O_2 \rightarrow 2\ NO_2$

c) absorption of nitrogen dioxide into water $4\ NO_2 + O_2 + 2\ H_2O \rightarrow 4\ HNO_3$

A few well-known processes are shown below:

	50-70% acid	98% acid
Bamag/Davy	x	x
Espindesa	x	
Grande Paroisse	x	x
Hercules Maggie		x
Montedison	x	
Stamicarbon	x	x
Sumitomo	x	x
Uhde	x	x

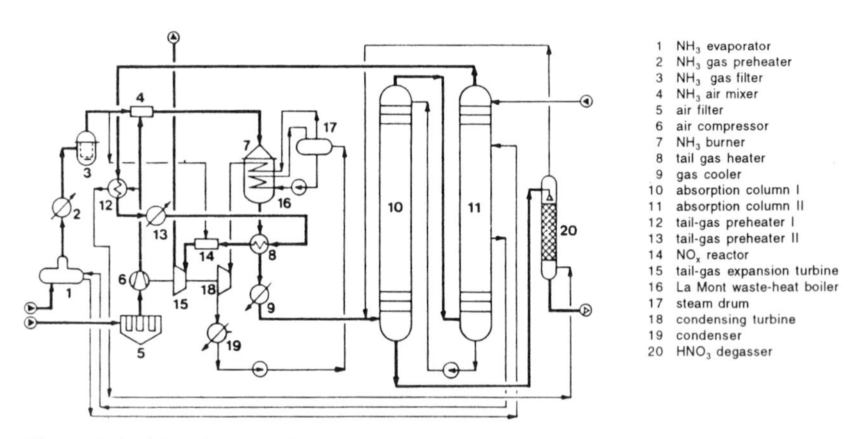

Figure 6.6: Manufacture of nitric acid using the Uhde medium-pressure process [4]

In the Uhde medium-pressure process, for example, conversion of ammonia to nitrogen monoxide takes place at a temperature of approximately 880°C and a pressure of approximately 5 bar. Acid concentrations of up 68% are achieved. Figure 6.6 shows a flow chart for such a medium-pressure plant. For a number of components which are in contact with hot nitric acid, such as preheaters and cooling coils for absorption columns or reheaters, normal stainless steels such as alloy 304 L (1.4306) or alloy 304 LN (1.4311) can be used. For more severe conditions, however, it is recommended to use alloy 304 L produced with controlled chemistry and remelted using the electro-slag remelting process [5]. Under very severe conditions, as exist in cooler-condensers or reheaters, the special nitric acid grade Cronifer 2521 LC - alloy 310 L (1.4335), with its controlled structure and chemistry, is often specified. It has been demonstrated that only by closely controlling the structure and chemical composition of alloys 304 L and 310 L can their good performance be ensured in long term service in nitric acid. The standard Huey test of 5 x 48 hour boiling periods in 67% nitric acid is not always selective enough to predict this. It is necessary to prolong the test to 20 or even 50 boiling periods to ensure that the material is in the correct metallurgical condition for such service. Figure 6.7 shows the effect of optimizing the processing parameters on the evolution of the corrosion rate of Cronifer 2521 LC (1.4335) in a prolonged Huey test [5].

High-temperature materials such as Nicrofer 3220 H/HT - alloy 800 H/HP (1.4958/ 1.4959), Nicrofer 7520 - alloy 75 (2.4951), or Nicrofer 5520 Co - alloy 617 (2.4663) are used for catalyst baskets. This contains the Pt-Rh wire gauze catalyst on which the combustion of ammonia takes place. Principal requirements for the catalyst basket materials are high creep strength, nitriding resistance, and metallurgical stability at service temperatures in the range of 900°C. Nicrofer 3220 H - alloy 800 H (1.4958) has also been used for the dome of waste heat boilers, where conditions are essentially similar.

Highly concentrated acids contain 95-99 wt.% HNO_3. During production, temperatures may rise to as high as approximately 100°C. As the flow chart in Figure 6.8 (SABAR process) shows, a portion of the nitrogen oxides formed is ab-

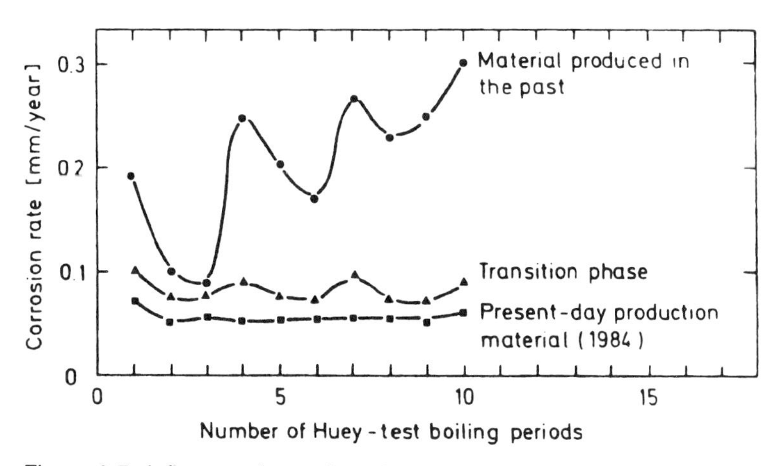

Figure 6.7: Influence of manufacturing conditions on the corrosion rate of Cronifer 2521 LC in the Huey test [5]

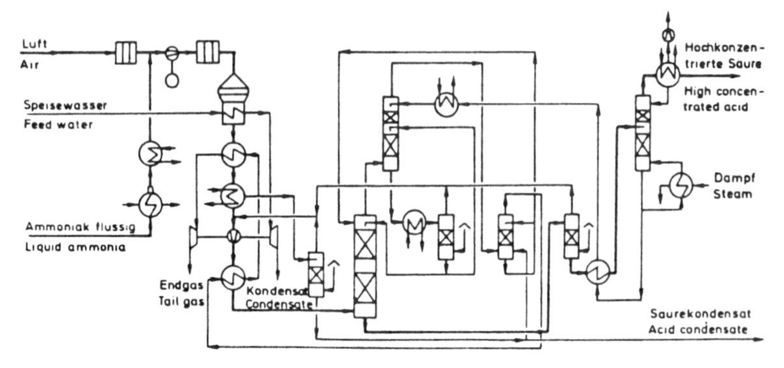

Figure 6.8: Production of highly concentrated acid using the SABAR process [6]

sorbed in approximately 68% acid in a column, with the formation of nitric acid of higher concentration. This superazeotropic acid is divided under vacuum in a rectification column into highly concentrated acid (head product) and azeotropic acid (bottom product). The highly concentrated acid is condensed in a cooler and a side-stream flows as reflux back into the column, the remainder leaving the plant as end product [6]. Constructions in pure aluminium were previously used for production, storage and transportation of the highly concentrated acid. However, this resulted in frequent failures of welded joints at higher temperatures. The flowing hot acid also caused erosion damage [7]. Furthermore, construction of major components was limited by the low strength properties of aluminium. Service life was considerably improved by the use of chromium-nickel steel containing approximately 4% silicon e.g. Cronifer 1815 LCSi - alloy 306 (1.4361). The silicon forms a protective layer on the surface and reduces corrosion-erosion to technically acceptable levels.

Figure 6.9 shows the corrosion rates of some chromium-nickel steels in boiling nitric acid of various concentrations. It can be seen that Cronifer 1815 LCSi - alloy 306 (1.4361) should be used for acid concentrations of above approximately 75%. Cronifer 1815 LCSi - alloy 306 (1.4361) is used for rectification and bleaching columns, coolers, reoxidizers, piping and vessels. Special regulations must be adhered to for the manufacture and working of this material. The following applies for the welding of Cronifer 1815 LCSi - alloy 306 (1.4361).

- Use of a welding filler material which produces a ferrite content of at least 4.5% in the weld metal.
- Cooling by water during welding.
- Heat treatment at approximately 1140°C after welding, with subsequent water quenching, for welded components intended for use at application temperatures above 40°C.

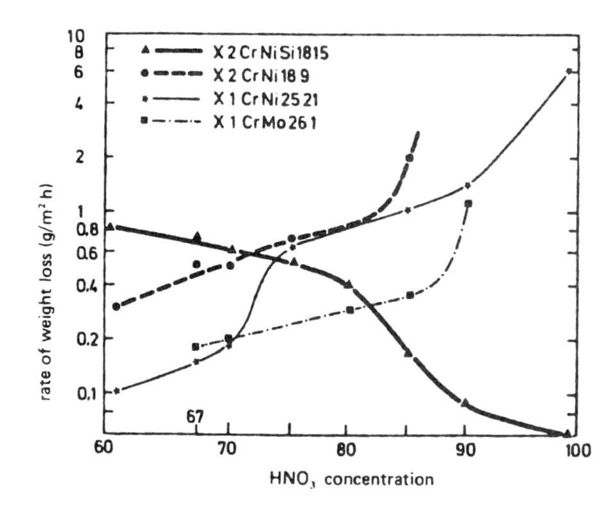

Figure 6.9: Area-specific rates of weight loss of austenitic chromium-nickel-molybdenum steels and a ferritic chromium-molybdenum steel in boiling nitric acid, acccording to Horn [7]

Amongst others, these precautions are necessary to ensure that the metal has a homogeneous structure free of precipitates which would prevent the formation of an uninterrupted layer of silica on the surface [8]. Figure 6.10 compares the behaviour of Cronifer 1815 LCSi - alloy 306 (1.4361) in a prolonged Huey test in three different metallurgical conditions; incorrectly heat treated [3], correctly heat

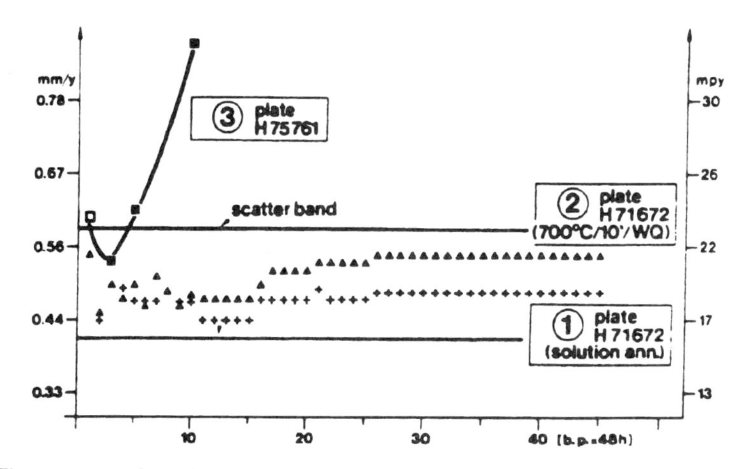

Figure 6.10: Cronifer 1815 LCSi: corrosion behaviour in 65% nitric acid in the extended Huey test (48 x 48 h) [8]

treated [1], and correctly heat treated and sensitized [2]. It can be seen that even material that has undergone a brief sensitization treatment shows no significant increase in corrosion rate over the duration of this extended test, provided that the initial solution treatment was correctly executed.

6.4 Materials for the production and handling of sulphuric acid

Sulphuric acid is by far the most important basic chemical produced worldwide. It is used as reagent or as catalyst in many chemical processes, and as a drying agent in others. Furthermore, sulphuric acid is a by-product in combustion processes of sulphur-containing fuels or waste. Thus the range of conditions in which sulphuric acid can be encountered as a corrosive medium in industrial processes is vast. Within the scope of this book it is impossible to cover all these areas. The following therefore is limited to a consideration of materials used in the production and handling of concentrated sulphuric acid. The essential process to obtain sulphuric acid is the production of sulphur dioxide (SO_2) by combustion of elemental sulphur, roasting pyrite ores or as a by-product of other processes. This sulphur dioxide is catalytically converted to sulphur trioxide (SO_3), then absorbed into concentrated sulphuric acid in one or two stages.
There is little use of nickel alloys or special stainless steels in the first two stages of the process. However, problems may be encountered in the combustion stage with materials in contact with sulphur dioxide at temperatures in the range of 700 - 800°C. Recent investigations by Krupp VDM have shown that alloys such as Nicrofer 45 TM - alloy 45 TM (2.4889) offer excellent resistance under such conditions [9].
In the third absorption stage, major use of special alloys is absolutely essential. The absorption of SO_3 into concentrated sulphuric acid takes place in packed towers. Most existing towers are lined with acid-resistant brickwork. Absorption occurs in a counter-current system with the SO_3-gas flowing upwards and the concentrated (98-100 %) sulphuric acid flowing downwards. The absorption of SO_3 into the acid is highly exothermic.
Until recently, few materials were available that offered adequate corrosion resistance to highly concentrated sulphuric acid at temperatures above about 120°C. This imposed severe limitations on downstream equipment such as heat exchangers, pumps and piping, and inlet temperatures of sulphuric acid into the absorption tower had to be limited. This led to a considerable loss in energy. For components operating at these lower sulphuric acid temperatures, special alloys had found application:

- Cronifer 1925 hMo - alloy 926 (1.4529) or high-silicon containing steels for acid distribution systems, replacing cast iron suffering from maintenance problems.
- Nicrofer 5716 hMoW - alloy C-276 (2.4819) and Cronifer 2521 LC - alloy 310 L (1.4335) for plate heat exchangers (acid coolers).

- Cronifer 1925 hMo - alloy 926 (1.4529) or Nicrofer 3127 hMo - alloy 31 (1.4562) for tubular acid coolers using sea- or brackish water for cooling.
- Cronifer 1925 hMo - alloy 926 (1.4529) for acid make-up tanks.

In recent years much work has gone into the development of materials for resisting concentrated sulphuric acid at temperatures above 120°C. The motivation was to replace brick-lined absorption towers by all-metal towers requiring far less installation time, and to recuperate more energy by operating heat exchangers and boilers at higher temperatures. A first significant breakthrough was achieved by the introduction of Cronifer 2803 Mo (1.4575), a super-ferritic stainless steel with extremely low corrosion rates in concentrated sulphuric acid at temperatures as high as 200°C (Figure 6.11) [10].

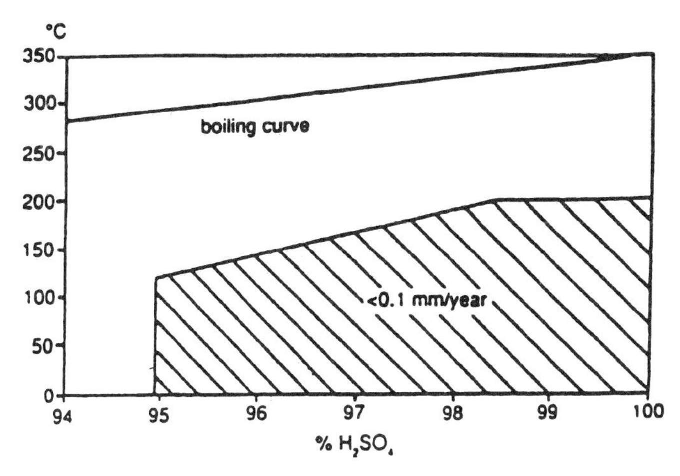

Figure 6.11: Application area of Cronifer 2803 Mo (1.4575) in concentrated sulphuric acid

This material is now successfully in service in heat recovery systems and associated applications in a number of sulphuric acid plants [10].

More recent research and development work on materials for use in hot concentrated sulphuric acid has been directed towards alloys with very high silicon contents. It has been found possible to produce special stainless steels with up to 7% silicon resistant to sulphuric acid over a wider range of concentrations and temperatures. Such a material is Krupp VDM's Nicrofer 2509 Si7 - alloy 700 Si (1.4390), with 25% nickel - 9% chromium - 7% silicon - bal. iron [11]. This alloy has proved to have an even better corrosion resistance in this medium than Cronifer 2803 Mo (1.4575).

An even more recent Krupp VDM development, Nicrofer 3033 - alloy 33 (1.4599) with nominally 31% nickel, 33% chromium and 32% iron, shows a unique combination of properties, which make it particularly suited for applications where no fully satisfactory metallic solution has yet been found. Not only does Nicrofer 3033 - alloy 33 (1.4599) offer outstanding resistance to hot concentrated sulphuric acid, it further combines excellent weldability with a high degree of resistance to corrosion in brackish or saline waters [12].

As mentioned at the beginning of this article, it is almost impossible to give a comprehensive picture of the behaviour of materials in sulphuric acid as encountered in industry: the range of conditions is simply too vast. It is worth remembering, however, that this apparently easy product can show a very complex behaviour: it may be either oxidizing or reducing, corrosion rates may rise by several orders of magnitude with just a few degrees Celsius increase in temperature, traces of impurities may radically affect the behaviour, as can flow rates. The selection of construction materials therefore requires a particularly close dialogue between the plant operator and the materials supplier [13].

6.5 Materials for phosphoric acid production

A large proportion of the world's fertilizer supply depends upon the production of phosphoric acid. Although two principal methods are used to produce the acid, it is the wet process which accounts for some 95% of the total. This process consists basically in digesting calcium phosphate in hot concentrated sulphuric acid, to form liquid phosphoric acid and solid calcium sulphate (gypsum). After filtration, the phosphoric acid, usually at a concentration of about 28% P_2O_5, is concentrated either to the standard commercial grade of 52% or 54% P_2O_5, or further to so-called "superphosphoric" acid at 70% P_2O_5.

Pure phosphoric acid is only mildly corrosive to metals. However, the presence of fluorides, chlorides and silica in the phosphate ore leads to the formation of corrosive fluosilicic acid and of aggressive acid chloride and fluoride conditions. The chloride content can be further increased should seawater be used to wash the ores, and in some cases the presence of sulphides further augments the aggressivity of the acid. This has an impact on materials selection at all stages of the production of phosphoric acid by the wet process.

Figure 6.12 is a schematised layout of one type of wet-process phosphoric acid plant [14]. The first stage at which special alloys are required is the digester. The vessel itself is usually of carbon steel with a non-metallic lining. This solution is also sometimes used for the mixers and pumps, but more frequently these are of special stainless steels such as alloy 904 L (1.4539), Cronifer 1925 hMo - alloy 926 (1.4529) or Nicrofer 3127 LC - alloy 28 (1.4563), or high-chromium castings. The use of increasingly impure ores, and higher temperatures, means that in some cases even these materials are no longer adequate to resist the extremely

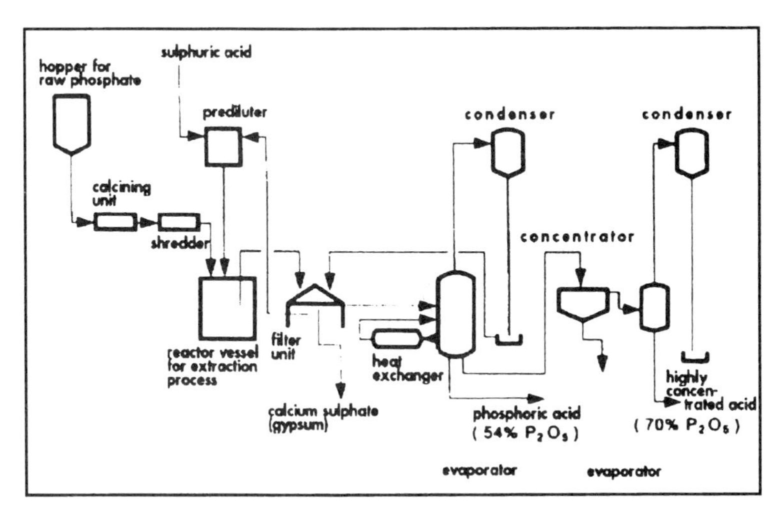

Figure 6.12: Schematic flow chart of the phosphoric acid process [14]

severe erosion-corrosion conditions in the digester. Extensive studies under real service conditions have shown that the VDM alloys Nicrofer 3127 hMo - alloy 31 (1.4562) and Nicrofer 6020 hMo - alloy 625 (2.4856) are particularly resistant to erosion-corrosion in phosphoric acid slurries. In fact Nicrofer 3127 hMo - alloy 31 (1.4562) was demonstrated to be comparable to, or even superior in performance to higher alloys of significantly higher price.
Interestingly, these studies also clearly demonstrated that in these conditions it is the corrosion resistance of the material which determines its behaviour, rather than its hardness [15]. They likewise confirmed that alloys for this service must have a high chromium content.

The second major unit where special alloys are required is the filter. Various designs exist, Ugeco and Prayon being amongst the most widely used. Here it is resistance to pitting and crevice corrosion which is required. In these less severe conditions somewhat lower alloys can be employed than in the erosion-corrosion conditions of the digester. Alloys used for filters include 317 LN (1.4439), 904 L, Cronifer 1925 hMo - alloy 926 (1.4529), Nicrofer 3127 LC - alloy 28 (1.4563) and Nicrofer 3127 hMo - alloy 31 (1.4562).

After filtration, the as-produced phosphoric acid, containing residual sulphuric acid, chlorides, fluorides, fluosilicic acid, etc., is concentrated from 28-30% to 54% P_2O_5. This is done in evaporators which may be either of the metallic type or of graphite.

Experience has shown that metallic exchangers give reliable service when hot water is used as the heating medium. This is because the deposits which form

on the walls are of gypsum ($CaSO_4$), which are easily removed during the cleaning cycle, so preventing concentration and overheating under the deposits, and consequently accelerated corrosion. In such cases the material which is frequently used for the heat exchangers is Nicrofer 3127 LC - alloy 28 (1.4563). When steam is used as a heating medium, however, the wall temperatures in the exchanger are significantly higher, sometimes exceeding 120°C. In this case adherent fluosilicate deposits form under the gypsum layer, and these do not readily dissolve in the washing cycle. Chlorides and fluorides can therefore concentrate to high levels under these deposits, and temperatures can rise, leading to accelerated corrosion of the metal. Graphite therefore still remains the most common material for steam-heated phosphoric acid evaporators, despite the maintenance problems it presents.
Some phosphoric acid is further evaporated to “superphosphoric” acid at 70-72% P_2O_5. In fact, this step is not particularly corrosive, as the previous evaporation stage has eliminated some of the corrosive fluorine compounds originally present. Metallic materials are thus used for evaporators. A standard alloy for this application is Nicrofer 6020 hMo - alloy 625 (2.4856).

Other applications for special stainless steels and nickel alloys in phosphoric acid production include plate heat exchangers in HF effluent systems, pumps, fans and ducting.

6.6 Materials for acetic acid production

Acetic acid (CH_3COOH) is a major raw material for many chemical processes, used to produce, for example, cellulose acetate, vinyl acetate monomer, acetic esters, pharmaceuticals, plastics, insecticides and dyes. Whilst it is naturally produced by the fermentative oxidation of ethanol, the principal production is by the catalytic reaction of methanol and carbon monoxide:

$$CH_3OH + CO \xrightarrow{\text{catalyst}} CH_3COOH$$

The most established process using this route is that developed by Monsanto, the licence now being held by BP Chemicals - Figure 6.13 [16]. Rhodium activated with iodide acts as catalyst. This forms a complex with the carbon monoxide and methyl groups, which transform by hydrolysis to acetic acid, with carbon dioxide and water as by-products. The exothermic reaction takes place at a temperature of 180 - 200°C and a pressure of 15 bar in a special reactor. The formed acetic acid is then collected in an overhead condenser from where the iodine compounds are recycled into the reactor. The raw acetic acid is then distilled in two stages, with the unconverted methanol being returned to the reactor.

Hydroiodic acid, acetic acid and water in the reactor, scrubber and columns produce extremely corrosive, highly reducing conditions. The material most fre-

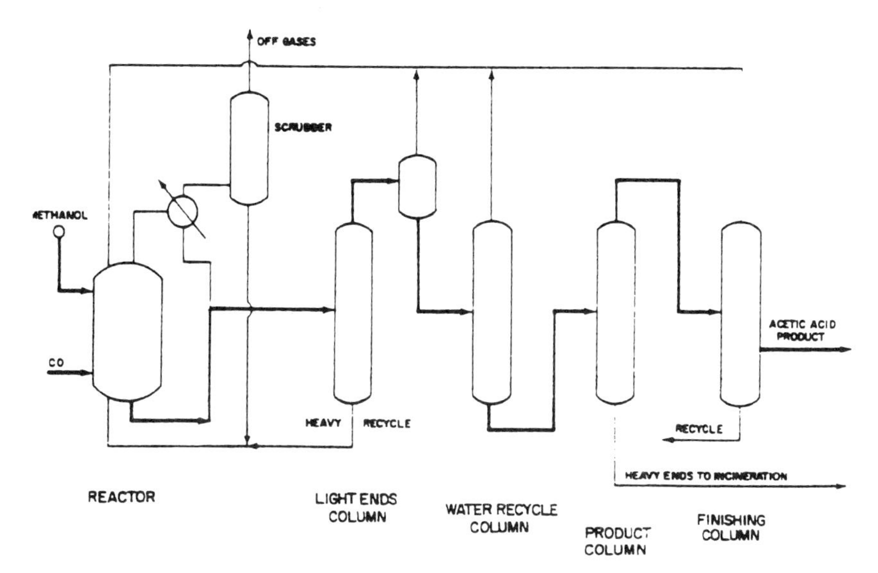

Figure 6.13: Schematic flow chart of the Monsanto/BP Chemicals acetic acid process [16]

quently used for these conditions is Nimofer 6928 - alloy B-2 (2.4617). The distillation columns may also be made of this material, or alternatively of Zirconium 702.

The high temperatures and pressures in the reactor require a very rigid construction. Because of the inherent metallurgical instability of Nimofer 6928 - alloy B-2 (2.4617), special precautions need to be taken for welding and fabrication. In particular, the complete reactor vessel has had to be solution treated and quenched after fabrication. After separation of the halogen ions, the acetic acid is less corrosive and materials such as Nicrofer 5716 hMoW - alloy C-276 (2.4819) and Nicrofer 4823 hMo - alloy G-3 (2.4619) are used for downstream equipment.

6.7 Materials for the production of hydrofluoric acid and aluminium fluoride

Hydrofluoric acid (HF) is produced by reaction between concentrated sulphuric acid and fluorspar (calcium fluoride):

$$H_2SO_4+CaF_2 \longrightarrow CaSO_4+2HF$$

Often the reaction takes place in two stages. In the first stage, or pre-reactor, the sulphuric acid and fluorspar are mixed together in a screw-feed mixer. The reaction is exothermic, and temperatures are of the order of 150°C. Materials for this component must therefore be able to resist corrosion by hot, concentrated sulphuric and hydrofluoric acids, and erosion by the fluorspar and calcium sulphate crystals. Nicrofer 5716 hMoW - alloy C-276 (2.4819), Nicrofer 5923 hMo - alloy 59 (2.4605) and Nicrofer 3620 Nb - alloy 20 (2.4660) have been used for this pre-reactor.

The second stage is a rotary kiln in which the reaction is taken to completion. Nicrofer 3620 Nb - alloy 20 (2.4660) is again used for the kiln or the kiln lining in some plants; others have used special stainless steels such as alloy 904 L. The crude hydrogen fluoride gas which leaves the kiln is contaminated with entrained sulphuric acid and silicon tetrafluoride. The next stage is therefore a scrubber to remove these contaminants. Materials selection for the scrubber will be determined by the concentrations of sulphuric acid and hydrofluoric acid in the scrubbing liquid. Nimofer 6928 - alloy B-2 (2.4617) is used for this application. The cleaned HF gas is next condensed to anhydrous hydrofluoric acid. This is not corrosive at room temperatures, so carbon steel can be used for the condensers.

The crude hydrofluoric acid is then distilled in two stages to remove impurities. In the first stage, volatile impurities are eliminated in a carbon steel column. In the second stage it is the heavier impurities which are eliminated by driving the hydrofluoric acid overhead, leaving a ternary mixture of hydrogen fluoride, sulphuric acid and water behind. In this stage the upper part of the distillation column can be in carbon steel, whilst Nicorros - alloy 400 (2.4360) is used for the lower part and the reboiler [17].

Aluminium fluoride is produced in fluidized bed reactors by reaction between aluminium oxide (Al_2O_3) and hydrogen fluoride at 650°C. Nicrofer 7216 - alloy 600 (2.4816) has been used for HF piping, reactor and product coolers, because of the excellent resistance of this material to high temperature corrosion by gaseous HF. Other candidate alloys are Nicrofer 6020 hMo - alloy 625 (2.4856) and Nicrofer 6616 hMo - alloy C-4 (2.4610).

6.8 Materials for the production of titanium dioxide by the chloride route

Titanium dioxide (TiO_2) is a white powder used in the production of special papers, plastics, porcelain and in paints. Two principal processes are used to produce titanium dioxide:

- the sulphate process, in which the ore is digested in sulphuric acid to form titanium sulphate, which is then calcined to titanium dioxide;
- the chloride process, in which the first stage is the production of titanium tetrachloride, which is subsequently oxidized to titanium dioxide.

The chloride process has become increasingly popular in recent years because it produces smaller quantities of waste products, especially sulphuric acid, which need recycling or special disposal.

Figure 6.14 is a schematised layout of the process. The raw materials, natural titanium oxide (ilmenite or rutile) and hot chlorine gas, react together in a ceramic-lined reactor in the presence of coke to form titanium tetrachloride gas and some iron chloride from the presence of iron in the ilmenite. The raw product is first condensed to liquid $TiCl_4$, then vaporized to pure titanium tetrachloride with the by-product iron chloride separated. The pure titanium tetrachloride is reheated and reacted with hot oxygen at 1050°C to form pure titanium dioxide. The chlorine gas is regenerated. The titanium dioxide is filtered, and the chlorine gas cooled in a submerged pipe coil before being returned into the production cycle.

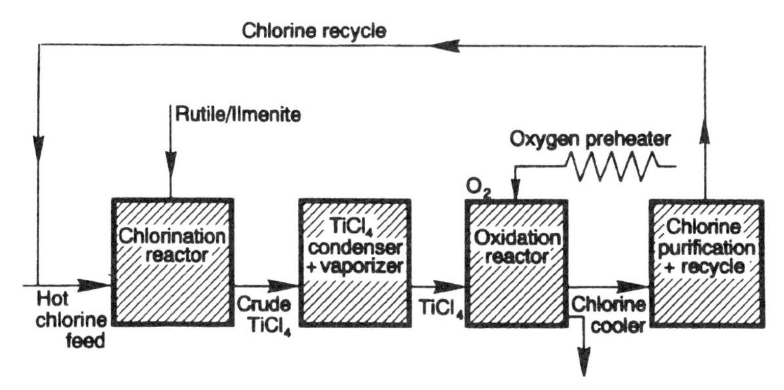

Figure 6.14: Schematised flow chart of the chloride process to produce titanium dioxide

Dry chlorine at high temperatures aggressively attacks many materials: steel will ignite at temperatures above about 200°C and titanium at temperatures as low as -18°C. Nicrofer 7216 - alloy 600 (2.4816) has long been known to have excellent resistance to corrosion by dry chlorine at temperatures as high as 500°C [18]. The alloy is also known to resist corrosion by other gaseous chlorinating compounds [19]. This corrosion resistance, together with the alloy's strength, ease of fabrication, and resistance to external corrosion from hot combustion gases, explains why it is used for almost all components in contact with hot chlorine or hot titanium tetrachloride.

The construction material for the oxygen preheater must be resistant to oxidation by pure oxygen at temperatures as high as 1050°C, whilst retaining sufficient creep strength to avoid its deformation. Nicrofer 6023 H - alloy 601 H (2.4851) is often used for such severe service conditions. A new Krupp VDM alloy, Nicrofer 6025 HT - alloy 602 CA (2.4633), offering higher creep strength and oxidation resistance, is expected to find increasing applications.

6.9 Materials for the production of vinyl chloride monomer

The production of vinyl chloride monomer is an interesting illustration of the very different types of corrosion which may be encountered in a single chemical process. Two basic reaction steps are involved:

1. Oxychlorination or direct chlorination of ethylene to dichloroethane.
2. Thermal cracking of dichloroethane to vinyl chloride monomer [20] - Figure 6.15

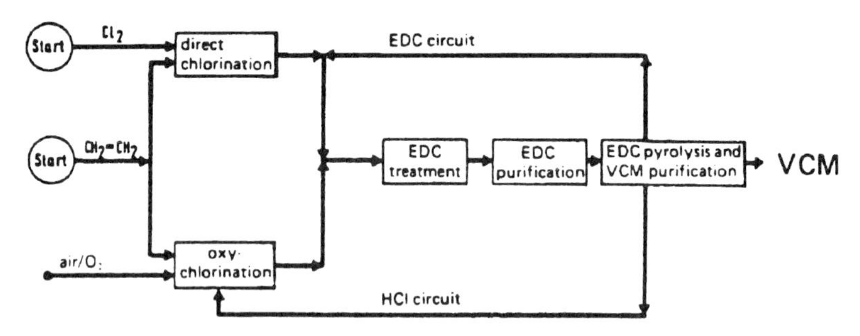

Figure 6.15: Typical VCM block flow diagram [20]

The direct chlorination process does not require special alloys, so only the more widely used oxychlorination process will be discussed. Ethylene, hydrogen chloride and air or oxygen react exothermically over a cupric chloride catalyst at temperatures in the range 225° - 325°C:

$$CH_2{=}CH_2 + 2HCl + 1/2\ O_2 \xrightarrow{CuCl_2} ClCH_2.CH_2Cl + H_2O$$

Reactors may be of the fluidized bed or fixed bed type. The fluidized bed reactor can be of carbon steel. Because both hydrogen chloride and cupric chloride are highly corrosive in the presence of traces of humidity, alloys such as Nicrofer 4221 - alloy 825 (2.4858) are often used for sparger tubes, and Nicrofer 5716 hMoW - alloy C-276 (2.4819) or Nicrofer 5923 hMo - alloy 59 (2.4605) may be used for the tube sheet and reactor lining in the catalyst zone.

In the fixed bed reactor the catalyst is contained in a number of vertical tubes. Temperature control is more difficult than with the fluidized bed reactor, and so the catalyst tubes are usually in nickel, Nicrofer 7216 - alloy 600 (2.4816) or even Nicrofer 6020 hMo - alloy 625 (2.4856).

After purification of the raw dichloroethane, it is thermally cracked at 425°C - 550°C to vinyl chloride and hydrogen chloride:

$$ClCH_2 \cdot CH_2\,Cl \xrightarrow{\text{Heat}} CH_2 = CHCl + HCl$$

Materials for the pyrolysis furnace tubing must be able to resist high temperature chlorination attack, carburization from coke formation, external corrosion by combustion gases, and have adequate mechanical properties at service temperatures. Nicrofer 3220 H - alloy 800 H (1.4958) is a standard material for this application. Nicrofer 7216 H - alloy 600 is also used in some plants.

In the dichloroethane purification section, the main corrosion problem is created by hydrochloric acid. Here alloys such as Nicorros - alloy 400 (2.4360), Nimofer 6928 - alloy B-2 (2.4617) or Cronifer 1925 hMo - alloy 926 (1.4529) find application in columns, reboilers, pumps and valves.

A common feature of many VCM plants has been the flare stack used to burn off waste gases. This must not only resist high temperature corrosion, but also wet corrosion by condensing hydrochloric acid. Nicrofer 5923 hMo - alloy 59 (2.4605) and Nicrofer 5716 hMoW - alloy C-276 (2.4819) are used in this application.

Environmental considerations are dictating that such waste gases will need to be recycled in the future. The corrosion problems created by this change in technology will be similar to those cited above, so that the same material considerations will apply.

6.10 Acrylic acid and acrylate esters

Acrylic acid, C_2H_3COOH, is of increasing importance as a starting material for the production of polyacrylates and acrylate esters. Polyacrylates find use in paints, water treatment chemicals and detergents. Acrylate esters such as methyl acrylate, ethyl acrylate, butyl acrylate and so forth represent an important class of monomers from which a wide range of polymers can be derived, with uses ranging from contact lenses to paints and fabrics.

The most important process for the production of acrylic acid is by the catalytic partial oxidation of propylene. In general, this process is considered as essentially non-corrosive, so that low-alloy steels or standard stainless steels are used. However, in recent years a number of serious corrosion problems have occurred in one particular part of this type of plant, the solvent extraction column. These problems have been attributed to the presence of traces of formic acid in the process stream. The varying intensity of the corrosion has meant that different materials have been specified to resist it. In one case Cronifer 1925 hMo - alloy 926 (1.4529) was used for the column and reboiler. In another, more severe case, Nicrofer 5716 hMoW - alloy C-276 (2.4819) was chosen. In a third unit the corrosion was limited to the reboiler. Following extensive comparitive testing, Nicrofer 3127 hMo - alloy 31 (1.4562) was specified for this reboiler.

Acrylate esters are produced by reacting acrylic acid with the corresponding alcohol, in the presence of an acid catalyst such as sulphuric acid or para-toluene sulphonic acid. Reaction temperatures are of the order of 130°C. An oxidizing or reducing stabilizing agent is added to prevent polymerization. It is obviously the catalyst which is the corrosion vector in this process. The choice of materials for the reactor, steam coil and mixer system will then be determined by their resistance to corrosion by sulphuric or sulphonic acids at high temperatures under either oxidizing or reducing conditions, according to the nature of the stabilizing agent. When reducing conditions are encountered, Nicorros - alloy 400 (2.4360) has been successfully employed. Under oxidizing conditions, chromium-containing alloys are required.

One major manufacturer of acrylate esters encountered rapid corrosion of his alloy 400 reactor and steam coil when a change was made from a reducing stabilizing agent to an oxidizing one. Extensive plant tests were made with a range of alloys including alloy 904 L (1.4539), Nicrofer 3127 LC - alloy 28 (1.4563), Nicrofer 4823 hMo - alloy G-3 (2.4619), Nicrofer 6020 hMo - alloy 625 (2.4856), Nicrofer 5716 hMoW - alloy C-276 (2.4819), Nicrofer 3127 hMo - alloy 31 (1.4562) and Nicrofer 5923 hMo - alloy 59 (2.4605). The only material to show no pitting or crevice corrosion, and which had an overall corrosion rate of below 0.01 mm/year, was Nicrofer 5923 hMo - alloy 59 (2.4605).

This alloy was therefore selected for the most severely-solicited component, the steam coil. When this showed negligeable corrosion after 3 years service, the decision was taken to use Nicrofer 5923 hMo - alloy 59 (2.4605) for the reactor itself at the next replacement.

6.11 Materials for the production of styrene

Styrene, $C_6H_5CH{:}CH_2$ is a major chemical used in the production of a range of important plastics: polystyrene, SBR, ABS and SAN resins, synthetic latexes, and polyesters. Although various processes are used to produce styrene monomer, the most important ones involve the same two basic steps of first producing ethylbenzene, then dehydrogenating the ethylbenzene to styrene (16, 21).

(1) $\underset{\text{Benzene}}{C_6H_6} + \underset{\text{Ethylene}}{C_2H_4} \xrightarrow{\text{Catalyst}} \underset{\text{Ethylbenzene}}{C_6H_5C_2H_5}$

(2) $\underset{\text{Ethylbenzene}}{C_6H_5C_2H_5} \xrightarrow{\text{Catalyst + Steam}} \underset{\text{Styrene}}{C_6H_5C_2H_3} + \underset{\text{Hydrogen}}{H_2}$

In the first reaction stage, in which ethylbenzene is produced, it is the catalyst which determines the types of materials required. The Monsanto process employs aluminium chloride as a catalyst. This is extremely aggressive to metals. The reactor is therefore made of acid-resistant brick, with the liners, heating coils and other internals in Nimofer 6928 - alloy B-2 (2.4617) and Nicrofer 5716 hMoW - alloy C-276 (2.4819). These same alloys are used in downstream heat exchangers and piping systems. Columns and exchangers handling trace chlorides are often in Nicorros - alloy 400 (2.4360); special stainless steels such as Cronifer 1925 hMo - alloy 926 (1.4529) could also be considered for such service.

Other ethylbenzene processes use non-corrosive catalysts, and only low-alloy materials are required.

The second stage, the dehydrogenation of ethylbenzene, takes place at very high temperatures (950° - 1000°C) in the presence of a large excess of superheated steam. This steam is produced in furnaces in which the tubes and manifolds may be in Nicrofer 3220 H or 3220 HT - alloy 800 H/800 HP (1.4958/1.4959). The steam transfer line from the superheater to the reactor may also be in this alloy.

Various materials are used for the dehydrogenation reactor. These include heat-resistant stainless steels, Nicrofer 3220 H/HT or high-performance alloys such as Nicrofer 5520 Co - alloy 617 (2.4663), according to the individual reactor design.

6.12 Materials for the production of MDI and TDI for polyurethanes

MDI (methylene di-para-phenylene isocyanate) and TDI (toluene di-isocyanate) are the basic materials for the production of polyurethanes. The processes for synthesizing MDI and TDI are essentially similar: they involve the production of diamines, which are then phosgenated to form the isocyanate; e.g.

$$CH_3C_6H_3\,(NH_2)_2 + 2\,COCl_2 \xrightarrow[170-180\,^\circ C]{} CH_3C_6H_3\,(NCO)_2 + 4HCl$$

In the synthesis of both MDI and TDI, hydrogen chloride is generated as a by-product. It is in the processing of this by-product HCl that special corrosion-resisting alloys are required. In TDI production, heat exchangers down-stream from the reactor have to handle mixtures of TDI + HCl + orthodichlorobenzene + traces of phosgene at up to 200°C. Nicrofer 7216 - alloy 600 (2.4816) is used for such exchangers; in the absence of humidity, service life is of the order of 10 years. Elsewhere in the systems, materials are required to handle the hydrochloric acid in aqueous solution. Depending on the individual process and the consequent severity of the process conditions, the materials used can range from Nimofer 6928 - alloy B-2 (2.4617) and Nicrofer 5716 hMoW - alloy C-276 (2.4819) through Nicorros -alloy 400 (2.4360) to the special stainless steel Cronifer 1925 hMo - alloy 926 (1.4529).

6.13 Materials for tube reactors

A number of hydrometallurgical reactions can only be carried out economically under conditions of high pressure and/or high temperature. Examples of such processes, employed on an industrial scale, include the digestion of bauxite concentrates in caustic soda solution, the pressurized leaching of lateritic nickel and uranium ores in sulphuric acid, as well as the precipitation of copper, nickel or cobalt from sulphuric acid or ammoniacal solutions using hydrogen sulphide.

An example in environmental engineering is the oxidation of organic substances in highly-contaminated waste waters.

A further example is the hydrothermal production of sodium metasilicate (water glass) from the raw materials caustic soda solution and quartz sand. The reaction takes place at a temperature of 175 - 190°C and a pressure of 15 bar.

These reactions are generally performed in spherical autoclaves equipped with mechanically driven agitators. Such units have the disadvantage of discontinuous operation and of substantial risk of damage to the moving parts and seals. On the other hand, the principle of processing with continuous material throughput represents a technological improvement. Development has led to a tubular reactor successfully marketed by Lurgi-VAW [22]. The tubular reactor has the following advantages over the spherical autoclaves described earlier:

- simpler technology, with only one moving mechanical unit (piston diaphragm pump)
- improved mass transfer and heat exchange, shorter reaction times
- lower investment costs
- rapid start-up and shut-down, and
- easier maintenance.

Because of the many advantages of the continuous process, development work is continuing to enable chemical processes to be performed in tubular reactors by modifying the processing parameters. In aluminium oxide extraction, the nature and composition of the bauxite available determines whether treatment can take place in discontinuously-operated individual autoclaves connected in series, or by using the tubular reactor process. In the latter there are two variations: the double tube plant (heat exchange: suspension/suspension) and the decompression plant (heat exchange: vapour/suspension) [23].

In the production of water glass, for example, ways are now being sought to perform the process described above in a tubular reactor. Because of the shorter residence time, considerably higher pressures and temperatures would have to be used.

The following requirements are generally made of materials for pressure vessels and tubular reactors:

- general corrosion resistance to the reagents used (acids, alkalis).
- erosion resistance.
- resistance to local corrosion and to stress-corrosion cracking.

These requirements may be met by nickel-based alloys such as Nicrofer 7216 - alloy 600 (2.4816), Nicrofer 4221 - alloy 825 (2.4858), Nicrofer 5923 hMo - alloy 59 (2.4605), and by some special stainless steels such as Nicrofer 3127 hMo - alloy 31 (1.4562), Cronifer 1925 hMo - alloy 926 (1.4529) or Cronifer 2419 MoN - alloy 24 (1.4566).

6.14 Conclusions

In addition to the processes described, a large number of other processes can only be carried out economically if nickel-containing materials are employed. Krupp VDM has an abundance of laboratory data and practical experience at its disposal and will be able to offer useful suggestions, particularly at the process development stage. In addition Krupp VDM operates an extensive computer-controlled information system which was developed together with the DECHEMA and leading material users from the chemical industry, and which is continuously upgraded [24].
The customer can equally rest assured that his interests will be safeguarded in all respects. This also applies in the case of fields of application which have not been covered or fully dealt with here - for instance shipbuilding and offshore engineering, liquefied gas and chemical products transportation, automotive engineering, mining and energy generation.

6.15 References Chapter 6

[1] "Corrosion Resistance of Nickel and Nickel-Containing Alloys in Caustic Soda and other Alkalies", NiDI Publication 281
[2] VDM Information No. 10, September 1975
[3] A.J. Sedriks: "Corrosion of Stainless Steels", John Wiley + Sons, New York-Chichester-Brisbane-Toronto, 1979
[4] Uhde Publication: "Nitric Acid from Ammonia"
[5] R. Kirchheiner, U. Heubner, F. Hofmann: "Increasing the Lifetime of Nitric Acid Equipment Using Improved Stainless Steels and Nickel Alloys", CORROSION 88, Paper

No. 318, National Association of Corrosion Engineers, Houston, Texas, 1988; Materials Performance 28 (1989), No. 9, 58 - 62
[6] Davy Publication, 1979
[7] E.M. Horn: "Korrosionsverhalten siliciumhaltiger nichtrostender Stähle in praktisch wasserfreier Salpetersäure", Werkstoffe und Korrosion 30 (1979) 723 - 732
[8] R. Kirchheiner, F. Hofmann, Th. Hoffmann: "A Silicon Alloyed Stainless Steel for Highly Oxidising Conditions", CORROSION 86, Paper No. 120, National Association of Corrosion Engineers, Houston, Texas, 1986
[9] U. Brill, J. Klöwer: "Corrosion behaviour of high silicon alloys in carbon-bearing and high-sulphur atmospheres", Materials at High Temperatures, Vol.11. No. 1 - 4, 1993
[10] H. Decking, W. Schalk: "Cronifer 2803 Mo - W.-Nr. 1.4575, Eigenschaften, Verarbeitung und Einsatz des superferritischen Stahles", ACHEMA 1994
[11] "Alloy 700 Si, a new corrosion resistant material for handling of hot, highly concentrated mineral acids", Werkstoffe und Korrosion 46, 18 - 26 (1995)
[12] M. Köhler, U. Heubner, K.-W. Eichenhofer, M. Renner: "Alloy 33, A New Corrosion-Resistant Austenitic Material for the Refinery Industry and Related Applications", CORROSION 95, Paper No. 338, National Association of Corrosion Engineers, Houston, Texas 1995
[13] "Behaviour of some metallic materials in sulphuric acid", VDM Report No. 22, August 1994
[14] "Korrosionsbeständigkeit nichtrostender Stähle und Nickellegierungen in Phosphorsäure", Druckschrift der Mannesmann Edelstahlrohr GmbH, Langenfeld 1988
[15] M. Aakka et al: "Corrosion-abrasion of nickel-base alloys in industrial phosphoric acid". Proceedings of "Corrosion in Natural and Industrial Environments: Problems and Solutions". Grado, Italy, May 23 - 25, 1995. NACE Italia, c/o OMECO, Monza (I) 1995
[16] "The Petrochemical Industry Handbook", Princeton Advanced Technology, Inc., Fourth Edition 1990 - 1991
[17] "Corrosion resistance of nickel-containing alloys in hydrofluoric acid, hydrogen fluoride and fluorine", NiDI Publication 443
[18] M.H. Brown, W.B. Delong, J.R. Auld: Ind. Eng. Chem., Vol 39 (No. 7), 1947, p 839
[19] U. Brill: "Proceedings of the Heat Resistant Materials Conference", Lake Geneva, Wisconsin, USA, 22 - 26 September 1991
[20] R.W. McPherson, C.M. Starks and G.J. Fryar: "Vinyl Chloride Monomer", Hydrocarbon Processing, March 1979, 75 - 88
[21] D.W. McDowell Jr: "Choosing materials for ethylbenzene service", Chemical Engineering, January 16, 1978, p 159 - 160
[22] "Metallurgische Drucklaugenpresse im Rohrreaktor", Lurgi Schnell-Information, 1984
[23] F.W. Kämpf: "Der Bauxitaufschluß im Bayer-Verfahren", Leichtmetalltagung Leoben/Wien, 1987, 30 - 37
[24] A. Hatzinasios, R. Eckermann: "CORIS - a computer based corrosion information system", Werkstoffe und Korrosion 42 (1991) 326 - 328

7 Nickel alloys and high-alloy special stainless steels in thermal waste treatment

J. Klöwer

7.1 Introduction

The falling number of suitable disposal sites, the need to treat hazardous industrial waste instead of dumping it and new laws which encourage the exploitation of waste, like the German "Kreislaufwirtschaftsgesetz" (Recycling Law) [1], have caused an increased demand for effective methods of thermal waste treatment. At the same time biomass and waste are becoming more and more important as a cheap fuel which can be used to produce electric power and heat .

To meet future requirements, different concepts for thermal waste treatment have been developed. The most common method is the burning of waste on a grate at a temperature between 850 °C and 1000 °C. A simple waste incineration plant is shown schematically in Fig. 7.1 [2]. It consists of the waste pit, the furnace with a grate, a boiler and the gas cleaning units. Heat recovery is possible by using the steam either in a steam turbine or in district heating systems. The residual slags and ashes, which amount to about 10 mass% of the original waste, can be used as construction material or disposed of in a landfill.

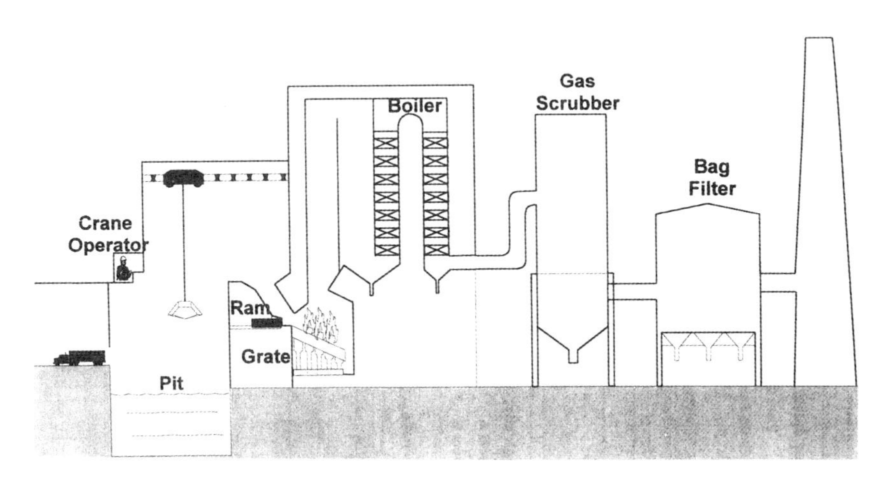

Figure 7.1: Schematic diagram of a conventional waste incineration plant consisting of the waste pit, the furnace with grate, the boiler, the gas scrubber, the filter and the stack (after Meadowcroft [2])

To increase plant efficiency and to reduce the concentration of harmful substances in the clean gases and the slags, conventional waste incineration systems are constantly being improved and new concepts are being designed (e.g. [3,4,5,6]). Some of the more recently developed systems are summarised in Table 7.1. Most of the processes comprise a two-stage system of thermal treatment with the first stage being pyrolysis (thermal decomposition in the absence of oxygen) and the second stage being either a burner (Siemens KWU), a gasifier (Noell, Veba, Thermoselect) or a rotary kiln (von Roll). In most cases the pyrolysis is carried out in a rotary kiln; exceptions are the Thermoselect process, where the waste is subjected to pyrolysis under high pressure in a non-rotating kiln, and the "von Roll" process, where the pyrolysis takes place on a grate.

Table 7.1: Summary of different methods of thermal waste treatment

	Process:						
	GRATE	**FBC**	**Schwel-brand**	**Conversion**	**Thermo-select**		**Duotherm**
	ABB, Martin,	**Lurgi**	**KWU**	**Noell**	**Thermoselect S.A.**	**Veba**	**von Roll**
Type of waste	household refuse	municipal waste, sewage sludge	household refuse, sewage sludge	household refuse, sewage sludge etc.	household refuse, (hazardous waste)	household refuse, sewage sludge	household refuse, industrial waste, sewage sludge
1st stage: pyrolysis	-	-	rotary kiln, 450-550 °C	rotary kiln, 500-550 °C	pyrolysis under pressure, 500-600 °C	rotary kiln,	pyrolysis on grate, 450-550 °C
2nd stage:	grate 850-1000 °C	fluidised bed, 850 °C	burner 1300 °C	gasifier	gasifier T=2000 °C	gasifier	rotary kiln
Fuel gas	air	air	air and pyrolysis gas	oxygen	oxygen		air and pyrolysis gas
Heat and gas recovery	district heating, steam turbine	district heating, steam turbine	district heating, steam turbine	gas turbine, gas engine, steam turbine		gas turbine, gas engine, steam turbine	
References	[2,3,4]	[3]	[3,7]	[8]	[9]	[10]	[11]

The concentration of noxious substances in the clean gases of waste incinerators can be kept low by means of careful process control, modern gas scrubbers and filter systems. However, the raw gases which are encountered in the furnace, the boiler and the scrubber contain high levels of corrosive components and corrosion problems in waste incineration and pyrolysis plants are frequently reported. The corrosion mechanisms in waste incineration have been extensively studied over the last decade; many of the reasons for corrosion failures are well understood and corrosion problems have been solved by choosing an appropriate material or even changing the plant design. Some of the more recent results are discussed below.

7.2 Corrosion mechanisms in plants for thermal waste treatment

There is general agreement that corrosion is the life limiting factor of most components in waste incineration and related plants. The major factors that govern corrosion are:

- the concentration of chlorine, sulphur, sulphates and heavy metals (Hg, Pb, Sn)
- the oxygen partial pressure of the gas (whether it is oxidising or reducing)
- the flue gas temperature, which determines the types of volatile species that can be transported in the gas and also determines the maximum temperature of any deposit [12]
- the metal temperature and
- the frequency of shut-downs to room temperature (danger of dewpoint corrosion).

Depending on the type of process, the process temperature, the concept of heat recovery and the composition of the waste, a variety of corrosion problems may be encountered. These include [12-20]:

a) Chlorination of the rotary kiln, the furnace walls and the heat exchanger tubes on the fire side
b) Sulphidation of the rotary kiln, the furnace walls and the heat exchanger tubes on the fire side
c) Corrosion by aggressive and/or low melting deposits on furnace or tube walls
d) Combined erosion-corrosion in the rotary kiln, and, to a lesser degree, of heat exchanger tubes
e) Wet corrosion, comprising
   - wet corrosion due to high concentrations of sulphuric acid and hydrochloric acid in the gas cleaning unit, especially in the scrubber
   - Aqueous corrosion in the boiler, superheater or evaporator during shutdown times (dewpoint corrosion)
   - stress-corrosion cracking on the water side of heat exchanger tubes

The type and degree of corrosion depends to an appreciable extent on the aggressivity of the raw gases, and therefore on the composition of the waste which is burnt. Table 7.2 shows the typical compositions of raw gases in waste incineration plants. It can be seen in the table that the concentration of chlorine can be as high as 4,000 mg/cm^3 in the raw gas of waste incinerators. After short periods of operation thick deposits consisting of fly ash, slags, alkali salts and heavy metals form on heat exchanger tubes. Under certain circumstances these deposits may protect the subjacent material against corrosion by the gases, but frequently they cause corrosive attack themselves. Table 7.3 summarises the analysis of deposits on heat exchanger tubes after operation in a waste incineration plant in Stockholm, Sweden. It can be seen that the concentrations of lead and chloride can vary by more than one order of magnitude.

Table 7.2: Typical concentrations of contaminants in the raw gas of waste incineration plants burning household refuse. The raw gases additionally contain sodium, potassium and calcium (after VDI 2114, June 1992 [21])

	Concentration in mg/m^3		Concentration in ppm	
	from:	to:	from:	to:
HCl	1000	4000	560	2240
SO_2	100	2000	35	700
SO_3	5	70	1.4	19.6
NO_2	150	450	71.5	214.5
CO	80	800	64	640
C_xH_y	10	400	16	640
Cd	0.3	2.5	0.06	0.5
Pb	10	60	1.08	6.49
Cu	10	50	3.56	27.78
Zn	10	150	3.45	51.69
Hg	0.4	0.7	0.04	0.08

Table 7.3: Composition of deposits found in boiler 2 and boiler 3 after inspection of the waste incineration plant in Stockholm, Sweden; after [22]

Element	**Concentration in mass%**			
	Boiler 3, analysis of January 1990		Boiler 2; analysis of June 1989:	
	1st measurement	2nd measurement	1st measurement	2nd measurement
Al	3.7	3.1	4.3	4.8
Si	6.3	4.6	5.0	
Ca	11	19	21	
Fe	1.2	0.91	0.76	0.89
K	5.2	4.7	3.4	
Na	2.7	3.3	3.1	
P	0.48	0.49	0.7	
S	15	15	6.7	7.5
Zn	4.0	2.2	0.83	0.82
Pb	4.1	1.5	0.045	0.052
Cl	0.24	2.5	12	11.9

Chlorination:
It is generally assumed that the aggressivity of the gases in the high temperature section of the plant is mainly dependent on the concentration of chlorine. Consequently, plants where waste with high chlorine concentrations is treated frequently suffer extremely severe corrosive attack. When detected, this was a surprising result, since chlorides were not expected to come into contact with the tube or furnace walls. According to the following reaction, referred to as "sulphation",

$$MeCl + SO_2 + O_2 \Leftrightarrow MeSO_4 + \tfrac{1}{2} Cl_2 \qquad (Me= K \text{ or } Na) \qquad (1)$$

alkali chloride should be converted to alkali sulphate. The sulphate $MeSO_4$, which is innocuous at moderate temperatures (T< 600 °C), condenses on tubes or furnace walls, while the volatile chlorine is transported with the gas stream into the quencher without coming into contact with the metal walls.

Recent investigations and calculations, however, have shown that the sulphation reaction does not happen if the temperature of the flue gas exceeds a certain value. The results of thermodynamic calculations carried out by Singheiser and Bossmann are shown in Fig. 7.2 [19]. It can be seen that up to a temperature of 600 °C the concentration of the alkali chlorides NaCl and KCl is low, while those

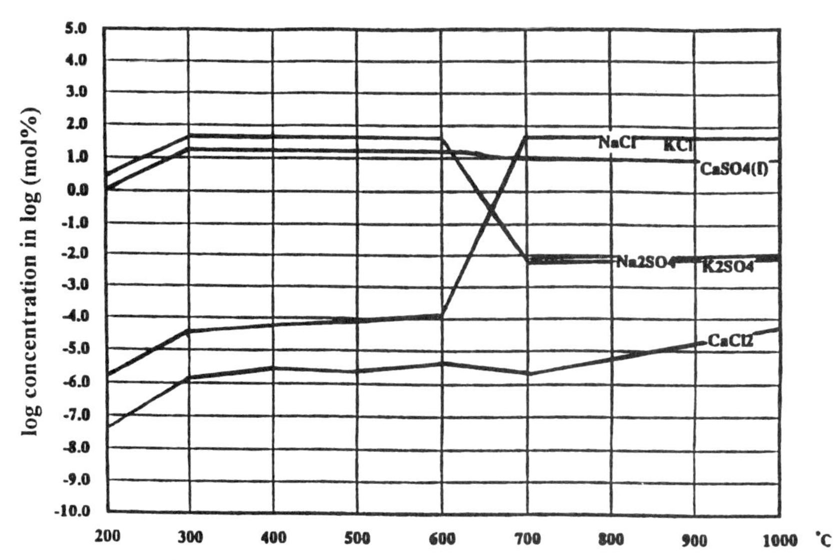

Figure 7.2: Calculated concentrations of alkali sulphates and alkali chlorides in flue gases as a function of flue gas temperature. Flue gas: 10 % O_2, 770 ppm HCl. 87 ppm SO_2, 5 ppm NaCl, 5 ppm KCl, 1 ppm $CaCl_2$, N_2 bal (after [19])

of alkali sulphates are high. At 600 °C and above, however, the equilibrium is shifted to NaCl + SO_2. On the basis of these considerations, the corrosion rate should increase with an increasing temperature of the flue gas, i.e., with an increasing concentration of NaCl and KCl and decreasing concentrations of gaseous Cl_2. Fig. 7.3 shows that the corrosion rate indeed increases with increasing flue gas temperature despite the constant metal temperature.

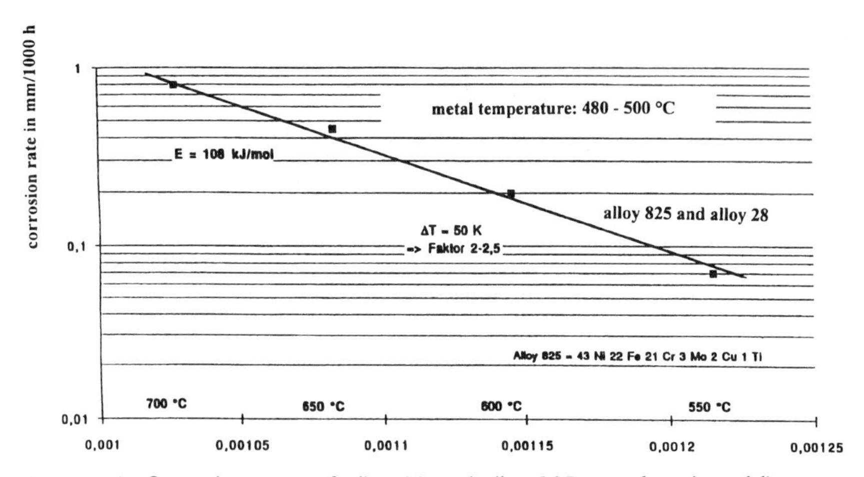

Figure 7.3: Corrosion rates of alloy 28 and alloy 825 as a function of flue gas temperature. The metal temperature has been kept constant at 480-500 °C (after [19])

The mechanism of chlorination in waste incineration has been studied extensively by Grabke and co-workers [23,24,25]. According to their investigations, the sulphation process takes place at the tube or furnace walls if it is prevented in the gas phase. The next steps are a build-up of chlorine in the deposit, the penetration of chlorine into the metal and the formation of metal chlorides (predominantly Fe_xCl_y). If oxygen is present, the chloride is converted to a metal oxide and chlorine is released for further reactions. Hence the only effect of chlorine is to catalyse the oxidation process and only small concentrations of chlorine/chloride are required to cause high corrosion rates.

If deposits are not present, chlorination may also occur by evaporation of volatile metal chlorides. It can be seen from Table 7.4 that most metal chlorides have a very high vapour pressure and consequently tend to evaporate even at moderate temperatures. Iron base alloys in particular are susceptible to this type of corrosion because of the high vapour pressure of iron chlorides.

Table 7.4: Temperature at which the vapour pressure of the oxide is 10^{-4} bar and significant evaporation has to be reckoned with (after [26])

Chloride	Tv in °C
$FeCl_3$	167
$FeCl_2$	536
$CoCl_2$	587
$NiCl_2$	607
$CrCl_3$	611
$CrCl_2$	741

Sulphidation is a much less significant corrosion problem in waste incineration plants. Highly sulphidising gases, however, may be encountered in plants for thermal treatment of waste containing coal, wood or paper and hence high concentrations of sulphur. In such plants material wastage may occur by a process of internal sulphidation (see chapter 4 in this book).

Corrosion by sulphate-containing salts, slags and ashes is widely known from industrial gas turbines. This mechanism of so-called "metal fluxing", which is a basic or acidic dissolution of metals, has been described by several researchers [27,28,29]. It is generally assumed that the process starts with the decomposition of sulphate:

$$Na_2SO_4 \Leftrightarrow Na_2O + SO_3. \qquad (2)$$

The same reaction applies to potassium sulphate. The subsequent dissolution of chromia-forming materials has been described as follows:

$$Cr_2O_3 + 2\ Na_2O \Leftrightarrow 2\ CrO_2^- + 2\ Na^+ \qquad (3)$$

in the case of basic disolution, and

$$Cr_2O_3 + 2\ SO_3 \Leftrightarrow 2\ CrSO_4 + 1/2\ O_2 \qquad (4)$$

in the case of acidic dissolution. Similar reactions apply to the dissolution of other metal oxides.

In pure sodium sulphate the prevailing reaction below 884 °C, i.e. below the melting point of sodium sulphate, is an acidic dissolution according to equation (4). In more complex deposits the type of dissolution depends on the basicity and the acidity of the salt mixture, which is given by the activity of Na_2O and the SO_3 partial pressure respectively. The corrosion rate, on the other hand, is determined by the solubility of the metal oxide in the liquid salt. The solubility of different metal oxides as a function of the basicity (Na_2O activity) is shown in Fig. 7.4. It can clearly be seen that most metal oxides exhibit a minimum of solubility - and hence a minimum of the corrosion rate - at the transition from basic to acidic behaviour. Even small additions of other components may shift the conditions from neutral to acidic or basic behaviour and cause high corrosion rates. Under certain circumstances even aggressive pyrosulphates or hydroxides may be formed [18,30]. It should be noted that silica scales show a behaviour which is different from that of other oxides; their solubility remains at a low level over a wide range of Na_2O activity.

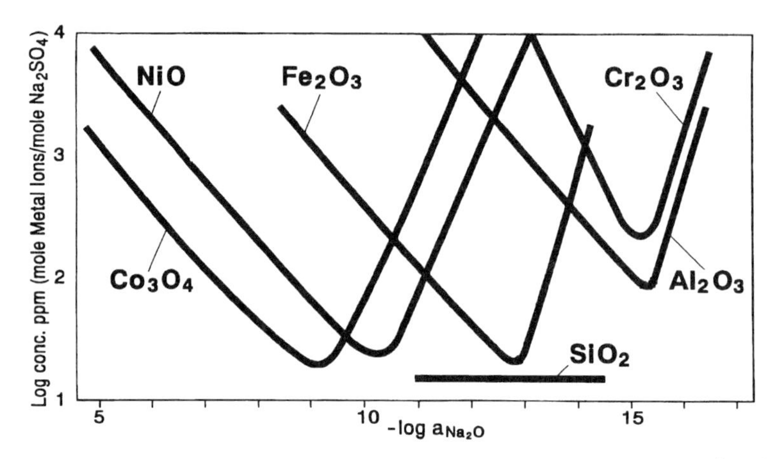

Figure 7.4: Solubility of metal oxides as a function of Na_2O activity (log Na_2O). A basic salt is represented by a high level of Na_2O activity, an acidic salt by a low level of Na_2O activity. The minimum of solubility corresponds to the minimum of the corrosion rate (after Rapp [29])

If complex salt mixtures are deposited, *low melting eutectics* may form and cause rapid corrosion due to dissolution of metal in the liquid phase. Corrosion rates of several mm/day have been observed in the presence of molten salts. It can be seen from Table 7.5 that combinations of alkali chlorides and heavy metal salts become liquid at very low temperatures. Fig. 7.5 shows a typical pitting corrosion attack which is caused by molten salts.

Table 7.5: Melting temperatures of some common eutectic metal chloride compounds (after Husemann [18])

Eutectic composition in mol%	Melting temperature in °C
25% NaCl / 75% $FeCl_3$	156
37% $PbCl_2$ / 63% KCl_3	175
60% $SnCl_2$ / 40% $FeCl_3$	176
70% $SnCl_2$ / 30% NaCl	183
70% $ZnCl_2$ / 30% $FeCl_3$	200
20% $ZnCl_2$ / 80% $SnCl_2$	204
55% $ZnCl_2$ / 45% KCl	230
70% $ZnCl_2$ / 30% NaCl	262
60% KCl / 40% $FeCl_2$	355
58% NaCl / 42% $FeCl_2$	370
70% $PbCl_2$ / 30% NaCl	410
52% $PbCl_2$ / 48% KCl	411
72% $PbCl_2$ / 28% $FeCl_2$	421
90% $PbCl_2$ / 10% $MgCl_2$	460
80% $PbCl_2$ / 20% $CaCl_2$	475
49% $PbCl_2$ / 51% $CaCl_2$	500

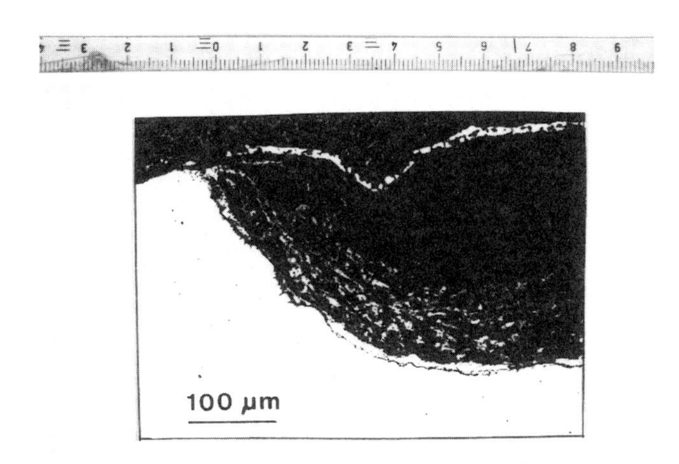

Figure 7.5: Typical high-temperature pitting corrosion in alloy 800H caused by a liquid salt after 168 hours exposure at 700 °C. Salt: 3.75 mol% $Fe_2(SO_4)$ added to a synthetic melt of 78 mol% Li_2SO_4+13.5 K_2SO_4 + 8.5 Na_2SO_4

While corrosion by salt/ash deposits is mainly encountered in the boiler section, combined *erosive-corrosive* attack predominantly takes place in the rotary kiln of pyrolysis plants. It is caused when abrasive particles, erosive gases or liquids remove the protective oxide scale of the metallic kiln walls and hot corrosive gases therefore have access to the base metal. The type and degree of erosion-corrosion depends on the hardness of the metal and the oxide, the shape and hardness of the particle, the angle of impingement, the velocity and the temperature [31-33]. The temperature dependence of the erosion-corrosion rate is typically given by a bell-shaped curve: above a certain temperature limit, which is about 500 °C in the case of ferritic steels and 750 °C in the case of nickel-base alloys, the rates of erosion-corrosion decrease significantly.

Aqueous corrosion is predominantly observed in the gas cleaning units of waste incineration plants. The corrosion problems are very similar to those encountered in flue gas scrubbers of power plants and the reader is referred to chapters 1 (by U. Heubner) and 8 (by R. Kirchheiner) of this book, where the corrosion problems of flue gas scrubbers and the selection of appropriate materials are discussed.

Stress-corrosion cracking caused by chloride-containing water is a frequent problem enountered with stainless steels. To avoid stress-corrosion cracking, steels with a high nickel concentration or even nickel-base alloys have to be used. At least 42 % nickel is required to ensure quasi-immunity to stress-corrosion cracking in boiler tubes [13].

7.3 Material selection

7.3.1 General remarks

Depending on the component for which they are to be used, materials for application in waste incineration and related plants have to meet different requirements: mechanical strength and creep resistance, ductility, corrosion resistance, weldability, approval by a pressure vessel authority and cost-effectiveness. It is clear from the preceding remarks that the selection of materials for furnace, boiler and scrubber has to made primarily with regard to corrosion resistance. Some materials used for the different components of waste incinerators are listed in Table 7.6.

The furnace walls in the lower part of both the furnace and the gasifier are attacked by corrosive combustion gases with temperatures of 1000 °C and above. Due to the very high temperature in this part, refractories or SiC tiles are usually used to line this area. To ensure a reasonable lifetime of the tiles, however, additional cooling is essential [34]. If studs are used to mount the tiles, these should be made from a high-chromium nickel-base material and/or covered with SiC caps.

Table 7.6: Summary of corrosive attack and material selection for different components of waste incineration and pyrolysis plants

Component	**Corrosive attack by**	**Material selection:**			
		Ceramic SiC tiles	Metals	Metal compounds	Reinf. plastics, rubber
Furnace:					
• Grate	• Hot slags and ashes		X		
• Furnace walls	• Hot gases, ashes, deposits, high temperature	X			
Pyrolysis stage:					
• Rotary kiln	• Erosion-corrosion, gases, deposits		X	X	
• Heating tubes	• Erosion-corrosion, gases, deposits		X	X	
Gasifier:					
• Walls	• Hot gases, ashes, deposits, high temperature	X			
Boiler, superheater, evaporator	• Chlorides, sulphates, low-melting deposits, erosion-corrosion		X	X	
Scrubber	• Sulphuric and hydrochloric acid, deposits		X	X	X

The grates of waste incineration plants are normally made of cast nickel-chromium alloys. High-chromium wrought materials may also be suitable.

Wrought alloys have to be be used for the boiler tubes, the rotary kiln and the scrubber. Table 7.7 summarises materials which are appropriate for these components.

Table 7.7: Selection of wrought iron and nickel-base alloys for application in the boiler, rotary kiln and scrubber of waste incineration and pyrolysis plants (RE=rare earth elements)

Alloy No.	DIN No.	VDM designation	Nominal composition in mass%					
		VDM designation	Ni	Fe	Cr	Mo	Si	Others
10CrMo910	1.7380		-	bal.	2.5	1	< 0.5	
AISI 314	1.4841	Cronifer 2520	19	bal.	24	-	1.9	
800H	1.4958	Nicrofer 3220H	31	bal.	20	-	0.5	
31	1.4562	Nicrofer 3127hMo	31	bal.	27	6.5	-	1.3 Cu
AC66	1.4877	Nicrofer 3228NbCe	32	bal.	27	-	-	0.8 Nb
45-TM	2.4889	Nicrofer 45-TM	bal.	23	27	-	2.7	0.1 RE
625	2.4856	Nicrofer 6020hMo	bal.	<4	22	9	-	3.4 Nb
59	2.4605	Nicrofer 5923hMo	bal.	-	23	16	-	

7.3.2 Alloys for the high temperature section of waste incineration and pyrolysis plants

10CrMo9 10 is an example of a conventional boiler steel which may be considered for low temperatures and low chlorine concentrations. Cronifer 2520 (AISI 314) is a stainless heat-resistant steel with excellent high-temperature corrosion resistance due to its high concentrations of chromium and silicon. Like all stainless steels, however, it is susceptible to stress-corrosion cracking and it tends to suffer ductility loss due to the formation of σ-phase at intermediate temperatures. Nicrofer 3220 H (alloy 800H) is a high-alloy special stainless steel widely used for applications in hot process gases. Under typical waste incineration conditions, however, its corrosion resistance is normally not significantly higher than that of ferritic boiler steels. The next alloy in the list, Nicrofer 3127hMo (alloy 31), has a higher resistance to both aqueous corrosion and high temperature corrosion. It is especially recommended for use in waste incinerators where dewpoint corrosion because of frequent shut-downs cannot be avoided.

Alloy AC66 has an excellent resistance to sulphidising/oxidising gas atmospheres. Its chlorination resistance, however, does not significantly exceed that of alloy 800H. If a high *chlorination resistance* is requested, nickel-base alloys like Nicrofer 45-TM (alloy 45-TM) or Nicrofer 6020 hMo (alloy 625) should be used. Nicrofer 45-TM has been designed especially for service in hot corrosive process gases. Besides a high chromium concentration, it contains 2.7 % silicon, which additionally protects the metal against corrosion by formation of a thin layer of silicon oxide [35,36]. Nicrofer 6020 hMo (alloy 625) is a typical Ni-Cr-Mo alloy designed mainly for wet corrosion applications. Because of its high strength and high chlorination resistance it is, however, also used in high temperature applications. The chlorination behaviour of alloys 625 and 45-TM in comparison to that of alloy AC66 and 800H after exposure to synthetic waste incineration gas is shown in Figs. 7.6 and 7.7. It can clearly be seen that the metal losses, which were caused by evaporation of volatile metal chlorides, are very low in the case of alloy 45-TM and alloy 625; while alloys 800H and AC66 already show significant metal losses at a temperature of 750 °C.

Both materials are indeed used in waste incineration plants. Alloy 625 is widely used for repairing corroded boiler tubes in waste incineration plants by overlay welding, especially in the United States. The alloy can be welded on ferritic boiler steels by common overlay welding methods. Nicrofer 45-TM is currently undergoing field tests in German and European waste incineration plants. Six 1000 mm long tubes were exposed in the evaporator of a waste incineration plant in Mannheim/Germany. An inspection after ten months revealed no significant evidence of corrosive attack. Tubes (dia. 6 mm) made of Nicrofer 45-TM have also been tested in the superheater section of a waste incineration plant in Bielefeld-Herford in Germany in comparison to 15Mo3 tubes of the same wall thickness. After 5,400 hours of operation the Nicrofer 45-TM tubes were still performing

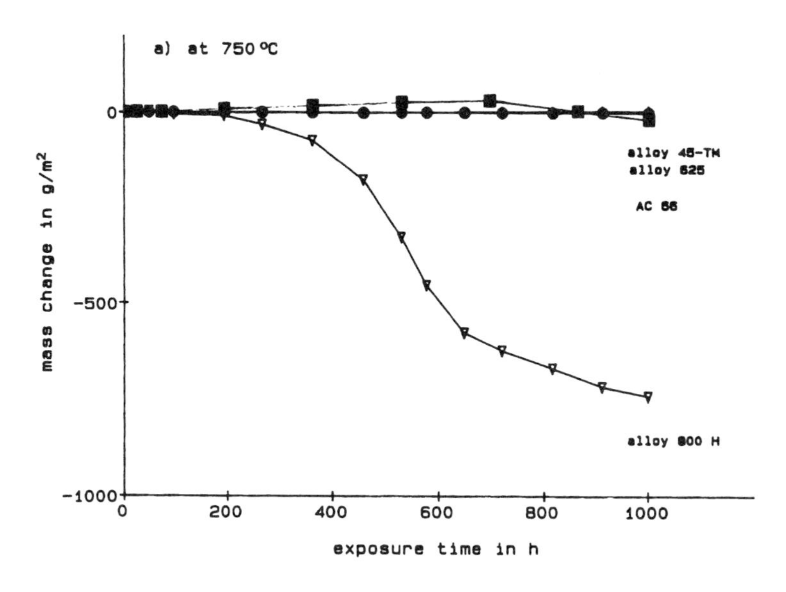

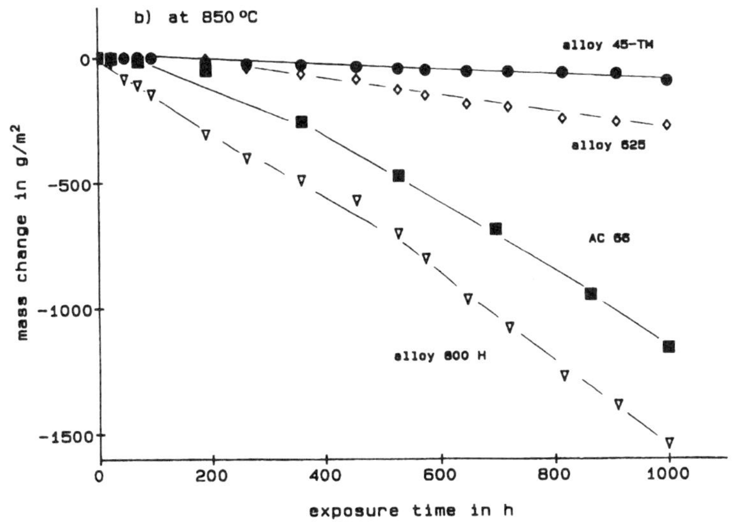

Figure 7.6: Mass loss of Nicrofer 45-TM (alloy 45-TM), Nicrofer 6020 hMo (alloy 625), Nicrofer 3228 NbCe (AC66) and Nicrofer 3220 H (alloy 800H) due to evaporation of volatile chlorides after 1,008 hours cyclic exposure in synthetic waste incineration gas (2.5 g/m^3 HCl, 1.3 g/m^3 SO_2, 9 vol% O_2, N_2: bal)
Figure 7.6a): Temperature T = 750 °C; 1 cycle = 24 hours
Figure 7.6b): Temperature T = 850 °C; 1 cycle = 24 hours

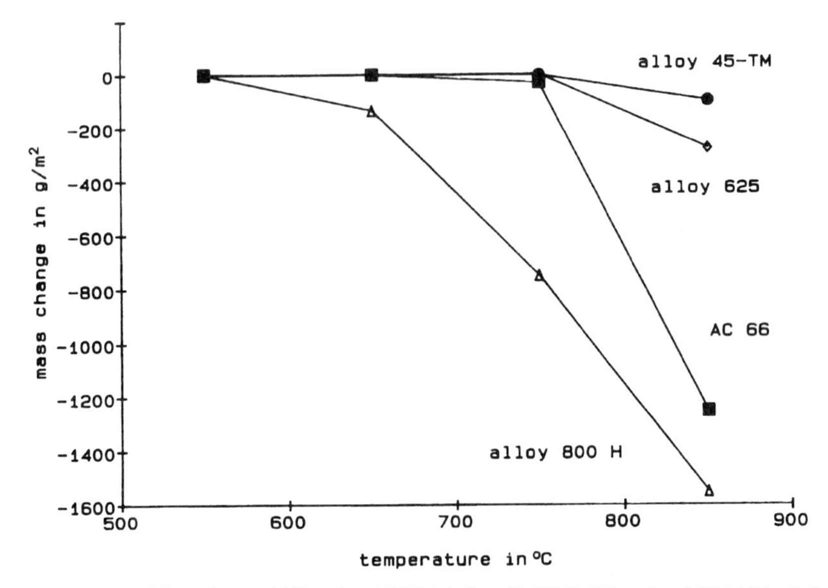

Figure 7.7: Mass loss of Nicrofer 45-TM (alloy 45-TM), Nicrofer 6020 hMo (alloy 625), Nicrofer 3228 NbCe (AC66) and Nicrofer 3220 H (alloy 800H) due to evaporation of volatile chlorides after 1,008 hours cyclic exposure in synthetic waste incineration gas (2.5 g/m^3 HCl, 1.3 g/m^3 SO_2, 9 vol% O_2, N_2: bal) as a function of time.
Maximum depth of corrosion attack after 1008 hours at 750 and 850 °C (in parentheses):
Alloy 800H: 880 µm (530 µm) + evaporated metal
AC 66: 210 µm (410 µm) + evaporated metal
Alloy 45-TM: 24 µm (266 µm)
Alloy 625: 110 µm (480 µm)

satisfactorily, whereas some of the 15Mo3 tubes had to be replaced after 3,900 hours because the wall thickness was reduced by corrosion to less than 3 mm [37]. Fig. 7.8 shows a cross-section of the superheater tubes after 5,400 hours of operation.

If *chloride-containing deposits* are encountered, all the foregoing comments regarding the chlorination resistance apply in full. Depending on the concentration of chloride in the deposit, alloys like Nicrofer 45-TM or Nicrofer 6020 hMo (alloy 625) should be selected. If waste containing very high concentrations of chlorides is treated (predominantly hazardous waste), an Ni-Mo-Cr-alloy like Nicrofer 6020 hMo (alloy 625) is recommended because of its higher resistance to chloride-induced internal oxidation.

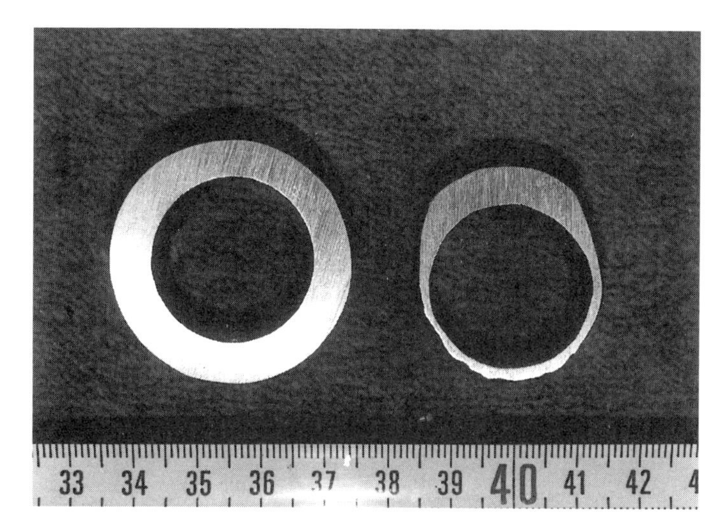

Figure 7.8: Tubes of alloy 45-TM and 15CrMo3 after 5,400 hours of operation in the superheater section of a waste incineration plant in Bielefeld/ Herford. The tube made of 15CrMo3 had to be replaced after 3,900 hours

If, on the other hand, *highly sulphidising gases* are encountered, as may be the case in plants for thermal treatment of wood and coal, Ni-Mo-alloys with 20 % Cr should not be used. According to laboratory tests, at least 25 % chromium is required to provide corrosion resistance in highly sulphidising process gases. As can be seen from Fig. 7.9, Nicrofer 45-TM with a chromium concentration of 27 % shows a low corrosion rate in a sulphidising/oxidising SO_2/O_2 gas even at high concentrations of SO_2. A comparably high sulphidation resistance was also found under sulphidising/oxygen-deficient atmospheres, which are typical of coal gasification plants [38].

If *sulphate-containing deposits* condense on boiler tubes and sulphate-induced corrosion has to be taken into account, molybdenum-containing alloys may suffer very high corrosion rates [39,40]. Fig. 7.10 shows the corrosion rates extrapolated from 1000-hours exposure of samples coated with a deposit of sodium sulphate and of sodium sulphate/potassium chloride. It can easily be seen from the bar chart that the molybdenum-containing alloys Nicrofer 6020 hMo (alloy 625), Nicrofer 3127 hMo (alloy 31), Nicrofer 6616 hMo (alloy C-4) and Nicrofer 5923 hMo (alloy 59) show high corrosion rates. The corrosion rate increases with increasing molybdenum concentration, as can be seen from Fig. 7.11. Molybdenum-free alloys like Nicrofer 45-TM do not suffer from this type of corrosion.

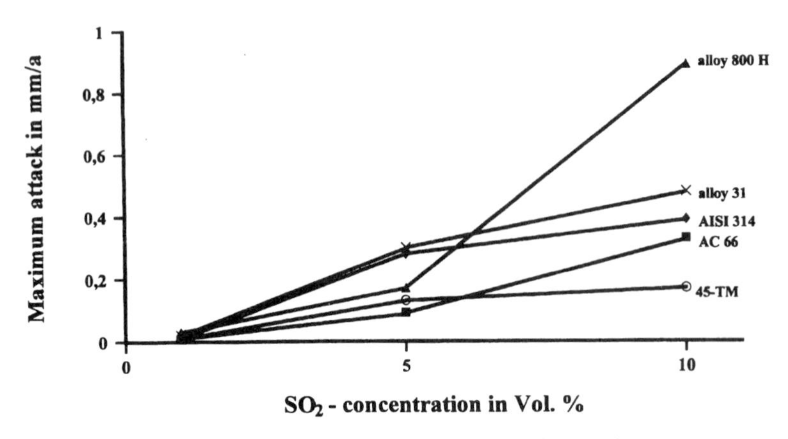

Figure 7.9: Extrapolated corrosive attack in mm/year after cyclic exposure at 650 °C in SO_2/air as a function of SO_2-concentration (linearly extrapolated from 1,008 hours, 1 cycle = 24 hours). The chromium content of alloy 800H is 20 %, all other alloys have a chromium content of 24 % or more)

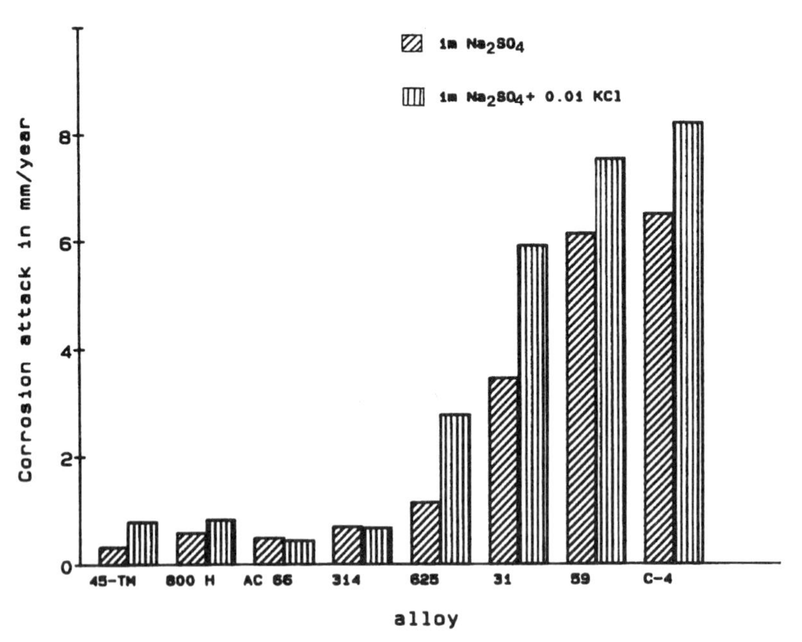

Figure 7.10: Extrapolated corrosive attack of samples coated with Na_2SO_4 and Na_2SO_4/KCl prior to cyclic exposure at 750 °C in air (linearly extrapolated from 1,008 hours, 1 cycle = 24 hours)

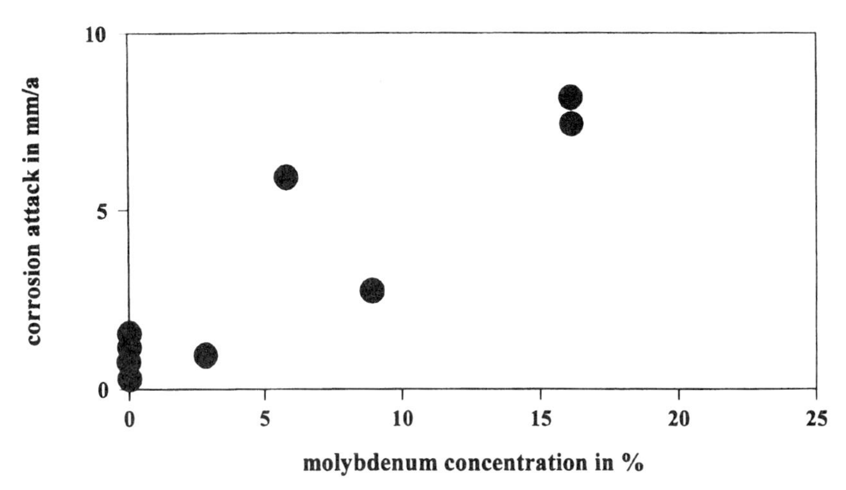

Figure 7.11: Effect of molybdenum on the corrosion rate of materials coated with Na_2SO_4/KCl prior to cyclic exposure at 750 °C in air. (Corrosion rate linearly extrapolated from 1,008 hours, 1 cycle = 24 hours)

Since the test temperature was below the melting point of sodium sulphate, acidic fluxing should be the predominant corrosion mechanism (see chapter 7.2). It is suspected that molybdenum additionally increases the acidity of the deposit according to the reaction:

$$MoO_3 + Na_2SO_4 \Leftrightarrow NaMoO_4 + SO_3 \qquad (5)$$

SO_3, which is released by this reaction, increases the acidity of the deposits and favours acidic fluxing of metal or oxide according to equation (4). A similar type of corrosion is the so-called "oil ash corrosion", which is caused by vanadium pentoxide. Vanadates, like molybdates, increase the acidity of deposits by releasing SO_3 [41].

Extremely high corrosion rates can be expected if the deposits become liquid. After exposure in a basic slag containing 45% Na_2S, 21% Na_2CO_3, 21% Na_2SO_4 and 13% NaCl, the corrosion rates of both stainless steels and nickel base alloys were found to be below 0.3 mm/year if the test temperature was 550 °C. If the test temperature was increased to 600 °C, the corrosion rates increased to levels of 1 mm/year (45-TM), 1.3 mm/year (alloy 625), 2.7 mm/year (alloy 800H) and 4.6 mm/year (AC66); at 650 °C corrosion rates between 1.8 mm/year (45-TM) and 27.2 mm/year (AC66) were measured [42]. The melting point of the slag was 580 °C [43]. Extremely high corrosion rates of high temperature materials were also detected in liquid 78 mol% $LiSO_4$/13.5 mol% K_2SO_4/8.5 Na_2SO_4 salt with additions of $Fe_2(SO_4)_3$ and NaCl respectively at 700 °C [44] and in 10% K_2SO_4/5% KCl/bal. Al_2O_3 salt at 600 °C [45].

Summarising these results, Nicrofer 45-TM can be recommended for use in the presence of both acidic and basic salts, whereas the use of Nicrofer 6020 hMo is restricted to less acidic salts. Liquid deposits should be avoided at all costs, since they can cause high corrosion rates in any metallic material. This can be done either by reducing the temperature or by shielding the affected tubes; both methods have been successfully applied [12,18,20]. Alloys like Nicrofer 45-TM or Nicrofer 6020 hMo, however, have a certain resistance to fluxing by liquid phases too.

It should be noted that a change in the gas composition may change the aggressiveness of the deposit significantly. It was found in laboratory tests that the corrosion rate of molybdenum-containing alloys was lowered when HCl/SO_2-containing gas was used instead of air [40] and that additions of SO_2 to HCl-containing test gas increased the resistance of stainless steels and nickel-base alloys to chloride-induced corrosion [25].

Combined attack by *erosion and corrosion* is predominantly suffered by rotary kilns in pyrolysis plants. Erosion by abrasive particles can affect the corrosion behaviour significantly, as can be seen from Figs. 7.12, 7.13 and 7.14. Fig. 7.12 shows the metal loss of different materials due to combined erosive-corrosive attack in a synthetic waste incineration atmosphere as a function of tempera-

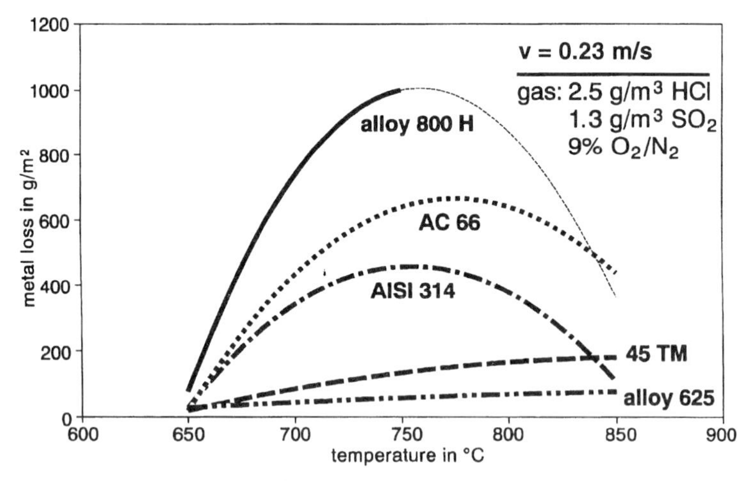

Figure 7.12: Metal loss of high temperature materials rotated for 1,008 hours in a bed of erosive particles in the presence of synthetic waste incineration gas as a function of gas temperature.
Erosive particles: SiO_2, d < 0.5 mm, relative velocity between particles and metal samples: 0.23 m/s, particle density: » 1.5 g/cm^3
Gas: (2.5 g/m^3 HCl, 1.3 g/m^3 SO_2, 9 vol% O_2, N_2: bal).

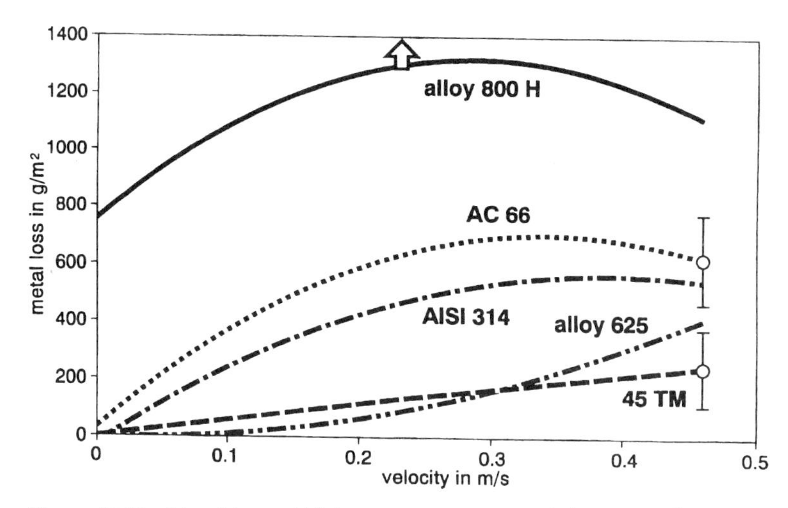

Figure 7.13: Metal loss of high temperature materials rotated for 1,008 hours in a bed of erosive particles in the presence of synthetic waste incineration gas as a function of the relative velocity of particles and metal samples.
Erosive particles: SiO_2, d < 0.5 mm, relative velocity of particles and metal samples: 0, 0.23 and 0.46 m/s, particle density: » 1.5 g/cm³
Gas: (2.5 g/m³ HCl, 1.3 g/m³ SO_2, 9 vol% O_2, N_2: bal).
Gas temperature: 750 °C

ture, while in Fig. 13 the metal loss is plotted against the relative velocity of the particle and the metal sample. Some typical edges of the specimens after the erosion-corrosion tests are shown in Fig. 7.14. The results clearly demonstrate that the resistance to erosion-corrosion is mainly dependent on the corrosion resistance to the respective gas (here: synthetic waste incineration gas). Hence the chlorination-resistant alloys Nicrofer 45-TM and Nicrofer 6020 hMo (alloy 625) exhibit high resistance even under combined erosive-corrosive attack. Nicrofer 3220H (alloy 800H), on the other hand, suffers high metal losses despite its high erosion resistance [46]. Consequently material selection for waste incineration plants should be carried out with a view to high corrosion resistance even if erosion has to be taken into account. Only if the temperature and corrosive attack are moderate can materials like Nicrofer 3220H (alloy 800H) be used. A higher corrosion-erosion resistance is obtained if corrosion-resistant materials like Nicrofer 3228 NbCe, or, with even higher resistance, Nicrofer 45-TM and Nicrofer 6020 hMo are used. Fig. 7.15 shows a rotary kiln made from AC66.

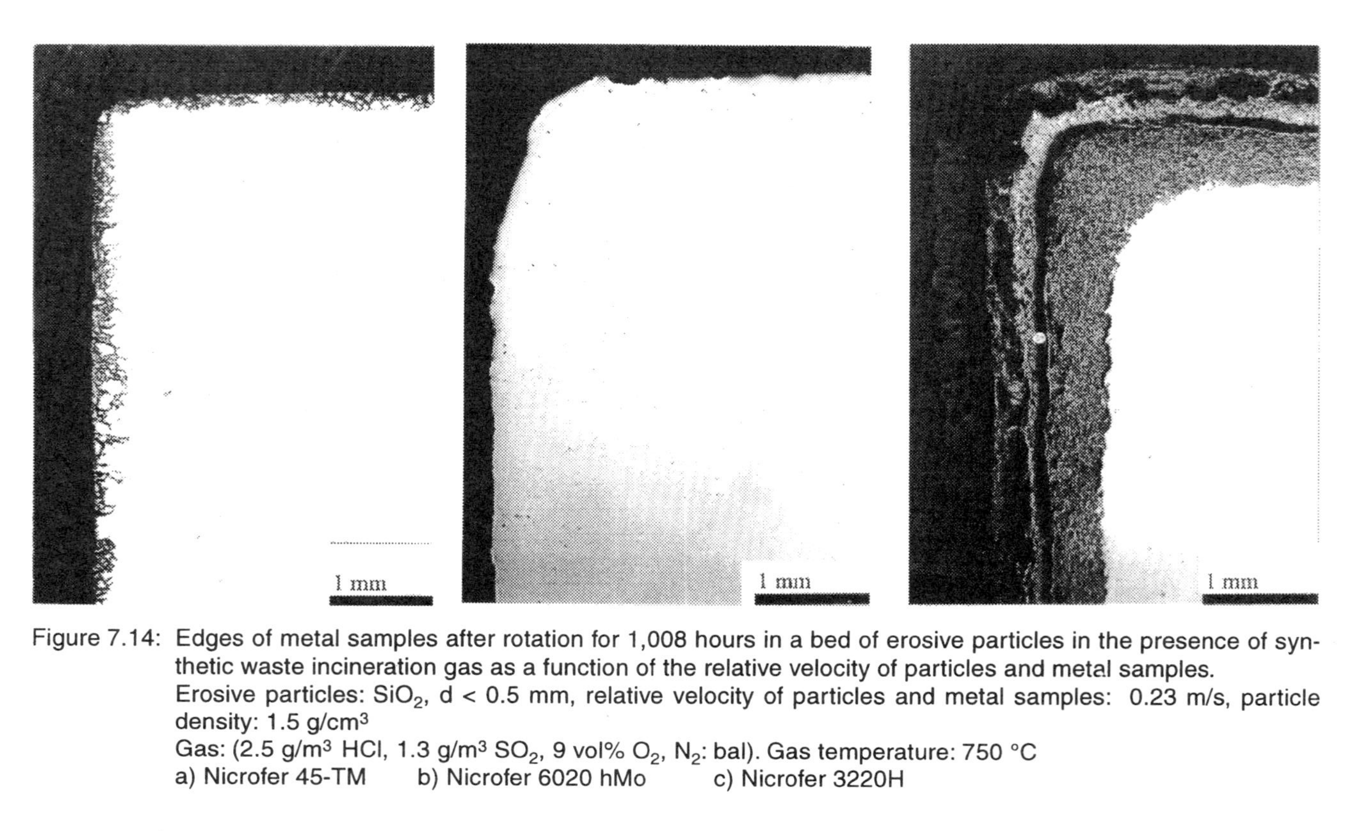

Figure 7.14: Edges of metal samples after rotation for 1,008 hours in a bed of erosive particles in the presence of synthetic waste incineration gas as a function of the relative velocity of particles and metal samples.
Erosive particles: SiO_2, d < 0.5 mm, relative velocity of particles and metal samples: 0.23 m/s, particle density: 1.5 g/cm³
Gas: (2.5 g/m³ HCl, 1.3 g/m³ SO_2, 9 vol% O_2, N_2: bal). Gas temperature: 750 °C
a) Nicrofer 45-TM b) Nicrofer 6020 hMo c) Nicrofer 3220H

Figure 7.15: Rotary kiln made from Nicrofer 3228 NbCe (AC 66) for a pyrolysis plant in Burgau prior to installation.
Technical data:
Type of waste: household refuse, industrial waste, sludge
Number of furnaces: 2
Quantity of waste: 35.000 t/year
Vapour pressure: 30 bar Vapour temperature: 400 °C
Power: 2.2 KW

7.3.3 Materials for the scrubber

The corrosion problems in the scrubber are generally similar to those encountered in scrubbers of flue gas desulphurisation plants and materials established for application in FGD plants can also be used in waste incineration scrubbers. One of these materials with high resistance to sulphuric and hydrochloric acid is Nicrofer 5923 hMo (alloy 59) with 16 % molybdenum and 23 % chromium. Experience with this alloy has already been obtained in the scrubber of the waste incineration plant at Essen-Karnap, Germany, where the scrubber was lined with sheets of Nicrofer 5923 hMo [47]. Originally the scrubbers were made from 4 mm carbon steel lined with rubber. After a short period, however, the rubber was damaged and the steel started to corrode. The plant was repaired in 1992 by cladding the walls with 2-mm sheets and the bottom with 3-mm sheets of Nicrofer 5923 hMo. It has been in operation since 1993; the last inspection after 26,000 hours showed no evidence of corrosive attack. Fig. 7.16 shows a schematic drawing of the plant.

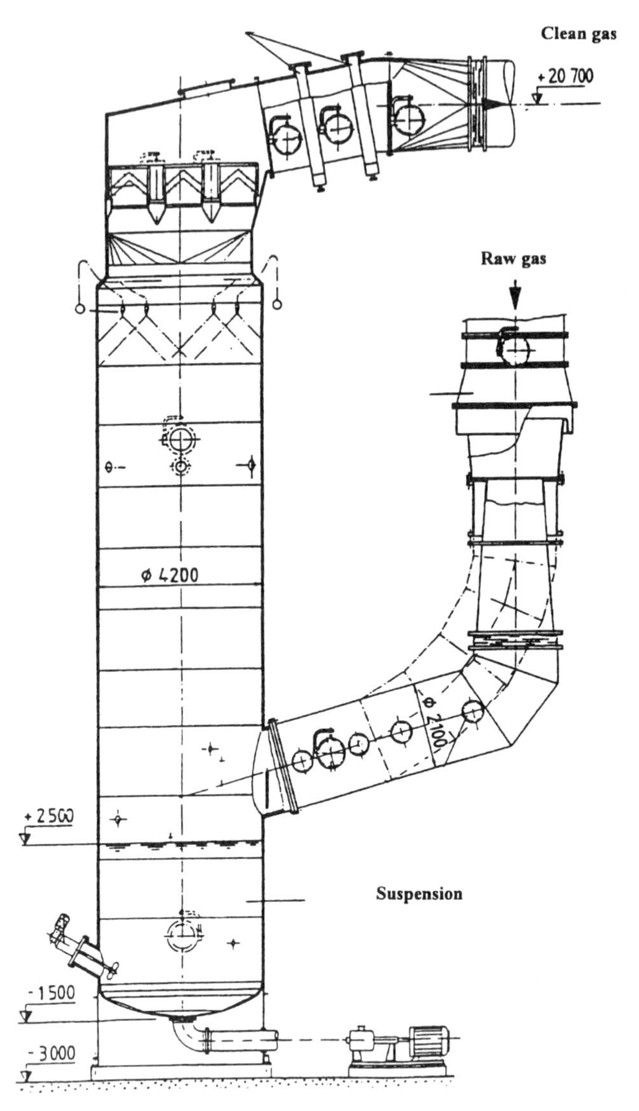

Figure 7.16: Schematic drawing of the gas scrubber used in the Essen-Karnap waste incineration plant. The scrubber has been completely lined with Nicrofer 5923 hMo (alloy 59), after [3,47].
Technical data:
Type of waste: household refuse, industrial waste,
Quantity of waste: 745.000 t/year
Number of incineration units: 4

7.4 Summary and concluding remarks

Highly corrosive atmospheres are encountered by materials in the rotary kiln, the furnace, the boiler and the scrubber of waste incineration and pyrolysis plants, with the type and degree of corrosion depending to an appreciable extent on the process, the temperature of the flue gas and the metal wall, the composition of the waste and the design of the plant.

The corrosion problems in the gas scrubber are similar to those encountered in FGD plants and consequently the same materials may be used. These are Ni-Cr-Mo-(Fe) alloys with high concentrations of chromium and molybdenum. Nicrofer 5923 hMo (alloy 59) has already been successfully tested in flue gas scrubbers in both power and waste incineration plants.

Corrosive attack of furnace and boiler walls occurs either by chlorination, sulphidation, attack by corrosive salt/ash deposits, by low-melting eutectics or by wet corrosion (dewpoint corrosion) during shut-down. Rotary kilns may additionally suffer combined erosion-corrosion. Since chlorination and/or chloride-induced oxidation is the predominant corrosion mechanism under most plant conditions, chlorination-resistant nickel-base alloys like Nicrofer 45-TM (alloy 45-TM) or Nicrofer 6020 hMo (alloy 625) should be used for these components. Nicrofer 6020 hMo (alloy 625) is suitable if very high chlorine concentrations are encountered, e.g. if chloride-containing hazardous waste is treated. It should be noted, however, that some national pressure vessel authorities have restricted the application of alloy 625 to low or medium temperatures (the max. temperature for alloy 625 in pressure vessel service laid down by the German VdTÜV is 450 °C). Nicrofer 45-TM has high corrosion resistance if chlorides are encountered in combination with sulphidising gases or sulphate-containing deposits. If dewpoint corrosion during frequent shut-downs cannot be avoided, selection of Nicrofer 3127 hMo (alloy 31) may also be considered.

High temperatures in combination with high chloride concentrations and/or low melting eutectics on the tube and furnace walls can cause extremely high corrosion rates of several mm/day. Under such conditions the corrosion rates can be controlled only by a combination of appropriate plant design, correct material selection and suitable process parameters. Reduction of the flue gas temperature, sufficient turbulence to enable sulphation, the avoidance of reducing conditions and the protection of special parts with shieldings can help to reduce the corrosion rates in waste incineration plants to a reasonable level.

7.5 References

[1] "Stellenwert der Hausmüllverbrennung in der Abfallentsorgung" Information brochure from the Federal Ministry of the Environment, Bonn

[2] B. Meadowcroft in: D. Coutsouradis et al. (Eds.) "Materials for Advanced Power Engineering 1994" Kluwer Academic Publ., Dordrecht 1994, p. 1413

[3] N.N.: "Thermische Abfallverwertung" Report by Krupp VDM GmbH, Werdohl, 1995

[4] J. Krüger: VGB Kraftwerkstechnik 74 (1994), p. 978

[5] J. Krüger: VGB Kraftwerkstechnik 76 (1996), p.134

[6] R. Scholz, R. Jeschar, N. Schopf and G. Kloppner: Chemie-Ing.-Technik 62 (1990), no. 11

[7] K.W.May, R. Langer, R. Kilian and N. Wieling: Ingenieur-Werkstoffe 3 (1991), no. 6, p. 35

[8] Energie-Spektrum June 1994, p. 28

[9] "Bezwinger des Feuers" umwelt und technik, December 1994, p. 16

[10] "Kombi-Angebot" Energie und Management 1-2/1995, p. 60-61

[11] EBB Nr. 318, German edition dated 28 October 1994

[12] I.G. Wright, H.H. Krause and R.B. Dooley: CORROSION '95, Paper No. 562, NACE International, Houston, TX, 1995

[13] G. Sorrell: Chemical Engineering Processing, March 1994, p. 49

[14] H.H.Krause, I.G.Wright and V.K.Sethi in: K. Natesan and D.J.Tillack (Eds.) "Heat Resistant Materials", Proc. Conf. ASM, Ohio 1991, p. 623

[15] K. Kautz and W. Berent: VGB Kraftwerkstechnik 59 (1979), p. 349

[16] P. Seifert and M. Born in: Proceedings of the VGB conference "Korrosion und Korrosionsschutz in der Kraftwerkstechnik 1995", 29/30 Nov. 1995 in Essen, Paper No. 9, p. 1-10

[17] H. Krause and I.G. Wright: CORROSION '95, Paper No. 561, NACE International, Houston, TX, 1995

[18] R.U.Husemann: VGB Kraftwerkstechnik 72 (1992), p. 918

[19] B. Boßmann and L. Singheiser: Proceedings of the VGB conference "Korrosion und Korrosionsschutz in der Kraftwerkstechnik 1995", 29-30 Nov. 1995 in Essen, Paper No. 11

[20] B. Meyer, O. Willmes and B. Röper: VGB Kraftwerkstechnik 75 (1995), p. 1043

[21] VDI-Richtlinie No. 2114, June 1992

[22] E. Häggblom and J. Mayrhuber in D. Coutsouradis et al. (Eds.) "Materials for Advanced Power Engineering 1990" PART I, Kluwer Academic Publ., Dordrecht 1990, p. 91

[23] E. Reese and H.-J.Grabke: Werkstoffe und Korrosion 43 (1992), p. 547

[24] D. Bramhoff, H.-J.Grabke, E. Reese and H.P.Schmidt: Werkstoffe und Korrosion 41 (1990), p. 303

[25] M. Spiegel and H.-J. Grabke: Werkstoffe und Korrosion 47 (1996), p. 179

[26] P. Elliot et al. in M.F.Rothman "High Temperature Corrosion in Energy Systems" AIME, Warrendale PA, (1985), p. 437

[27] R. Bürgel, H.W. Grünling and K. Schneider: Werkstoffe und Korrosion 38 (1987), p. 549

[28] A. Rahmel: Mat. Sci. and Eng. 87 (1987), p. 345

[29] R.A. Rapp: Mat. Sci. and Eng. 87 (1987), p. 319

[30] L.D. Paul et al. TAPPI Journal 76 (1993), p. 73

[31] I. Finnie in K.Natesan (Ed.) "Proc. Conf. Corrosion-Erosion Behaviour of Materials" AIME, Warrendale, PA, (1980), p 118

[32] K. Natesan and Y.Y. Liu: Mat. Sci. and Eng. A121 (1989), p. 571
[33] V.K.Sethi and R.G.Corey in: J.E. Field and J.P. Dear (Eds.) "Proc. 7th Conf. on Erosion by Liquid and Solid Impact", Cambridge (1987), p. 73
[34] V. Köhne, H. Knorth and H.-D.Pötsch: VGB Kraftwerksechnik 68 (1988), Issue 12, p. 1279
[35] U. Brill and Klöwer: Z. Metall 46 (1992), No. 9
[36] U. Brill and Klöwer: Materials at High Temperatures 11 (1993), p. 151
[37] D.C. Agarwal, J. Klöwer and G. Grossmann: CORROSION '97, Paper No. 155, NACE International, Houston, TX, 1997
[38] N. Ullrich, W. Schellberg and W.T.Bakker: 12th EPRI Conf. on Gasification Power Plants, October 1993, San Francisco, CA
[39] J. Klöwer in: K. Natesan, P. Ganesan and G. Lai (Eds.) "Heat Resistant Materials", (Proc. Conf.), ASM, Materials Park, OH, (1995), p. 245
[40] J. Klöwer: CORROSION '96 Paper No. 173, NACE International, Houston, TX, 1996
[41] P. Kofstad: "High Temperature Corrosion", Elsevier Applied Science, London, New York, 1988
[42] J. Klöwer and F. White in: "Proc. Conf. Int. Symp. on Corrosion in the Pulp and Paper Industry" Stockholm, Sweden, 14-16 May 1995, NACE, TAPPI (1995), p. 179
[43] O. Moberg: "Recovery Boiler Corrosion in Pulp and Paper Industry", Corrosion Problems, Vol 1, NACE, Houston, 1974, p. 125
[44] U. Brill, J. Klöwer, E. Maasen, H. Richter, W. Schwenk and J. Venkateswarlu in Ref. [2] p. 1585
[45] J. Mayhuber and H. Cerjak: Paper presented at the VGB conference "Forschung in der Kraftwerkstechnik 1993" 24-25 February 1993, Essen
[46] J. Klöwer in: K. Natesan, P. Ganesan and G. Lai (Eds.) "Heat Resistant Materials", (Proc. Conf.), ASM, Materials Park, OH, (1995), p. 147
[47] P. Prefke, W. Scharnberg, W. Schliesser, E. Schwab: Paper presented at the VGB conference "Werkstoffe und Schweißtechnik im Kraftwerk 1994", 15-16 March 1994

8 The use of nickel alloys and stainless steels in environmental engineering

R. Kirchheiner

8.1 Introduction

Since air pollution control occupies a leading position in the market for environmental technology, this chapter concentrates on the engineering and material concepts in flue gas desulphurisation in fossil-fuel-fired power stations. Starting from the developments in the USA and Japan from about 1975 the use of metallic materials of construction to accomplish these air pollution control tasks is described, using Germany as an example. Measures to reduce the pollutants in question were given a legal basis in Germany in 1974 with the Pollution Control and Noise Abatement Act (BlmSchG). In the same year the Technical Instructions for Air Pollution Control (TA-Luft) was passed, from which the regulations for large combustion plants (GFAVO) were later derived [1]. On this basis, from 1983 to mid 1988 a power station capacity of 37,000 MW at 72 sites in West Germany with 165 flue gas desulphurisation scrubbing lines and a total throughput of 135 million standard cubic metres of flue gas was desulphurised [2].

8.2 Processes and materials for the desulphurisation of flue gas in fossil-fired power stations

8.2.1 Chemical technology of desulphurisation plants

The subject of this chapter is the removal of oxidation products of the sulphur and of the accompanying HCl/HF gases from the combustion processes of organic materials (coal, lignite, oil) by chemical gas scrubbing. This process is generally known as 'flue gas desulphurisation' and the associated plants as 'flue-gas desulphurisation plants'.

The wide range of scrubbing processes developed can be divided into scrubbing processes with regeneration of the scrubbing solution, spray absorption processes, and scrubbing processes without regeneration of the scrubbing solution. Whereas the ecologically very worthwhile scrubbing processes with regeneration of the scrubbing solution have only been used to a very limited extent because of process engineering difficulties, of the scrubbing processes without regeneration the so-called limestone process is used throughout the world in about 90% of all installed flue-gas desulphurisation plants. There are about 200 different versions of this process alone [1, 2, 3]. The systematically best presentation of all process engineering questions in the field of limestone scrubbing is given in [4]. In this section, by taking into account process engineering and design problems, a concept is finally developed for the application of highly corrosion-resistant metallic materials. Fig. 8.1 presents a schematic view of a flue gas desulphurisation plant. After passing through an electrostatic precipitator, the dust-free raw gas is sprayed with a scrubbing solution in a quencher, whereby it is cooled and saturated with water vapour. In plants with a heat recovery system, the raw gas is cooled from about 200 - 300 °C (with/without NO_x reduction) to 150 °C before entering the scrubber by means of a so-called gas preheater. The gas flow and the circulation of the scrubbing solution are adjusted so that the temperature of the scrubbing solution is about 60 °C. The pH value is adjusted at 4.5 to 6. In the scrubber sulphur dioxide/sulphur trioxide (SO_2/SO_3) are transformed by the addition of limestone into calcium sulphite ($CaSO_3$) and subsequently transformed by injection of air or oxygen into essentially insoluble $CaSO_4 \cdot 2H_2O$, i.e. gypsum. This essentially insoluble gypsum precipitates out and thus promotes the very high SO_2/SO_3 conversion rate of of over 90% because of law of the mass balance [1, 2]. A suspension of quicklime (CaO), hydrated lime ($Ca(OH)_2$) or limestone ($CaCO_3$) is used as the scrubbing solution. Chemical buffer reactions provide a constant pH value in the absorber after the start-up phase of a flue gas desulphurisation plant. Secondary reactions that take place after entry into the quencher or the bottom section of the integrated absorber are the almost 100% removal of hydrogen chloride and hydrogen fluoride (HF), which are produced by chemical reactions in the combustion process and carried over with the flue gas. In counter-current systems the flue gas is again sprayed with the scrubbing suspension in the upper region of the scrubber, whereby the limestone excess in the added suspension results in a further increase in the degree of desulphurisation. A pH shift towards a pH of approx. 6 is obtained by the buffer reaction of hydrogen carbonate. The main reaction in flue gas desulphurisation therefore consists in the transformation in a weakly acidic medium of sulphur oxides into inert insoluble end products (gypsum). The resultant stress on the material from this process is in principle not critical. Practice shows, however, that particular zones within the plant are more sensitive with regard to materials than others, apparently because of secondary reactions or non-equilibrium conditions. Effects such as local temperature variations, catalytically acting surface conditions and the tendency towards formation of incrustations and concentrations of individual harmful ions contribute to this. Those zones are indicated in Fig. 8.1 and roughly classified according to the aggressiveness encountered there [5].

Dew point corrosion.

In the rankings of harmful effects on any material, the formation of a condensed acidic liquid phase from flue gases due to the temperature dropping below the dew point is at the top. The conditions for the formation of condensation are

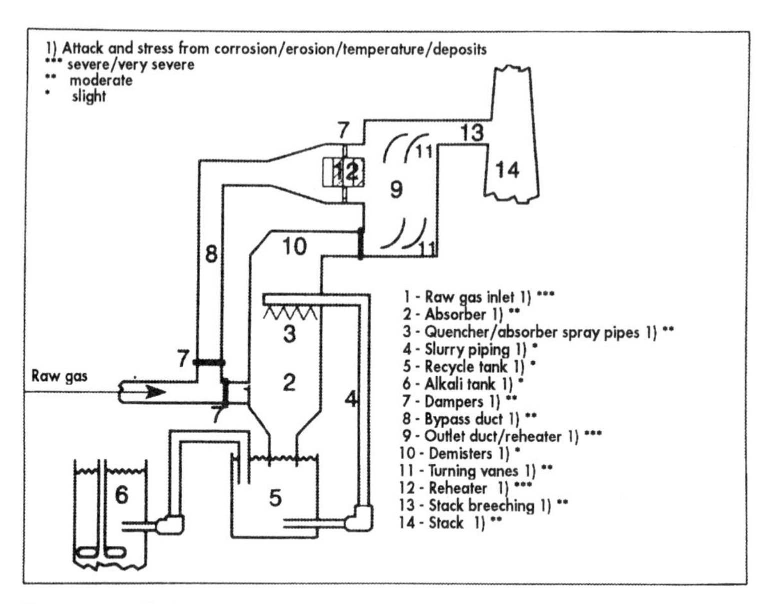

Figure 8.1: Schematic diagram of a flue gas desulphurisation plant (FGD) based on alkaline spray absorption scrubbing

described here in more detail because of the far-reaching importance of this mechanism for the use of metallic materials. Sulphur from the fuel is initially converted to sulphur dioxide (SO_2) with the oxygen present in the air in a strongly exothermic reaction. Sulphur trioxide forms to a much lesser extent, whereby the sulphur dioxide/sulphur trioxide equilibrium is displaced markedly towards sulphur dioxide as the final product because of the substantial quantity of which is released. At about 600 °C the thermal decomposition of the sulphur trioxide formed to sulphur dioxide commences, which also severely reduces the proportion of SO_3. In contrast to the inert sulphur dioxide, the sulphur trioxide, which is only present in traces, immediately undergoes a reaction with the water vapour always present in the raw gas through sulphurous acid up to sulphuric acid [2]. Because of the low initial quantity of SO_3, the flue gas thus only contains small quantities of gaseous sulphuric acid (H_2SO_4). Contents of 1 to max. 15 vol.ppm of gaseous sulphuric acid in the raw gas are typical. When the raw gas is cooled, certain gaseous constituents of the raw gas then start to condense out when the saturation pressure for these constituents, which decreases with temperature, attains the partial pressure of these constituents in the raw gas. The temperature at which this occurs is known as the dew point. In flue gases sulphuric acid, sulphurous acid, nitric acid, hydrogen chloride and hydrogen fluoride as well as water vapour are constituents that can condense. Sulphuric acid has the highest dew point of all these constituents, so that when one speaks of the acid dew point, the dew point of sulphuric acid is always meant. SO_2 therefore always condenses out first from the high raw gas temperatures (>350 °C) on cooling. With the typical sulphuric acid partial pressures in coal combustion of 10^{-6} to 10^{-5} bar and the associated water vapour partial pressure of 0.02 to 0.5 bar, practical measurements have shown dew point temperatures of about 100 to 150 °C, maximum 180 °C.

The processes of condensate formation due to the temperature falling below the dew point and their effect on materials are described in great detail in [6]. Because of the extreme differences in the boiling points of water (H_2O)/BP =100 °C) and sulphuric acid (H_2SO_4/BP = 338 °C) there is a marked separation of both constituents during boiling and condensation. This means for condensation under dew point conditions that high concentrations of sulphuric acid in the condensate can occur even with very low concentrations of sulphuric acid in the raw gas, as shown in Fig. 8.2. The condensing sulphuric acid can vary in concentration between 65 and 95 wt.% [2]. The concentrated sulphuric acid formed at this high condensation temperature has a large influence on the selection of possible materials in such zones. The inverse process, i.e. the heating of cold wall surfaces covered with condensate films, also results in extreme acid concentration.

The factor that determines resistance in the use of metallic materials for desulphurisation plants is therefore the reaction of the material under consideration to sulphuric acid. A close relationship of corrosion to temperature is caused by this sulphuric acid corrosion.

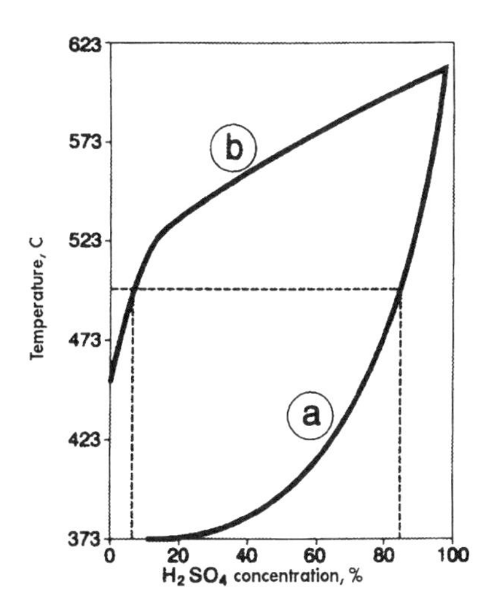

Figure 8.2:
Boiling diagram of a mixture of water and sulphuric acid to illustrate the formation of a highly concentrated sulphuric acid condensate from the raw gas [5a]

If the temperature only falls slightly below the dew point, the quantity of highly concentrated sulphuric acid precipitated is initially relatively small in the range T > 150 °C to T < 50 °C. Only with further continuous cooling does the quantity of condensate increase drastically. With the surface film formed for example on vertical walls, consisting of hot, highly concentrated sulphuric acid, the corrosion increases enormously, as is shown schematically in Fig. 8.3. Only later, according to the boiling point/dew point diagram in the system „water: sulphuric acid", does dilution of the condensing sulphuric acid due to simultaneous condensation of water take place, so that the attack on high-alloyed materials, i.e. usually those with a high chromium and molybdenum content accompanied by a high nickel content, is again reduced. Bäumel has calculated [2] that the maximum aggressiveness of the sulphuric acid condensates is about 30 °C below the observed sulphuric acid dew point.

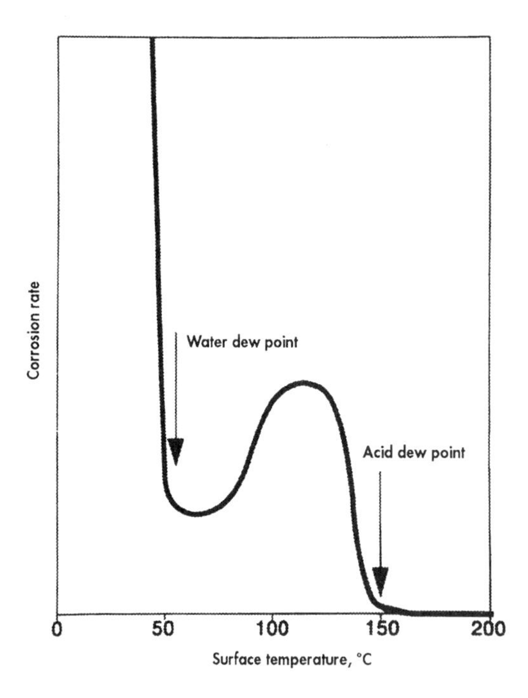

Figure 8.3:
Schematic diagram of a corrosion maximum below the sulphuric acid dew point of combustion gases

Additional attack is caused by halides (chlorides and fluorides) that are contained in the impurities of the fossil fuels (dead rock/salts); however, it can also be caused by the use of scrubbing suspensions with contaminated natural water and contaminated starting products. Because of the high rate of conversion of hydrogen chloride (HCl) and hydrogen fluoride (HF) with limestone, calcium chloride ($CaCl_2$) and calcium fluoride (CaF_2) are formed immediately. Calcium chloride dissociates instantly into calcium ions and chloride ions, whereas calcium fluoride is largely insoluble in the absorber suspension and immediately precipitates out. Whereas the quantity of chloride ions stemming both from the combustion of the coal and from the feed waters used for the limestone suspension can easily be as high as 1% to 5% in the absorber, the free fluoride ions in solution are limited to substantially less than 0.02%. As well as the danger from sulphuric acid corrosion in the active state of stainless steels, there also arises the risk of localized types of corrosion in acidic to neutral chloride-containing solutions. In addition, heavy metals and their oxides as well as nitrates drastically change the chemistry of their immediate surroundings by the formation of high redox potentials.

The corrosive nature of the surrounding conditions can only be influenced to a limited extent by selection of the feedstocks/fuels and control of the combustion. On the level of process engineering, however, pH value adjustment, temperature limitation and water management can be used to influence the corrosiveness of the conditions and thus the long-term resistance of the construction materials used. Fig. 8.4 summarizes these factors of influence and describes their effect on the corrosion behaviour of metallic materials. Metallic materials, in particular those that protect themselves by passivation of their surface, e.g. stainless steels and titanium, can be damaged by localized types of corrosion including pitting, crevice and stress-corrosion cracking. Erosive corrosion by solid particles and contact corrosion by incorrect material couples are further causes of premature failure of wrongly selected materials [7]. In addition to the reaction mechanisms there are also fuel-specific differences. In lignite-fired power stations a raw gas inlet temperature of 170 to 230 °C is measured because of the different calorific value and the impurities in the lignite. This is in contrast to desulphurisation plants for coal-fired power stations, in which the raw gas reaches an absorber inlet temperature of about 130 to 160 °C. The zones of greatest material attack, both thermal as well as chemical, occur here for both fuels. In the comparison of coal-fired and lignite-fired power stations it is also apparent that the adiabatic saturation temperature of 65 to 70 °C in the lower scrubber cylinder in lignite-fired power stations is about 20 °C higher than the 45 to 50 °C

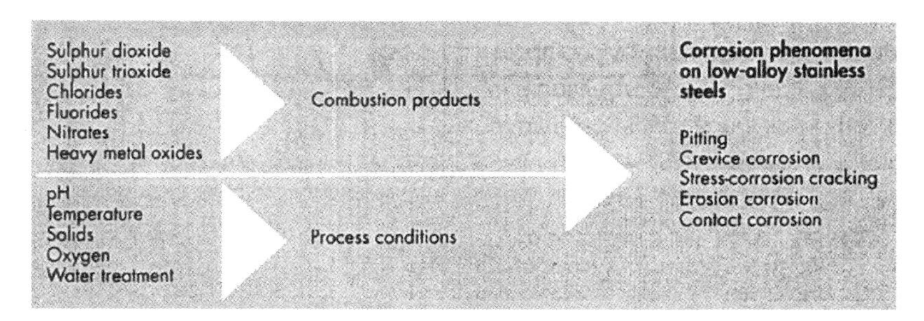

Figure 8.4: Types of corrosion on low-alloy stainless steels initiated by combustion products and process conditions

in coal-fired power stations [8]. In the absorber itself the fluids have a constant acid strength on account of the buffer system. Organically based materials, e.g. rubber linings, have accordingly been damaged to a great extent by steam permeation [8]. Downstream of the flue gas scrubber, i.e. in the clean gas section, the desulphurisation plants for lignite-fired and coal-fired power stations differ in that in the latter there is relatively high droplet entrainment with precipitation of large quantities of particulates in the 6 o'clock positions of clean gas ducts. On the other hand, desulphurisation plants for lignite-fired power stations are relatively clean but exhibit massive condensate formation depending on the mode of operation and the insulation of the clean gas lines. pH values < 2.0 are not uncommon. The aggressiveness of desulphurisation plant media is thus largely determined by the equilibrium between sulphuric acid, sulphurous acid, calcium sulphite and calcium sulphate as well as chlorides and fluorides in every compound form. The qualification of metallic materials must be related closely to these conditions, which differ from location to location.

8.2.2 Metallic materials for corrosion protection in flue gas desulphurisation (FGD) plants

The description above clearly shows that construction materials to protect components in flue gas desulphurisation must withstand extreme corrosive attack. These can be summarized as attack by sulphuric, hydrochloric and hydrofluoric acids as well as by acid- and halide-contaminated water at, in some cases, extremely high temperatures and with the simultaneous effect of particulate erosion and deposit formation. Whereas normal engineering steels cannot withstand attack by reducing mineral acids, of the metals that can be passivated titanium is damaged by extreme corrosion due to attack by acidic fluoride-contaminated water. The chemical industry has successfully used high-alloy stainless steels and nickel-base alloys in such corrosive conditions for many decades. These count as spontaneously passivating alloys in the classification of metallic materials. In the meantime, experience from over twenty years of using metallic materials in FGD plants as well as many individual investigations have shown that high-alloy steels and nickel alloys have outstanding resistance for use in desulphurisation plants. These are described in more detail below.

8.2.3 Application range of stainless steels and nickel-base alloys

In the early stages of designing flue gas desulphurisation plants, usually for coal-fired power stations, large quantities of the materials 316 LN (1.4429), 317 LN (1.4439) and 904 L (1.4539) were used, based on experience in the USA. These three materials are susceptible to corrosion in normal FGD service [9] and today are only occasionally used because of more demanding conditions in FGD plants. The materials Cronifer 1925 hMo (1.4529) and Nicrofer 3127 hMo - alloy 31 (1.4562) are being used to an increasing extent in zones of FGD absorbers that are subject to slight or moderate attack. In addition to high nickel contents, these materials have significantly higher molybdenum contents than ordinary stainless steels, as shown in Table 1.1. The high chromium content in conjunction with molybdenum ensures their outstanding resistance.

Of the nickel alloys shown in Table 1.1, the Nicrofers 6020 hMo - alloy 625 (2.4856), 5716 hMoW - alloy C-276 (2.4819) and 5923 hMo - alloy 59 (2.4605) are used in FGD plants.

Corrosion behaviour in pure sulphuric acid:

Material service in FGD plants basically requires resistance to sulphuric acid containing media. The isocorrosion diagrams of the material Cronifer 1925 hMo and Nicrofer 3127 hMo - alloy 31 in sulphuric acid are therefore shown in Fig 8.5. and Fig. 8.6. In these diagrams the limiting lines of the resistance in pure, slightly

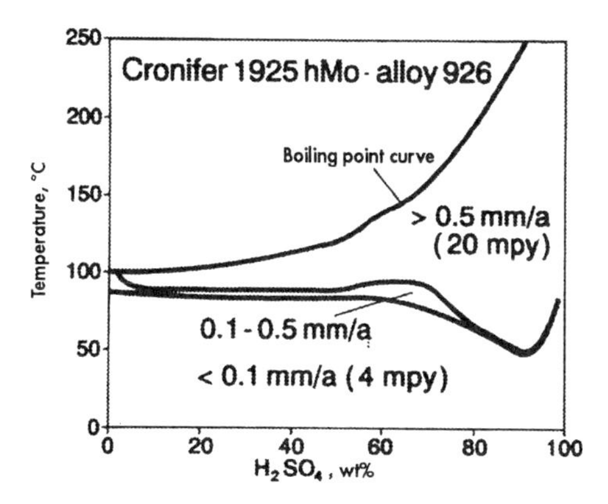

Figure 8.5:
ISO-corrosion diagram of the material Cronifer 1925 hMo in sulphuric acid; corrosion rate in mm/a, calculated from immersion tests over at least 120 h

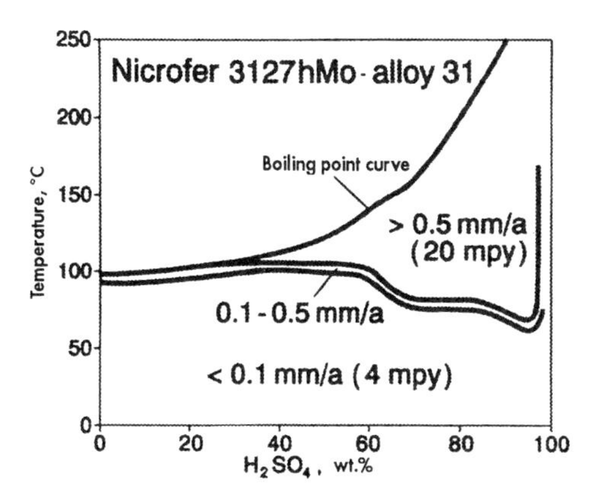

Figure 8.6:
ISO-corrosion diagram of the material Nicrofer 3127 hMo - alloy 31 in sulphuric acid; corrosion rate in mm/a, calculated from immersion tests over at least 120 h

agitated and aerated sulphuric acid in the concentration range 0 to 100 wt.% H_2SO_4 were determined by immersion experiments according to DIN 50905 [10]. Both the isocorrosion lines for 0.1 mm/a and 0.5 mm/a are plotted as a function of temperature and concentration. The area below 0.1 mm/a gives the concentration-temperature points in which the material can be used without restriction, whereas in the region between 0.1 mm/a and 0.5 mm/a limited use is possible. Above 0.5 mm/a, engineering applications of these materials are basically no longer possible. It should be noted that this is for pure sulphuric acid with no impurities or additions, so these diagrams can only represent a rough guide to the selection of specific material groups. This also applies to the NiCrMo materials Nicrofer 5716 hMoW - alloy C-276 (2.4819) and Nicrofer 5923 hMo - alloy 59 (2.4605) described below.

Cronifer 1925 hMo can be described as having unlimited resistance in pure sulphuric acid up to a concentration of approx. 60 wt.% and temperatures up to 80 °C. The region of unrestricted application (< 0.1 mm/a) is only separated from the region of no practicable application of this material by a very narrow temperature band. In an initial approximation these lines follow the same path, so it is possible to speak of a strict temperature limit of 80 °C up to 60 wt.% sulphuric acid. From 60 to about 90 wt.% sulphuric acid line this resistance line decreases to about 50 °C.

Under the same conditions the material Nicrofer 3127 hMo - alloy 31 shows, according to Fig. 8.6, a parallel displacement upwards of the isocorrosion lines of about + 20 °C on average. At a sulphuric acid concentration up to > 30 wt.% the unrestricted resistance is exceeded when the boiling point is reached at atmospheric pressure. In the region > 60 wt.% sulphuric acid the resistance limit clearly falls to lower temperatures, but even for 80% sulphuric acid it is still just above 80 °C.

The isocorrosion diagram in Fig. 8.7 for the alloy Nicrofer 5716 hMoW - alloy C-276 (2.4819) exhibits for the range > 5 and < 70 wt.% sulphuric acid, a larger division between the completely resistant region (< 0.1 mm/a) and the non-resistant region (> 0.5 mm/a). The absolute resistance limit (> 0.5 mm/a) is however, identical when compared to Nicrofer 3127 hMo - alloy 31 up to 60 wt.% sulphuric acid. Only over 60 wt.% sulphuric acid does the higher nickel content have a positive effect as the range of limited resistance is clearly displaced to higher temperatures with, however, the exception of the concentration range above 80% sulphuric acid. The isocorrosion diagram of the alloy Nicrofer 5923 hMo - alloy 59 (2.4605) produced under identical criteria is shown in Fig. 8.8. [10, 11]. As with the high-chromium materials Cronifer 1925 hMo and Nicrofer 3127 hMo - alloy 31, the region of restricted application (0.1 to 0.5 mm/a) is very narrowly defined. In comparison to the material Nicrofer 5716 hMoW - alloy C-276 this narrow band of limited resistance falls approximately in the upper third of the region of restricted application for the last named material up to 60 wt.% H_2SO_4. Above 60 wt.% sul-

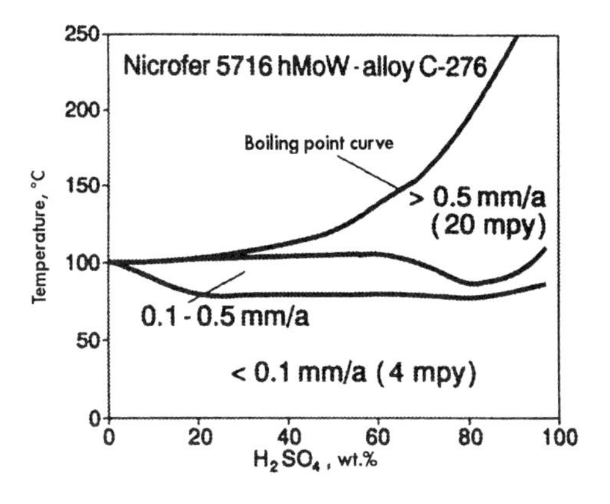

Figure 8.7:
ISO-corrosion diagram of the material Nicrofer 5716 hMoW - alloy C-276 in sulphuric acid; corrosion rate in mm/a, calculated from immersion tests over at least 120 h

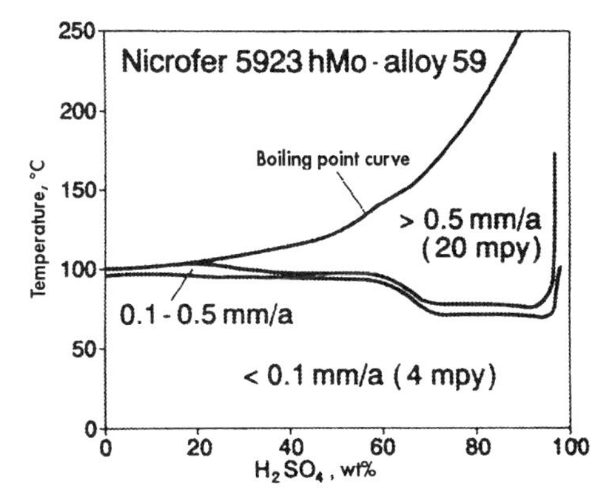

Figure 8.8:
ISO-corrosion diagram of the material Nicrofer 5923 hMo - alloy 59 in sulphuric acid; corrosion rate in mm/a, calculated from immersion tests over at least 120 h

phuric acid, however, the line of unsuitability is clearly below 80 °C, i.e. it is still below the limiting line given for Nicrofer 5716 hMoW - alloy C-276.

Taking the FGD plant conditions into account, it should be remembered that the following acid concentration relationships occur:

H_2SO_4 concentration 2 wt.% = pH approx. +4
H_2SO_4 concentration 5 wt.% = pH approx. -0.02
H_2SO_4 concentration 10 wt.% = pH approx. -0.4.

Reliable pH measurements in absorber slurries of flue gas desulphurisation plants vary between pH $\geq$4 and pH $\leq$7. These pH values valid for the absorption zones thus reflect extremely dilute aqueous sulphuric acid containing media and are typically not included in the isocorrosion diagrams. Only in the zones of condensate formation, as described above, are sulphuric acid concentrations within the measurement range of an isocorrosion diagram formed. These observations can only be utilized as an initial guide for use in FGD components subjected to condensate attack, in which condensation of moderately concentrated sulphuric acid is to be expected.

The influence of heavy metal ions.

The condensates of fossil fired power stations always contain heavy metal ions from the impurities in the fuels used. The contact between very acidic condensates and simple carbon steel, e.g. in raw gas ducts, also produces large quantities of ferric ions in the condensate. The influence of these on the higher-valency oxidizing heavy metal ions has been investigated in more detail. A non-standardized test developed in the USA to determine the critical local corrosion temperatures using the time lapse effect is called „Green Death" because of its very aggressive nature and deep green colour.

This test is based on observations in flue gas desulphurisation systems in coal-fired power stations. Wall condensates in the gas ducts were collected, in particular during the shut-down phase of the power stations, and analysed to determine the nature of these very aggressive solutions. It was established that there was a wide band of dilute sulphuric acid solutions with a measurable fraction of dissolved hydrogen chloride as well as a fraction of heavy metal ions. The resultant model solution contained 7 vol. % sulphuric acid, 3 vol. % hydrochloric acid and, for intensification, 1 wt.% copper (II) chloride and 1 wt. % iron (III) chloride. The time lapse effect is used in a test to determine the critical pitting or crevice corrosion temperature of nickel-base alloys by using material specimens at a defined starting temperature and 5 °C temperature step increases every 24 hours. The critical pitting or crevice corrosion temperature is that at which local corrosion first becomes clearly evident. To clarify the influence of the heavy metal ion content on the aggressive nature of dilute sulphuric/hydrochloric acid solution the results from three selected solution compositions are given in Fig. 8.9.

Solution 1
7 vol.% H_2SO_4, 3 vol % HCl, 0 wt.%, $CuCl_2$, 0 wt.% $FeCl_3$

Solution 2
7 vol.% H_2SO_4, 3 vol.% HCl, 0.1 wt.% $CuCl_2$, 0.1 wt.% $FeCl_3$

Solution 3
7 vol.% H_2SO_4, 3 vol.% HCl, 1 wt.% $CuCl_2$, 1 wt.% $FeCl_3$.

Green Death corresponds to solution 3.

Whereas in the reducing solution 1 with no heavy metal ion additions neither pitting nor crevice corrosion occurred in either nickel-base alloy up to 135°C as

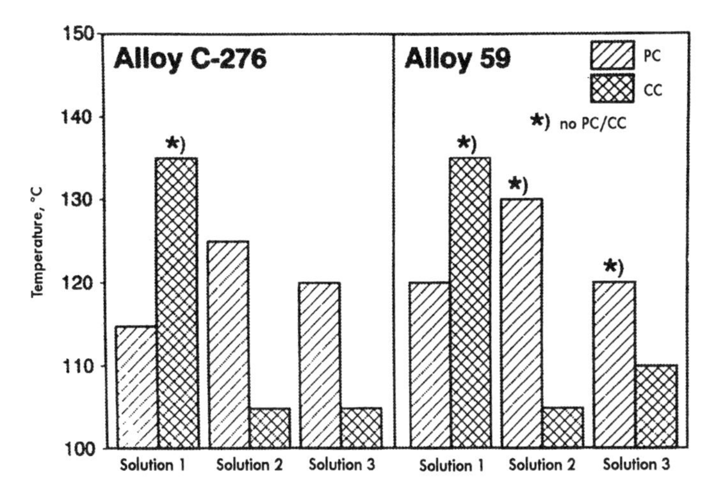

Figure 8.9: Influence of different heavy metal ion contents (Cu^{3+}/Fe^{3+}) on the local corrosion behaviour of NiCrMo materials in the Green Death test

expected, pitting (T = 125 °C) and crevice corrosion (T = 105 °C) occurred with a solution content of only 0.1 wt. % CuCl2, particularly with Nicrofer 5716 hMoW - alloy C-276.

As shown in Fig. 8.9 the same values displaced upwards by about 5 °C applied to the start of pitting on the alloy Nicrofer 5923 hMo - alloy 59. In solution 2 no pitting was observed up to 130 °C, whereas crevice corrosion already started at 105 °C. The original test solution with 1 wt.% $CuCl_2$ and 1 wt.% $FeCl_3$ produced on Nicrofer 5716 hMoW - alloy C-276 pitting at 115 - 120 °C and, as expected, also crevice corrosion at temperatures 15 °C lower. The alloy Nicrofer 5923 hMo - alloy 59 remained free of any localized corrosion up to 110 °C and then suffered isolated crevice corrosion. Pitting was not observed.

The behaviour of practically relevant welded joints was also investigated in this short-time test. Table 8.1 shows, using the alloy Nicrofer 5923 hMo - alloy 59 as an example, that within the scatter band of this test ($\pm$ 5 °C) the welding processes used in practice, tungsten inert gas (TIG), plasma keyhole and manual metal arc, do not impair the generally good local corrosion resistance of this material in the welded condition.

Only in the case of plasma keyhole welding with no filler material is the critical crevice corrosion temperature in the root region slightly lower than in the unwelded condition (see case a). A sufficiently high alloy content (Cr + Mo) is thus also necessary in the weld to withstand the aggressive effect of dissolved heavy metal ions.

Table 8.1: Influence of welding on the critical pitting and crevice corrosion temperatures (CPT and CCT) of alloy 59 (2.4605) in the Green Death solution a) 1st test series b) 2nd test series

State	Unwelded	Welded		
		TIG	Plasma keyhole	MMA
CPT	a) b) > 120	a) 115 HAZ + weld b) > 120	a) > 120	b) > 120
CCT	a) 110 b) > 120	a) 110 b) > 120	a) 105 weld-root	b) > 120 weld-root

Effect of oxygen in sulphuric acid solutions:

The influence of another, oxidizing, addition is dependent on the process technology. To achieve the highest possible SO_2 removal efficiency, the absorber suspension is given an excess of atmospheric oxygen. The formation of insoluble gypsum results in a large amount of the SO_2 being removed from the mass balance. This oxygen introduction, i.e. the addition of an oxidizing agent, causes an increase in the redox potential of the medium in a similar manner to the presence of heavy metal ions. The local corrosion susceptibility of insufficiently alloyed steels may be drastically increased as a result. Air-saturated, weakly acidic absorber solutions have a redox potential of approx. + 600 mV against the standard hydrogen electrode. The risk of localized types of corrosion increases still further if there are zones with restricted material exchange. These include dead areas resulting from the design as well as zones of extreme crevice formation, either from sealing materials or due to solid adhesions which form in large quantities in the operation of limestone based absorbers.

For these reasons, all FGD-relevant laboratory tests are carried out with air/ oxygen-saturated solutions. Crevice corrosion blocks enable oxygen-impoverished zones to be formed and thus local element formation.

The influence of halide ions.

A practically orientated differentiation within the group of nickel-base alloys is possible at the chloride-pH-temperature graph intersection point Cl^- 7% / pH = 1 / T = 105 °C. This laboratory test can give, after a minimum duration of 20 days, information on the behaviour of metallic materials in the raw gas inlet zone directly preceding the absorber sump and, to some degree, details on the behaviour in the clean gas outlet line with droplet entrainment and deposits. The results are shown in Fig. 8.10 [10].

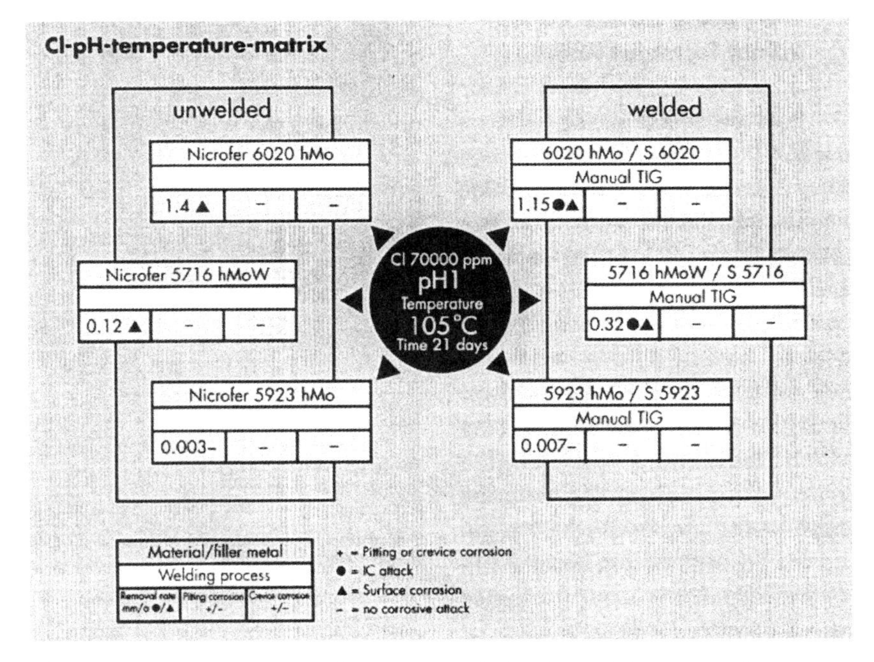

Figure 8.10: Comparison of the corrosion behaviour of NiCrMo materials in a diluted sulphuric-acid-containing solution with 7% chloride addition; base materials unwelded and base materials welded with identical material using the TIG process; corrosion rate given in mm/a, see legend for local corrosion assessment

Whereas the material Nicrofer 5923 hMo - alloy 59 exhibited uniform surface corrosion rates < 0.01 mm/a both in the unwelded and also in the manual TIG welded state with the simultaneous absence of any local corrosion, the material Nicrofer 6020 hMo - alloy 625 had corrosion rates up to 1.4 mm/a under these conditions. The base material Nicrofer 5716 hMoW - alloy C-276 suffered a mass loss rate of 0.12 mm/a, which was increased to 0.3 mm/a by welding. The use of this material under such conditions thus necessitates the meeting of extremely strict criteria from the point of view of the quality of the welding process.

It can be clearly recognized that the material Nicrofer 6020 hMo is not suitable for ensuring prolonged corrosion resistance under the conditions frequently encountered in FGD as a thin wallpaper cladding. Only the new alloy Nicrofer 5923 hMo - alloy 59 exhibits, with the absence of any localized corrosion, a resistance that makes many decades of low-maintenance use in extreme FGD conditions possible. Table 8.2 again clearly shows for Nicrofer 5923 hMo - alloy 59 that practical welding processes do not reduce the outstanding resistance under these conditions.

Table 8.2: Influence of welding on the corrosion behaviour of alloy 59 (2.4605) in a sulphuric-acid-containing solution with chloride addition: H_2SO_4, pH = 1, 7 wt% Cl^-, 105 °C (boiling), 21 days

State	Unwelded	Welded		
		TIG	Plasma keyhole	MMA
Rate of removal mm/a	0.003	0.007	0.003	0.002
Local corrosion	no	no	no	no

The limits to the application of high-alloyed nickel-based materials can be clearly illustrated by the following examples. Table 8.3 gives the corrosion rates of the three materials Nicrofer 6020 hMo - alloy 625, Nicrofer 5716 hMoW - alloy C-276 and Nicrofer 5923 hMo - alloy 59 in two extreme simulated solutions of possible FGD condensates. Whereas in the 20% sulphuric acid solution with 1.5% chloride addition the resistance is still very good at 80 °C, this limit is exceeded even at 50 °C if for the same chloride content the sulphuric acid content is increased to 50 wt.%. This means for practical FGD operation that in the wet/dry inlet zones of absorbers in lignite-fired power stations, for example, it is possible to use fully metallic semi-finished products in nickel-base alloys in the continuous presence of a condensate containing 20 wt.% sulphuric acid and 1.5% chloride impurity. If, on the other hand, cyclic attack by hot 50% sulphuric acid has to be allowed for because of the installation of heat recovery systems in the raw gas duct - this attack always occurs in the early stage of condensation - then nickel-base alloys with increased wall thickness have to be considered.

Table 8.3: Corrosion behaviour of alloy 59 in moderately concentrated sulphuric acid with chloride addition compared to other nickel-chromium-molybdenum-alloys

Alloy	Corrosion rate mm/a	
	20% H_2SO_4 + 1.5% Cl^- 80 °C	50 % H_2SO_4 + 1.5% Cl^- 50 °C
625 (2.4856	0.003	0.75
C-276 (2.4819)	0.007	0.42
59 (2.4605)	0.003	0.38

Table 8.4: Behaviour of the material Nicrofer 5716 hMoW - alloy C-276 (2.4819) in testing in highly chloride-enriched media; $CaCl_2$ solution saturated + 75 g/l fly ash + 2.5 ppm F^-, 0-40 days aeration with compressed air, 41-60 days without areation; test time 60 days; pH: 2.0 (HCl)

Material State	**Local corrosion assessment and rate of removal (V_k)**
As delivered (annealed/shot blasted/ pickled)	No crevice corrosion V_k < 0.01 mm/a
120 grit ground	Sample 1 no crevice corrosion Sample 2 slight crevice corrosion V_k < 0.01 mm/a
As delivered + welding manual TIG	No crevice corrosion V_k < 0.01 mma
milled, 120 grit ground + welding, manual TIG	Slight crevice corrosion in base metal, heat-affected zone, capping pass and root V_k 0.01 mm/a

For laboratory simulation of the extreme conditions in the raw gas inlet and the absorber sump region which may occur with combustion of coal with high salt contents, a saturated approx. 62% calcium chloride solution was used, with the pH adjusted to a mean value of 2.

In addition, 5 wt.% of pure fly ash was added, such as is collected in the electrostatic precipitator as a result of the combustion in the coal-fired power station. The test temperature was set at 80 °C. The total testing period was 60 days with intermediate evaluations every 20 days. The material Nicrofer 5716 hMoW - alloy C-276 was tested in the unwelded and welded states after fitting slotted Teflon blocks on both sides. Only slight localized corrosion of the material Nicrofer 5716 hMoW - alloy C-276 was found after 60 days in all four states, i.e. unwelded and manual TIG welded, with and without K 120 grinding. The removal rate was in all cases < 0.01 mm/a, thus qualifying this material for these conditions.

In a final stage of approximation to FGD practice, the FGD simulation solutions A to C were developed, based on our own experience. These model solutions are suitable for distinguishing resistant from non-resistant materials with high nickel, chromium and molybdenum contents within an acceptable time (max. 3 x 30 days). The chemical compositions of these FGD simulation media are :

Medium A
pH 1 (H_2SO_4), 3% Cl^- 0.05% F^- (total) + 15% FGD gypsum ($CaSO_4$, unpurified)

Medium B
pH 1 (H_2SO_4), 7% Cl^- 0.01% F^- (total) + 15% FGD gypsum ($CaSO_4$, unpurified)

Medium C
pH1 (H_2SO_4), 15% Cl^- 0.01% F^- (total) + 15% FGD gypsum ($CaSO_4$, unpurified).

All media are aerated and gently stirred throughout the experimental period. An additional time lapse effect is achieved by using these media with Teflon crevice blocks, whereby a crevice block is attached to each specimen side with a defined torque. The temperatures are usually set to between 60/65 and 80 °C, but also to 150 °C in individual cases.

Experiments in model medium A have shown, as recorded in Table 8.5 [10], under electrochemically controlled conditions that after 10 days the material Cronifer 1925 hMo is resistant under these conditions up to 60 °C but at only 80 °C goes into solution with a corrosion rate of 0.7 mm/a and suffers severe pitting. The material Nicrofer 3127 hMo - alloy 31 is significantly more stable under precisely similar conditions. A temperature-dependent reduction of the breakthrough or repassivation potential is not found within the accuracy of measurement (see Table 8.5). No localized corrosion attack was observed. The corrosion

Table 8.5: Corrosion behaviour of the materials Cronifer 1925 hMo and Nicrofer 3127 hMo - alloy 31 in the FGD simulation solution A: pH 1 (H_2SO_4), 3% chloride, 0.05% fluoride + 15% FGD gypsum ($CaSO_4$, contaminated, aerated, stirred), measurement of the breakthrough and repassivation potential using the potentiodynamic method, measurement of the corrosion rate after 10 days testing

Material No.	Material designation	Test temperature °C	Breakthrough potential mV	Repassivation potential mV	Corrosion rate mm/a	Surface assessment
1.4529	Cronifer 1925 hMo	60	1150	1065	< 0.001	no corrosive attack
		80	575	< 375	approx. 0.7	severe pitting
1.4562	Nicrofer 3127 hMo - alloy 31	60	1140	995	< 0.001	no corrosive attack
		80	1110	935	<0.001	no corrosive attack

rates are less than 0.001 mm/a for both temperatures. Without drawing premature conclusions as to the long-term corrosion behaviour of this latter material, it can be assumed that the material will have a significantly greater tolerance to the conditions simulated by the model medium A. The situation in the case of an increase in aggressiveness in FGD model medium B is similar, as shown in Table 8.6. With Cronifer 1925 hMo the temperature increase from 60 °C to 80 °C results in a drastic reduction in the breakthrough and repassivation potential as well as an increase in the corrosion rate to over 1 mm/a because of severe pitting. In contrast, the material Nicrofer 3127 hMo - alloy 31 is still largely stable even under these more aggressive conditions. Increasing the testing period from 10 to 30 days, however, resulted in localized corrosion even in this case with crevice corrosion depths up to 0.5 mm. This means for the application of Cronifer 1925 hMo that it can be specified up to 60 °C in acidic halogen-containing FGD media of type A, but massive failures must be anticipated with a temperature increase to 80 °C. The application range of Cronifer 1925 hMo (1.4529) therefore is more in the direction of weakly acidic to neutral solutions with high halide ion contamination (pH > 4, Cl^- < 5%). In contrast, the material Nicrofer 3127 hMo - alloy 31 (1.4562) is resistant in medium A up to 80 °C whereas it apparently cannot be used in a medium of type B over 60 °C. This also applies for typical TIG welds with higher-alloyed filler materials, e.g. with filler rods of the type Nicrofer S 6020 (2.4831).

Table 8.6: Corrosion behaviour of the materials Cronifer 1925 hMo and Nicrofer 3127 hMo - alloy 31 in the FGD simulation solution B: pH 1 (H_2SO_4), 7% chloride, 0.01% fluoride + 15% FGD gypsum ($CaSO_4$, contaminated, aerated, stirred), measurement of the breakthrough and repassivation potential using the potentiodynamic method, measurement of the corrosion rate after 10 days testing

Material No.	Material designation	Test temperature °C	Breakthrough potential mV	Repassivation potential mV	Corrosion rate mm/a	Surface assessment
1.4529	Cronifer 1925 hMo	60	1140	1030	< 0.001	no corrosive attack
		80	460	< 335	approx. 1.0	severe pitting preferred SL
1.4562	Nicrofer 3127 hMo - alloy 31	60	1155	1000	< 0.001	no corrosive attack
		80	1070	955	<0.001	no corrosive attack

Accordingly, there remain only the nickel-chromium-molybdenum alloys for use in FGD media of type B and C above 60 °C.

Table 8.7 shows that the material Nicrofer 5923 hMo - alloy 59 in medium B at a temperature of 80 °C is not affected by localized corrosion after 30 days of testing. Even under artificially extremely aggressive conditions (teflon crevice block) a crevice corrosion depth of less than 0.1 mm is measured. The specimen welded

Table 8.7: Local corrosion behaviour of Nicrofer 5716 hMoW - alloy C-276 and Nicrofer 5923 hMo - alloy 59 in the unwelded and welded condition in the FGD simulation solutions B and C after 30 days at 80 °C

Test medium	**Material** **1) Base material** **2) Manual TIG welding with filler** **3) Plasma welding**	**Corrosion rate** **mm/a**	**Crevice corrosion depth** **(mm)**
Medium B 7 % Cl^- 0.01 % F^- pH 1 (H_2SO_4) 15 % FGD gypsum derated stirred	1) Nicrofer 5923 hMo - alloy 59 (2.4605)	not measurable	< 0.1
	2) Nicrofer S 5923 (2.4607)	not measurable	-----
	3) without filler	not measurable	SG > 0.3
Medium C 15 % Cl^- 0.01 % F^- pH 1 (H_2SO_4) 15 % FGD gypsum derated stirred	1) Nicrofer 5923 hMo - alloy 59 (2.4605)	not measurable	> 0.1 < 0.5
	2) Nicrofer S 5923 (2.4607)	not measurable	> 0.1 < 0.5
	3) without filler	not measurable	> 0.1 < 0.5
	1) Nicrofer 5716 hMoW - alloy C-276 (2.4819)	not measurable	> 0.1 < 0.5
	2) Nicrofer S 5716 (2.4886)	not measurable	> 0.1 < 0.5
	3) without filler	not measurable	> 0.1 < 0.5

using TIG and matching filler is completely free of any localized corrosion. In the specimen welded using the plasma keyhole process with no filler, localized corrosion attack occurred in a third of the possible crevice corrosion sites, and in one case a site greater than 0.3 mm deep was found in the weld metal. A measurable corrosion rate was, however, not measured on any specimen.

The limits of application of the material Nicrofer 5716 hMoW - alloy C-276 (2.4819) and Nicrofer 5923 hMo - alloy 59 (2.4605) are attained in the FGD simulation solution C. With the continuous action of a high-halide sulphuric acid solution containing 15% Cl^- and the simultaneous presence of extreme crevice conditions, the occurrence of crevice corrosion regions up to 0.5 mm deep must be expected according to Table 8.7. Based on these laboratory investigations, however, this only applies to the continuous imposition of such conditions over a period of more than 6,000 operating hours out of a possible total of 8,000 FGD plant operating hours per year. This extreme case is in all probability rarely encountered during the life of a power station. Based on the marked repassivation ability of the high-nickel materials as shown by Table 8.8, it can be assumed that a corresponding increase in wall thickness largely guarantees security against corrosion when using these materials. The walls of raw gas inlet ducts are therefore frequently fabricated from solid plates more than 6 mm thick.

The behaviour of the passivatable materials described here in typical FGD media is thus determined by the strength of the sulphuric/hydrochloric acid mixture

Table 8.8: Local corrosion behaviour of Nicrofer 5716 hMoW - alloy C-276 (2.4819) and Nicrofer 5923 hMo - alloy 59 (2.4605) under electrochemically controlled conditions in the FGD simulation solution C; test method see Table 8.7

Material designation	Test temperature °C	Break-through potential mV	Repassivation potential mV	Corrosion rate mm/a	Surface assessment
Nicrofer 5716 hMoW - alloy C-276 (2.4819)	60	1125	935	---	---
	80	745	800	< 0.001	no corrosive attack
Nicrofer 5923 hMo - alloy 59 (2.4605)	60	1155	1015	< 0.001	no corrosive attack
	80	925	850	< 0.001	no corrosive attack

formed, the halide content and the valency and quantity of the dissolved oxidizing agents, e.g. heavy metal ions and/or atmospheric oxygen. Crevice conditions lead to the formation of local galvanic cells. These processes based on complex electrochemical reactions must be taken into account in the selection of materials for extremely severely attacked FGD plant components. They can be derived, depending on the location, from the composition of the fuel or the raw gas, the temperature profile and the operating principle of the desulphurisation plant.

8.2.4 Design of FGD plants using metallic materials under fabrication and cost aspects

The chronological development of the use of metallic materials for wet desulphurisation plants is described in detail in [12]. The sum of experience relating to the application limits of nickel-based alloys and high-alloy stainless steels is shown schematically in Fig. 8.11. This two-dimensional representation should not be used for individual decision making in the design of FGD components. It only gives an example of the conditions found in the real world of FGD and serves merely as a guide to the necessary discussion between the end-user of the power station/FGD plant, the engineering consultant, the equipment fabricator and the supplier of the mill product. The lower alloyed stainless steels of type 1.4435 up to 1.4539 can - as a simplification - only be used at a maximum temperature of 70 °C in the region above the mist eliminator and the clean gas outlet of the absorber, i.e. in the neutral to weakly acidic region, if the chlo-

A final material selection can only be made jointly with the material manufacturer and the plant operator, as many other parameters such as particulate contents (deposits), service temperatures or plant design and method fo operation have to be taken into account.

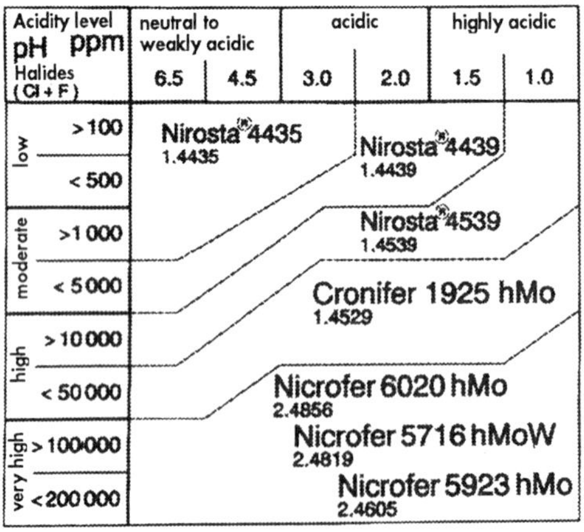

Acidity level pH ppm Halides (Cl + F)
neutral to weakly acidic 6.5 4.5
acidic 3.0 2.0
highly acidic 1.5 1.0
low >100 <500
moderate >1000 <5000
high >10000 <50000
very high >100000 <200000
Nirosta® 4435 1.4435
Nirosta® 4439 1.4439
Nirosta® 4539 1.4539
Cronifer 1925 hMo 1.4529
Nicrofer 6020 hMo 2.4856
Nicrofer 5716 hMoW 2.4819
Nicrofer 5923 hMo 2.4605

Figure 8.11: Application limits of nickel materials and high-alloy special steels in flue gas desulphurisation (FGD) in the temperature range 50-70 °C

ride concentration is adjusted to << 0.5%. The operating range of the 6% Mo steels, Cronifer 1925 hMo (1.4529) and Nicrofer 3127 hMo - alloy 31 (1.4562), commences in weakly acidic solutions and extends to strongly acidic solutions with moderate chloride ion attack. Nickel-based alloys are usually used at chloride concentrations > 0.5% and acidic to very strongly acidic conditions as well as where temperature increases to well over 70 °C are expected. This simplified schematic list leads, in conjunction with information on the behaviour of individual materials under specific conditions, to the development of a material concept for flue gas desulphurisation plants employing wet scrubbing processes.

This enables the individual materials to be classified according to their suitability for the routinely encountered FGD components, as shown in Fig. 8.12. It is economically sensible to produce in metal all components for the heat transfer system, the pre-scrubber with raw gas inlet, the absorber, the ducting, the second heat exchanger and the stack. Nickel-base alloys are used for the components that will be subjected to the most aggressive media, see Fig. 8.1. Heat transfer systems and pre-scrubbers are usually subjected to high condensate attack (see Section 8.2.1.) and must therefore be made of the most resistant material. In recent years Nicrofer 5716 hMoW - alloy C-276 (2.4819) and Nicrofer 5923 hMo - alloy 59 (2.4605) have been used to a large extent for this application.

The raw gas inlet in the absorber must also be specified in this material. If a stable chemical system with pH values in the range $\geq$ 4 and $\leq$ 6.5 is formed in

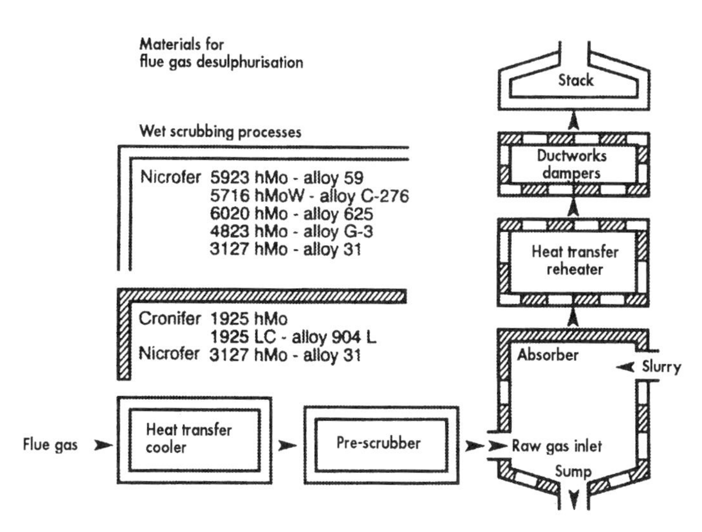

Figure 8.12: Material concept for the all-metal design of flue gas desulphurisation plants (FGD)

the absorber sump and thus the aggressiveness can be calculated, Nicrofer 6020 hMo - alloy 625 (2.4856) can be used and, given an exact knowledge of the operating conditions, even Nicrofer 3127 hMo - alloy 31 (1.4562). The condition for the use of the latter material however, is the avoidance of gypsum deposits by designing the agitating units accordingly. If there are any doubts, it is better to design the absorber sump region in Nicrofer 5923 hMo - alloy 59 (2.4605), e.g. in clad form (10 + 4 mm). In the transition zone from raw gas inlet to the wet/dry interface of the sump, certain areas have to be constructed with a greater amount of Nicrofer 5923 hMo - alloy 59 (2.4605) or Nicrofer 5716 hMoW - alloy C-276 (2.4819) depending on the design.

This also applies to the quencher nozzle system up to the process specific transition into the absorption region, e.g., bowl in dual-circuit scrubber. The first absorption planes, i.e. the first spray tube layers above the quencher nozzle system, can be constructed of Nicrofer 3127 hMo - alloy 31 (1.4562) and, on proceeding further into the upper zones of a counterflow absorber, in Cronifer 1925 hMo (1.4529). The head of an absorber may be made of Nirosta 4439 depending on the effectiveness of the mist eliminator. The heat transfer system, which is coupled with the raw gas inlet, must be fabricated from nickel-based alloys of the type Nicrofer 5716 hMoW - alloy C-276 or higher. It is necessary to make a distinction between coal-fired and lignite-fired power stations where the clean gas duct connections to the stack are concerned. Whereas in lignite-fired power stations with heat transfer systems very clean treated gas duct conditions usually occur and thus limited deposits form, the situation in coal-fired power stations is the opposite. In particular, if there is no heat transfer system and the mist eliminator is not operating effectively or is not rinsed out frequently, thick deposits form in coal-fired power stations which can give rise to an extremely high risk of corrosion with simultaneous condensate formation, particularly in the 6 o'clock position and in flow deviation areas and at design related flow interruption points. Clean-operating treated gas ducts can in some cases even be fabricated from the alloy Cronifer 1925 hMo (1.4529). With all other methods of operation, higher alloyed nickel-base alloys must be used, particularly in the 6 o'clock position and at the other flow interruption points, to achieve a corresponding increase in corrosion resistance. Stack inlet pipes or stacks built of solid metal usually require the alloys Nicrofer 5716 hMoW - alloy C-276 (2.4819) and Nicrofer 5923 hMo - alloy 59 (2.4605), since it is not possible to calculate accurately over the stack height and cross section where the condensation zone occurs as a function of the external climate. Massive damage to coating systems (resin coatings with flake filling) as well as serious rotting of unprotected concrete stacks have recently increased the application of metallic nickel-chromium-molybdenum materials in stack construction. Despite some fluid-mechanics-related requirements, materials for shut-off fittings, i.e. dampers and gates, must naturally be assessed according to their location in the gas stream. The material selection is obviously based on the above remarks.

On the basis of past experience, the selection of suitable materials specific to the component can guarantee better long term reliability and thus higher plant availability. If costs dictate thin-walled application of highly corrosion resistant materials, this can be produced by:

Cladding of existing or new carbon steel structures with thin large-size sheets of nickel-chromium-molybdenum materials (wallpapering)

The use of explosively or roll-bond clad composite materials with highly corrosion resistant cladding materials.

The concept of wall paper cladding:

Only materials with the highest corrosion resistance can be used because of the thin wall thickness of such claddings - usually 1.6 to 2.0 mm. Nicrofer 5716 hMoW - alloy C-276 (2.4819) is so far the most widely used material for the ongoing refurbishment by wallpapering of damaged FGD components. The switch to the modern material Nicrofer 5923 hMo - alloy 59 (2.4605) combines outstanding resistance under FGD conditions with simplified fabrication, in particular the application of thin mill products by welding. The 6 main steps in the corrosion-proofing and thus long-life cladding of large-volume components (ducts, absorber, stacks) are as follows:

1. Specification of large format sheets, e.g. 8 x 2.5 m, to minimize the welding and thus the fabrication costs.
2. Specification of welding parameters, selection of welding personnel and equipment after discussion with the manufacturer of the semi-finished product.
3. Preparation of the sheets according to the engineers' design up to the point of being ready for installation on the construction site, i.e. cutting, punching and preforming.
4. Attaching the thin sheets by welding using tack welds at the edges and spot welds on the surface.
5. Finishing with overlapping seal welds and
6. Inspection of all seal welds using a vacuum box and/or dye penetrant testing.

Details of the wallpapering technique are given in Section 3: Welding of nickel alloys and high-alloy special stainless steels.

For rapid fixing of the sheets over a large area, VDM has optimized spot welding by reducing it to a single operation. This technique is based on mechanically strong joining using two-pass build-up welding in precut holes in one operation. By controlling the process parameters a low-segregation, end-crater-free spot weld is obtained [13]. Testing of specimens with approx. 10-15% weld metal gave the results in Table 8.9. In the FGD simulation solutions A and B corrosion rates at the lower end of the accuracy of measurement were obtained from 30-day immersion at 80°C, regardless of the welding position. The relatively uniform

Table 8.9: Testing the corrosion behaviour of the materials Nicrofer 5716 hMoW - alloy C-276 and Nicrofer 5923 hMo - alloy 59 after welding using the technique of wall-papering by plug-spot welding; welding process: metal inert gas (MIG) build-up welding; two passes, in precut holes; tested over 30 days at 80 °C, corrosion rate in mm/year

Welding process: MIG	Pulsed arc	Spray short arc and pulsed arc	Pulsed arc	Spray short arc and pulsed arc
Base metal:	Nicrofer 5716 hMoW - alloy C-276 (2.4819)		Nicrofer 5923 hMo - alloy 59 (2.4605)	
Filler:	Nicrofer S 5923 (2.4607)			
Welding position:	S	W	S	W
FGD Simulation A	0.0007 0.0008	0.0007 0.0006	0.0008 0.0008	0.001 0.001
FGD Simulation B	0.002 0.002	0.002 0.002	0.001 0.001	0.001 0.01
FGD Simulation C	0.05 0.07	0.05 0.06	0.02 0.02	0.03 0.04

rate of corrosion was almost without exception less than 0.005 mm/a and the specimens were free from local corrosion. Even the much more aggressive FGD simulation solution C with its higher (15%) chloride content merely resulted in a slightly increased corrosion rate, which was nevertheless well below 0.1 mm/a for both materials regardless of the welding position.

The use of composite materials:

Whereas the wallpapering technique is mainly used for refurbishment of existing FGD plant components, clad materials with corrosion-resistant layers are being specified to an increasing extent for the construction of new FGD absorbers and ducts. These composite materials are produced by either explosive or roll-bond cladding.

The construction of FGD absorber components using explosively clad components is described in [14]. The details of the fabrication technique and in particular the problems of joint welding depending on the position are discussed in this article. Realistic FGD simulation tests have demonstrated the complete feasibility of this concept, in this case with the nickel-base alloy Nicrofer 6020 hMo - alloy 625 (2.4856).

In addition to explosive cladding, the use of hot roll-bond cladding has recently been growing [15, 16]. Welding of clad semi-finished products is described in detail in Chapter 3. Table 8.10 shows, as an example, the corrosion results of a Nicrofer 5716 hMoW - alloy C-276 cladding layer in FGD simulation solutions. The FGD media A and C with their widely differing chloride content were used for the assessment. The test was carried out over 30 days at a maximum absorber saturation temperature of 80 °C in an artificially aerated and stirred medium.

Table 8.10: Corrosion behaviour of a Nicrofer 5716 hMoW - alloy C-276 cladding in FGD simulation solution
FGD medium A: 3% Cl^- + 0.05% F- + 15% FGD gypsum, pH = 1 with H_2SO_4
FGD medium C: 15% Cl^- + 0.01% F^- + 15% FGD gypsum, pH = 1 with H_2SO_4
Test conditions: 30 days, 80 °C, aerated and stirred

Medium	Specimen	Corrosion rate (mm/a)
A	Cladding material	MW 0.016
A	Specimen with weld	MW 0.040
C	Cladding material	MW 0.042
C	Specimen with weld	> 0.1*

* Increase in the rate of weight loss by local corrosion on an incorrectly carried out sealing weld without filler material.

In both media a corrosion rate for the cladding < 0.1 mm/a was established on the unwelded cladding material. Specimens with welded joints confirmed these results, with the exception of one specimen which suffered a higher rate of removal due to local corrosion in the region of a test seal weld that had not been prepared correctly (no filler metal). Tests in coal-fired and lignite-fired power stations have confirmed the good corrosion behaviour of these composite materials. The ageing of composite material sheets clad with Nicrofer 5716 hMoW - alloy C-276, for example in the FGD plant of a coal-fired power station, has shown that no measurable weight loss occurs after 300 days in the position immediately above the first quencher nozzle plane at a temperature of about 60 °C and a chloride content > 20%, the pH varying between 2 and 5. Ageing tests in the 6 o'clock position of clean gas ducts of lignite-fired power stations have also shown no measurable weight loss rates after more than 1 year in operation.

8.2.5 Nickel-base alloys and stainless steels in FGD plant components

Sections 8.2.1 and 8.2.4 describe in detail the reasons for selecting materials in line with the corrosion conditions. In this concluding section the commercial implementation of this material concept is described using case studies. The latter are arranged by individual components; for each component, e.g. pre-scrubber/ absorber, several characteristic application examples have been selected from various power stations, which fully cover the spectrum of the actual operating conditions. This taking out of context should not detract from the overall view of the parameters specific to the power station and the process, within which such a scrubber or absorber has to be installed. For example, the upstream installation of a DENOX plant also changes, as a result of the higher reaction temperatures required, the operating conditions in the pre-scrubber because of the significantly higher transfer temperatures of the raw gas after passing through the DENOX plant. The installation of a heat transfer system (raw gas/clean gas) has the opposite effect, that is, a reduction of the raw gas inlet temperature. In addition to the case studies, a very detailed list is given in [1], which details the process technology along with the name of the power station, process and materials. A trend can be recognized from this. Based on experience with ordinary stainless steels and with the behaviour of organic-based corrosion protection systems (rubber coatings, GRP, vinyl ester) the tendency in material selection is increasingly towards more expensive completely and partly metallic solutions. The ordinary stainless steels initially preferred have been displaced by "superaustenitics" with more than 6% molybdenum and higher nickel contents. Where nickel-base materials are concerned, the change has been from materials such as Nicrofer 4823 hMo - alloy G-3 (2.4619) and Nicrofer 6020 hMo - alloy 625 (2.4856) to the high-alloyed nickel-chromium-molybdenum materials Nicrofer 5716 hMoW - alloy C-276 (2.4819) and Nicrofer 5923 hMo - alloy 59 (2.4605). Extensive field trials in working FGD installations were used in the development of the metallic refurbishment concepts which were in full use at the date of publication of this book.

8.2.5.1 Pre-scrubber/absorber

Case 1.
Raw gas inlet and absorber material concept in lignite-fired base-load power stations.

The wide variations in the composition of the lignite fuel requires a material concept for the design of the raw gas inlets, which are subjected to severe corrosion conditions, that also allows for the peak SO_2 loads up to several 1000 mg/m^3 (NTP) in the raw gas as well as the increased raw gas temperatures. Most of the total of 37 FGD absorbers fitted to lignite-fired West German power stations with a total energy generation output > 9000 MW have been built using the materials Nicrofer 6020 hMo - alloy 625 (2.4856) or Nicrofer 5716 hMoW -

alloy C-276 (2.4819) in the raw gas inlet, using either the wallpapering technique or solid metal sheets. These highly corrosion resistant constructions usually commence just before or at the shut-off damper to the raw gas inlet/absorber. The damper itself and the walls are subjected to extremely corrosive conditions because of the rebounding absorber/quencher suspension and the simultaneous mixing with hot incoming raw gas. The deposition of residual fly ash, rust particles from the upstream carbon steel ducts and the concentration and precipitation of salts from the quencher suspension rebound result in conditions which rapidly overtax all organic coatings and also some low-alloyed stainless steels. Nicrofer 5716 hMoW - allow C-276, in particular, selected and fabricated by several consulting engineers/process equipment manufacturers, has proved successful under these extreme operating conditions down to wall thicknesses of 1.6 mm. Apart from individual design and process related defects at welds, these materials have proved themselves in fabricated condition for more than 30,000 operating hours (4 years).

The actual absorber with or without integrated quencher depending on the FGD process, is connected directly to the raw gas inlet. In lignite-fired power stations in the Rhineland three types of scrubber are used.

Type 1 - single circuit countercurrent scrubber, in which the exhaust gas is passed into the lower part of the absorber and scrubbed by the scrubbing suspension from above in countercurrent.

Type 2 - combination scrubber, operating as a co-current scrubber in the quencher zone and as a countercurrent scrubber in the absorber zone.

Type 3 - dual circuit countercurrent scrubber, in which the quencher and absorber zones are separated from each other into two regions by a characteristic bowl. The separation enables each circuit to operate at a different pH, thereby permitting specific process parameters to be optimized.

The important components for the metallic design in these scrubbers are, besides the wall casing from the bottom to the top, the oxidation air systems (tubes and nozzles), the quencher tube planes, the bowl (depending on the process) the actual absorber spray lances and nozzles, the support trays of the mist eliminators and the flue gas exit segment of the ductwork. In addition, even with plastic mist eliminators the fixing elements - bolts and threaded bars - have to be manufactured from highly corrosion-resistant metallic materials. Stirring units and their components that come into contact with the medium such as bolts and bars also have to be made from metallic materials.

In the case of a 4 x 600 MW power station with a total flue gas volume of 2 million m^3/h (NTP), the raw gas inlet was fabricated from Nicrofer 5716 hMoW - alloy C-276 (1.6/2 mm cladding thickness) and the sump region of the absorber from Nicrofer 6020 hMo - alloy 625 sheet/plate with a thickness, depending on

Alloy	Sheet surface	Crevice block region	Weld and heat-affected zone
Cronifer 1812 LC (1.4435)	Pitting up to 0.6 mm deep at 0.7 mm diameter corr. rate: 0.07 mm/a	Very severe attack on all contact surfaces	Increased crevice corrosion
Cronifer 1925 hMo (1.4529)	Slight pitting up to 0.2 mm deep corr. rate: 0.0003 mm/a	1 of 24 test surfaces attacked	Uniform attack
Nicrofer 5716 hMoW - alloy C-276 (2.4819)	corr. rate: 0.001 mm/a	Slight crevice corrosion attack	No attack

Table 8.11: Long term corrosion behaviour of Cronifer 1812 LC, Cronifer 1925 hMo and Nicrofer 5716 hMoW - alloy C-276 under clean gas conditions in a lignite-fired large power station; condensate: pH 1.4 to 2, 70 °C, <0.1% Cl^-, particulates; ageing time: 10 000 operating hours

the location, of 1.6 to 8 mm. Spray systems and fixing elements were made of the 6% Mo steel Cronifer 1925 hMo. Mill products of 3 to 10 mm sheet/plate thickness were used. Agitators and flanges were produced from Nicrofer 6020 hMo - alloy 625 bar and cold-formed plates. Bolts, nuts and washers in corrosion prone-regions were also manufactured from Nicrofer 6020 hMo - alloy 625.

Case 2.
All-metal FGD absorber connected to a 420 MW coal-fired power station.

In 1980 the decision was taken in a coal-fired power station in Southern Germany to install a twin-line FGD system based on a limestone wet scrubbing process. The flue gas volume is 1.6 million m^3/h (NTP) and carries 2,700 mg SO_2/m^3 (NTP). Desulphurisation reduces this to below 400 mg/m^3 (NTP). The raw gas inlet temperature is about 130 °C, the clean gas outlet temperature at the scrubber head 45 to 50 °C. The pH in the scrubber sump is in the range 4 to 6, the Cl^- content varies around 4%. By separating the raw gas inlet/quencher zone from the absorber zone in this dual circuit scrubber, the chance was seen of realizing a progressive material concept. The raw gas inlet connections, quencher zone and scrubber sump as well as the oxidation air system in the absorber sump were manufactured from solid plates of Nicrofer 6020 hMo - alloy 625 (2.4856) with a thickness of about 6 mm, as shown in Fig 8.13. The absorber middle region (wall) and the bowl were fabricated from 1.4539 plates. The spray system in the actual absorber upper region was manufactured from 1.4435. The support tray in the upper region of this absorber was also made from this material. External stiffeners for both these towers, which are approx. 30 m high and 10 to 11 m in diameter, were made of the material 1.4462, i.e. a high strength austenitic-ferritic material. It was found after about 2 years of service that the

Figure 8.13: Installation of an all-metal dual circuit FGD absorber in a coal-fired base load power station.
Raw gas inlet, absorber sump, and oxidation air system in Nicrofer 6020 hMo - alloy 625 with welding filler Nicrofer S 6020

material 1.4539 was not sufficiently highly alloyed to withstand the unexpectedly high chloride content in the region of the bowl i.e. the middle part of the absorber, because of process technology problems. Large deposits of solid particles from the gas flow were a further cause of the localized corrosion resistance of this material being exceeded. Therefore the underside of the bowl, which was prematurely damaged by corrosion, was covered with Nicrofer 5716 hMoW - alloy C-276 (2.4819). The application of sheets of the latter material to the prematurely damaged and to some degree very rough plate surface presented no problems. This also confirmed the rightness of the concept of rapid repairability in cases of limited damage in the all-metal corrosion protection concept.

Small areas of shallow pit formation in the raw gas inlet fabricated from Nicrofer 6020 hMo - alloy 625 (2.4856), which subjected to heavy corrosion attack have meanwhile also been permanently repaired using this method of wallpapering with Nicrofer 5716 hMoW - alloy - C-276 (2.4819) along with additional process engineering modifications. The rightness of selecting the all-metal material concept was confirmed by the small amount of maintenance over a period of 6 years.

Case 3.
550 MW combined heat and power station with dual circuit scrubber.

Similar concepts were realized in a coal-fired 550 MW power station with a limestone scrubber for generation of electricity and district heating. With a total gas throughput of 1.4 million m³/h (NPT) with an SO_2 content of 2,600 mg/m³, here there was simply a deviation in the expected scrubber sump pH of 3.5 to 4.5. Based on the experience from the previous case, so-called deflection plates were installed in the annular gap between the bowl and the absorber wall, i.e. in the inlet to the actual upper part of the absorber of the scrubber. These were intended to prevent carry-over of chloride-containing foam into the absorber circuit. The raw gas inlet and the scrubber sump were again made of Nicrofer 6020 hMo - alloy 625 (2.4856). Immediately above the absorber sump liquid level up to the mid-point of the absorber region, at a height of approx. 15 m, the material 1.4539 was used. The absorption and mist eliminator zone was fabricated from the material 1.4435. Agitators and absorber oxidation spray systems were made from Nicrofer 6020 hMo - alloy 625 (2.4856). A special feature of this power station is that the clean gas outlet duct extends right up to the flue gas mixing zone. The stainless steel 1.4439, which is more resistant because of its higher molybdenum content, was used here because of the risk of SO_3 formation from clean gas aerosols.

After 5 years of low maintenance operation, the all-metal material concept has proved its capabilities, as in the previous case.

Case 4.
700 MW coal-fired combined heat and power station with a pre-scrubber.

The combined heat and power station had a total output of almost 1200 MW, composed of 4 smaller 100 and 125 MW units and a new 700 MW unit. The two FGD lines were built by a leading German engineering company under licence from Japan. The total volume of raw gas is 2.5 million m³/h (NTP) laden with 2,300 mg SO_2/m³ (NTP). The scrubber sump was designed for pH values between 3.5 and 4.5 and chloride contents of max. 4%. The inlet temperature of the raw gas is 110 °C and the outlet temperature is about 45 °C. The scrubber for the smaller power station units (100 to 125 MW) operate with a conventional prescrubber concept. The severest corrosive attack occurs here because of the precipitation of all the aggressive constituents. Consequently, the raw gas inlet connection and the wall regions of the prescrubber were fitted with a wallpaper cladding of Nicrofer 5716 hMoW - alloy C-276 (2.4819). The sump region of this precipitator is of rubber-lined carbon steel. The spray systems in the prescrubber are of all-metal construction in Nicrofer 5716 hMoW - alloy C-276 (2.4819). In contrast, the oxidation air systems in all three absorbers were manufactured from Nicrofer 6020 hMo - alloy 625 (2.4856).

The two scrubbers of the 700 MW unit have been constructed with 3 mm thick Nicrofer 6020 hMo - alloy 625 (2.4856) sheets in the most severely attacked region for corrosion protection using the wallpapering technique. The five spray planes are also of metal. The bottom two contact zones consist of Nicrofer 6020 hMo - alloy 625 (2.4856) tubes and the three above of Cronifer 1925 Mo (1.4529) tubes. The inspection carried out in the first four years of service suggest that the entire system, i.e. the metallic components, in particular, will achieve a very long service life.

Case 5.
Coal-fired peak load power station.

The technique of corrosion protection by wallpapering has been successfully practised since 1979 in a coal-fired medium-load power station in Northern Germany. The modified Japanese fgd-process, which operates with limestone and magnesium and produces gypsum as the end product, desulphurises the 1.3 million m^3 (NTP) of raw gas from 2,500 mg SO_2/m^3 to values less than 400 mg/m^3. The power station's function as a moderate to peak load electricity provider results in numerous start-up and shutdown procedures during a working year. This results in the frequent occurrence of flue gas condensates with pH values less than 1 with the simultaneous presence of high chloride and fluoride contents and severe attack of the material selected. For this reason, 12 years ago the raw gas/absorber inlet in this severely attacked region was clad with Nicrofer 6020 hMo - alloy 625 (2.4856), as were the oxidation lances in the oxidation zone. Over 10 years of low maintenance and repair operation have fully confirmed the initial material selection. Due to some localized corrosion, retrofitting of that area was engineered in Nicrofer 5716 hMoW - alloy C-276.

8.2.5.2 Clean gas ducts including reheating

Clean gas ducts transport scrubbed and therefore moist clean gases from the scrubber outlet to the stack. Clean gases from wet scrubbing always carry scrubbing suspension droplets in the form of aerosols. Depending on the flow rate and the possible presence of flow discontinuities, this moisture is either deposited in the clean gas duct, or enters and is discharged from the stack. To prevent residual acid condensation of these fine aerosols, the clean gas is frequently heated to above the calculated sulphuric acid dewpoint, thus giving the clean gas a greater impetus on leaving the head of the stack. The methods of clean gas reheating consist of direct addition of hot gases from combustion or indirect heating of the clean gas steam using large tubular, plate or hybrid heat exchangers. A particularly interesting form of clean gas heating is the use of the heat content of the raw gases from the power station to heat rotating storage stacks which then give their heat to the clean gas on the other side (gas/air preheater). Depending on the fuel and the process technology, massive deposits of the particulate content of the aerosols frequently occur, particularly in coal-fired power stations with defective demisters. In lignite-fired power stations these plant segments are relatively clean. Common to both types of power station, however, is the precipitation of condensate over the full length of the clean gas ducting. This is aggravated by climatically unfavourable conditions, i.e. cold winter weather or high external wind velocities, which may cause, for example, massive continuous condensate precipitation in poorly insulated clean gas ducting. Based on a wide range of experience over the last ten years, here there is again an increasing tendency to adopt durable, reliable all-metal solutions with suitable insulation.

Case 1:
Coal-fired power station

The 740 MW unit of this power station is desulphurised by 3 single circuit scrubbers using slaked lime. The clean gas leaving the top of the absorber is reheated with raw gas and then fed to the stack in a collecting line. Condensate samples in this clean gas duct show a large variation in pH, chloride content and also fluoride content:

Residual SO_2 content	< 400 mg/m^3 (NTP)
pH	1.9 to 4.0
Cl	186 to 6,292 mg/l
SO_4^{2-}	4,386 to 6,487 mg/l
F^-	10 to 36 mg/l
Fe	13 to 121 mg/l

Analyses made 3 years later in the stack condensate and thus directly related to the clean gas composition gave a pH value of 1.9, a chloride content of 2 g/kg and a sulphate content of 40 g/kg. The total fluoride content had meanwhile increased to 1.75 g/kg. This data shows that large location-and also time-dependent variations in the clean gas composition can be measured. The clean gas ducts of these three scrubbers were constructed of Cronifer 1925 LC (1.4539). Surprisingly, no indication of large-scale corrosion was found in these ducts after 10 years in service. Because of the heavy deposits in the clean gas duct as well as the detection of damage to the clean gas dampers made of 1.4439, approval was obtained two years after the commissioning of power station unit A to carry out prolonged field tests in this plant. The materials used were Cronifer 1925 LC (1.4539), Cronifer 1925 hMo (1.4529), Nicrofer 6020 hMo - alloy 625 (2.4856) and Nicrofer 5716 hMoW - alloy C-276 (2.4819). After a period of about 10,000 hours (approx. 2.5 years of operation with 4,000 annual operating hours), crevice corrosion and active pitting were detected on the materials Cronifer 1925 LC and Cronifer 1925 hMo in the region of very strongly adhering deposits and in the region of the specimen clamping. Under these extreme conditions of pH $\leq$ 2, 50 to 60°C, $\leq$ 0.7% Cl^- and adhering gypsum deposits only the materials Nicrofer 6020 hMo (2.4856) and Nicrofer 5716 hMoW -alloy C-276 (2.4819) exhibited total resistance. The corrosion rate on these specimens was less than 0.05 mm/a after 10,000 hours, which agrees well with the almost complete absence of any local corrosion.

The apparent contradiction between the behaviour of the clean gas duct constructed over 10 years ago of Cronifer 1925 LC (1.4539) and the relatively rapid corrosion of the test specimens is related to the location of these specimens. Whereas the duct floor made of Cronifer 1925 LC is almost completely covered with several centimetres of particulate deposits and thus has a certain amount of protection, the ageing specimens projected vertically into the clean gas flow about 0.5 m from a condensate discharge point in the 6 o'clock position. These specimens were therefore subjected to both wetting by the clean gas condensate and, in the lower zone, the high-chloride encrustation of the particulate deposits. A further intensification was the result of the selected screw connections with plastic crevice formers, i.e. Teflon washers. The interaction of these factors has evidently resulted in the accelerated formation of local corrosion with correspondingly high weight loss rates. Since the protective action of massive gypsum deposits cannot be relied on over a long period, refurbishment of these critical areas is today carried out in higher quality metal alloys, e.g. Nicrofer 5923 hMo - alloy 59 (2.4605), using the wallpapering technique.

Case 2:
Refurbishment of damaged synthetic resin lined clean gas ducts with an NiCrMo alloy liner.

In lignite-fired large power stations west of Cologne, the extremely large clean gas outlet ducts (several hundred metres long and 7 metres in diameter) are protected against corrosion by a synthetic resin coating system with a polyester and vinylester base. After 3,000 hours of service it was found that the multi-layer coating which contained glass flakes as the filler material had been damaged by massive water vapour permeation resulting in blistering [8]. The damage pattern discovered, i.e. water blisters statistically distributed over the entire clean gas duct, necessitated an immediate, durable refurbishment. Attack of the underlying carbon steel could no longer be ruled out.

Apart from other concepts it was decided that as soon as possible qualification of metallic cladding materials should be assessed by ageing tests in various clean gas duct segments. Three test series were carried out from 1988 to 1990 with more than 100 different ageing specimens in several clean gas ducts. The typical average analysis of the numerous condensate samples obtained in the clean gas ducts is: pH 1.4 to 2, 70 °C, < 2,000 ppm Cl^-, particulates.

In spite of small, time-dependent variations in the chemical composition of the condensate, the net result of all the field trials exhibited a clear trend. Simple stainless steels of types 1.4435 and 1.4439 up to the so-called sulphuric acid resistant steels like 1.4539 are damaged by the massive occurrence of localized corrosion after only a short period. Even the high-alloy austenitic 6% Mo steel exhibited isolated localized corrosion up to 0.2 mm deep, especially under deposits and the artificially applied cleavage blocks. Only the classical NiCrMo alloys with at least 56% nickel, 16% chromium and 15% molybdenum proved completely resistant under these conditions. Based on these ageing trials, a clean gas duct segment 35 m long was refurbished by means of a self-supporting inner liner. The materials used ranged from Cronifer 1925 hMo (1.4529) with its limited resistance up to Nicrofer 5716 hMoW - alloy C-276 (2.4819), which was totally resistant in the trials. After extensive calculations and intensive discussion regarding the working of the thin metal mill products with a wall thickness of 1.5 to 2.0 mm to be used, a self-supporting metal inner liner in the form of a tube made of the various material segments was placed in the dismantled duct segment using a purpose-built clamping unit. The narrow time frame allowed by the utility company for this refurbishment work was easily accommodated using this technology. Meanwhile, the third inspection after about 8,000 operational hours has revealed that the technique of fitting a self-supporting inner liner enabled damaged clean gas ducts to be given a lasting refurbishment without removing the rotted organically based corrosion protection system (synthetic resin, etc.) No corrosion or other material changes have so far been detected at the freely accessible surfaces at predetermined measurement points.

Case 3:
Clean gas re-heating in two combined heat and power stations in Berlin.

In two coal-fired combined heat and power stations, a special clean gas reheating technique was used to increase the impetus at the mouth of the stack. A hot gas is produced by combustion of heavy fuel oil or natural gas and is fed into the upward-flowing clean gas in the mixing chamber shown in Fig. 8.14a. A delta vane located in the flow, as shown in Fig. 8.14b, brings about mixing of the clean gas (T = 50 °C) and the hot gases (T = 750 °C) within a few metres. Because the corrosive conditions were not precisely known at the time of planning, the chamber and the delta vane were specified and manufactured in Nicrofer 6020 hMo (2.4856) welded with matching filler. Numerous inspections have shown that very corrosive conditions developed in the chamber with centimetre-thick deposits of soot from oil combustion as well as sublimed sulphur and the simultaneous presence of condensate moisture from the clean gas. After more than 7 years, the material has exhibited no indication of premature failure due to local corrosion, even in its welded form. It is to be expected that the mixing chamber will retain its full functionality over the life cycle of the power station, even with further variations in the operating conditions.

Figure 8.14: Static gas mixer for heating desulphurised clean gases for a Berlin combined heat and power station; Nicrofer 6020 hMo - alloy 625 (2.4856)
a) mixer chamber

8.2.5.3 Fittings and shut-off valves (dampers)

Flue gas dampers are control elements that direct the gas flow before and after flue gas scrubbing. This function enables complete FGD plant sections to be shut down or bypassed, e.g. for inspection or maintenance. There are three basic types depending on the application: guillotine as shown in Fig. 8.15, butterfly ("barn doors") and single or double louvre dampers. In this series the geometrical complexity increases as does the degree of corrosion prevention measures needed. Regardless of the design, all these valve types are manufactured from metallic materials, usually stainless steels because of the reliability demanded. These valves are subjected to a wide range of corrosive conditions depending on the location (raw gas/clean gas/reheating). This is described with reference to case studies:

Figure 8.14b: View of the mixing elements (impact plates and delta vane) from the outlet side

Case 1:
Clean gas duct shut-off damper in a coal-fired 740 MW power station

The power station discussed previously in Section 8.2.5.2 „Clean gas ducts" under Case 1 was provided with a clean gas guillotine damper made from Cronifer 1713 LCN (1.4439). After a relatively short time incipient local corrosion of an unspecified depth was found on this guillotine damper. This damper was manufactured using a hollow profile design and did not receive any trapped air heating

Figure 8.15: Valve for large flue gas ducts in desulphurisation plants: flue gas gate valve (guillotine damper); material: Nicrofer 5716 hMoW - alloy C-276 (2.4819)

before the commencement of FGD plant commissioning. For this reason, within a short period of a few months complete penetration of the flat section of the hollow frame chambers with permanently discharging condensate occurred. Under the clean gas duct conditions described previously in 8.2.5.2, i.e. condensate pH values below 2 and chloride contents up to 6.3 g/l, the material mentioned above 1.4439 basically cannot cope. In the search for a suitable corrosion-resistant material for sealing bars, 100-day trials were carried out with different metallic materials. Table 8.12 clearly illustrates that the spectrum of non-

Table 8.12: Long term corrosion behaviour of selected metallic materials in the exhaust gas duct of a coal-fired 740 MW power station; condensate: pH <<2, <0.63% Cl^-, < 0.004% F^-, 60 °C

Alloy	Corrosion rate mm/a	Surface change
Corten steel	1.8	Severely pitted
Cronifer 1713 LCN (1.4439)	0.5	100 % pitted surface 0.3 mm deep pitting
Cronifer 1925 hMo (1.4529)	no data	80 % under encrustration 0.3 mm deep pitting
Nicrofer 5716 hMoW - alloy C-276 (2.4819)	0.004	Hardly any corrosive attack visible

resistant to limited-resistance materials extends from weather-resistant Corten steel via the material 1.4439 to the material type 1.4529 (Cronifer 1925 hMo). Resistance only commences with the nickel-base alloys - in this case, for example, Nicrofer 5716 hMoW - alloy C-276 (2.4819). If there is no possibility of carrying out trapped air heating for space or economic reasons, this guillotine damper and the corresponding sealing segments can be refurbished with thin sheet cladding of, for example, Nicrofer 5716 hMoW - alloy C-276. In new installations it is advisable to use solid construction.

Case 2:
Clean gas duct louvre damper in lignite-fired large power stations.

In spite of different FGD processes, different designs, different clean gas duct routings and different reheating technologies, similar massive corrosion damage occurred on shut-off dampers made of the material 1.4439 after a relatively short period of operation in two lignite-fired large power stations. In one power station it was decided to refurbish four units with a total of 12 bypass louvre dampers upstream of the stack inlet. The materials used were Nicrofer 6020 hMo - alloy 625 (2.4856) and Nicrofer 5923 hMo - alloy 59 (2.4605) as reliable alternatives. The fabricator praised in this context the reliable, defect-free welding performance of the filler metal Nicrofer S 5923 compared to the older filler metal Nicrofer S 6020. The full area refurbishment required, per valve, a total of 30 m^2 of sheet/plate in the wall thicknesses 6.4 and 3 mm, with a total weight of 16 tonnes per damper. In addition to the already known clean gas duct conditions in lignite-fired power stations, the unreservedly favourable experience with the material Nicrofer 6020 hMo - alloy 625 in sealing strip elements helped to determine the material to be used for the refurbishment. In a similar type of power station, sealing strips made of 2.4856 had been successfully used for many years as 0.5 mm cold-rolled spring-hard strip. The damper frame, contact strip and the adjacent wall connection, which are located very close to each other and are made of ordinary Corten steel, had suffered enormous damage specifically in this power station. The stabilizer sheets of the sealing strips, probably made of ordinary type 1.4571 stainless steel, also exhibited uniformly distributed pitting. The fact that in this case both the contact strips made of 1.4539 as well as the bolts and stabilizer sheets made of 1.4571 had not also been extensively damaged is attributable to cell formation in the immediate vicinity of the Corten steel (wall). The Corten steel has acted as the sacrificial anode, thus causing the simpler stainless alloy steels to survive in this aggressive environment.

8.2.5.4 Stacks

Clean gas outlet stacks are - regardless of their design - the last place in the power station where the residual contents of the cleaned flue gases can precipitate. The height of the stack is governed by local conditions and depends amongst other things on the emission values permitted by the local authorities. The height of stacks is continually increasing because of the recently tightened requirements and the associated higher clean air dilution and wider dispersion. Multiple-flue stacks over 200 m tall are no rarity, particularly with newly erected plant. Whenever there is a risk of condensation or the deposition of sulphurous acid from the clean gas, it is necessary to line the chimney partly or completely with corrosion-resistant material. All-metal flues (tubes) within the concrete case are also conceivable.

Case 1:
300 MW coal-fired power station with a 150 m stack following the FGD plant.

The coal-fired power station mentioned above burns so-called soft coal (4.8% sulphur) and uses a double alkali process for wet desulphurisation. The clean gas outlet temperature after the absorber is 55 °C. The unit needing refurbishment had suffered a great many maintenance problems between the FGD plant and the stack in the first year of operation [17]. The use of an oil-fired clean gas reheater between the absorber outlet and the stack inlet failed to prevent the existing stack from being periodically attacked by wet clean gases because of the poor availability of the hot gas generator. Both the unprotected C steel in the interior of the stack as well as the lining of the top of the stack with a low-alloy stainless steel had been locally severely attacked. Further considerations led to the conclusion that the stack which had an excessively large diameter and consequently a low flow velocity, should be operated as, so to speak, the final "wet scrubber". The realization of these ideas led to the concept of wallpapering the complete stack with Nicrofer 5716 hMoW (2.4819), as shown in Fig. 8.16. The criteria for this corrosion-resistant cladding with large-format sheets were:

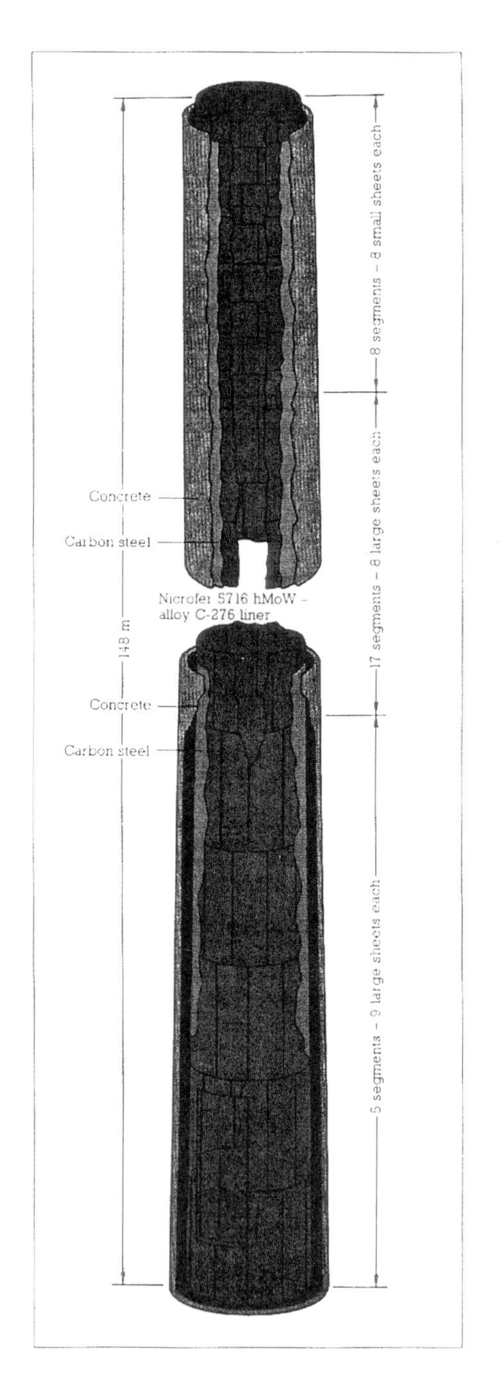

Figure 8.16:
Refurbishment of a concrete stack with a C steel inner tube (flue) by wallpapering with large format sheets; material: Nicrofer 5716 hMoW - alloy C-276 (2.4819)

1. A smooth, cold-rolled surface was preferred for reasons of aerosol precipitation.
2. In this context, the welds had to be minimized as potential flow obstructions. The use of large cold-rolled sheets made it possible to minimize the circumferential welds as well as saving on fabrication costs.
3. Temperature excursions, which may briefly peak at 180 °C during breakdowns, necessitate the selection of a highly corrosion-and temperature-resistant cladding material.
4. Inspections in the stack can only be carried out when the plant is completely shut down. The resultant requirement for process reliability automatically necessitates the use of a metallic cladding material.

The specifed material Nicrofer 5716 hMoW - alloy C-276 (2.4819) was produced in a thickness of 1.6 mm and a format of about 6000 x 2000 mm. In total, 190 of these large-format sheets with a total weight of 37 tonnes were supplied. The area of the chimney to be clad was about 2,700 m^2. The use of these cold-rolled large-format sheets enabled the weld length to be reduced by almost 50% in contrast to small sheet segment designs. As described in [18], the lined stack has been in operation for many years with no problems.

8.3 Summary and outlook

Flue gas desulphurisation plants are legally prescribed measures to safeguard the environment and must retain their functional efficiency in both new and refurbished plants for many decades. The chemistry of the highly efficient wet scrubbing process requires materials that can resist the ubiquitous attack of halogen-contaminated sulphuric acid for long periods. Numerous examples show that high-alloy stainless steels and nickel-base alloys are capable of meeting these requirements. The fuel- and process-specific features of each flue gas desulphurisation plant require a separate study and material selection. Whereas in the early eighties material selection for new construction was made relatively rashly because of legislative requirements, the tendency now is towards long-term reliable and thus economic metallic solutions for both the refurbishment of existing plants and the construction of new plants. The short life and repair proneness of certain organically based corrosion protection systems has contributed to this.

The advantages of metallic materials for the construction and operation of flue gas desulphurisation plants are summarized again as:

impermeable to gases and blister-free
resistant to swelling and free of porosity
mechanically and technologically resistant to ageing and embrittlement

corrosion resistant
non-combustible
environmentally friendly due to full recyclability
simple to work and repair

Thus the metallic material concept is a good solution considering all the evaluation criteria. The main advantages are reliability, economy and a long service life.

8.4 References Chapter 8

[1] W. Römer: Rauchgas-Entschwefelung - Verfahren, Werkstoffkonzepte, Case histories -, Krupp VDM GmbH (1991) available by Krupp VDM GmbH, Plettenberger Str. 2, D-58791 Werdohl

[2] A. Bäumel: Werkstoffauswahl für Bauteile in Rauchgasentschwefelungsanlagen, Haus der Technik, Essen, Fachtagung Nr. T-30-916-111-9 (1989)

[3] H. Kaiser: Langfristig mehr Chancen als Risken, Impressum ENVITEC FORUM, a publication of the Verlagsgruppe Handelsblatt, Düsseldorf

[4] D. S. Henzel, B. A. Laseke, E. O. Smith, D. O. Swenson: Handbook for Flue Gas Desulfurization Scrubbing with Limestone, Noyes Data Corporation, New Jersey, USA (1982)

[5] W. L. Silence, R. F. Miller, N. Ostroff, A. Kirschner: Simultaneous Design and Materials of Contruction Selection for FGD Systems, CORROSION 84, Paper No. 299, The National Association of Corrosion Engineers, Houston, Texas, 1984

[5a] U. Sander, U. Rothe, R. Kola, in: Ullmanns Encyklopädie der technischen Chemie, 4. Auflage, Band 21, Page 118, Weinheim (1982)

[6] D. R. Holmes: Dewpoint Corrosion, Institution of Corrosion Science and Technology / Ellis Horwood Limited, England (1985)

[7] Korrosionsbeständige VDM-Werkstoffe für Rauchgasentschwefelungsanlagen, VDM report No. 18 (1991), available by Krupp VDM GmbH, Plettenberger Str. 2, D-58791 Werdohl

[8] A. Möllmann: Betriebserfahrungen mit Innengummierungen und Beschichtungen in Rauchgaswäschern und Reingaskanälen, VGB-Fachtagung Korrosionsschutz für Stahl und Stahlbeton im Kraftwerksbau, 1989

[9] J. A. Beavers, G. H. Koch, Review of Corrosion Related Failures in Flue Gas Desulfurization Systems, Materials Performance 10 (1982), Page 13 - 25

[10] recherched by E. Altpeter, Central Laboratory of Metallgesellschaft AG

[11] R. Kirchheiner, M. Köhler, U. Heubner: Nicrofer 5923 hMo, ein neuer hochkorrosionsbeständiger Werkstoff für die chemische Industrie, die Umwelttechnik und verwandte Anwendungen, Werkst. Korros. 43 (1992) 388 - 395

[12] D. C. Agarwal, A. Chronological Survey: Solution to FGD Corrosion Problems via new and old Alloys, Air Pollution Control Equipment Conference 1990, Paper No. 5

[13] Th. Hoffmann, D. C. Agarwal: "Wallpaper" installation guidelines and other fabrication procedures for FGD maintenance, repair and new construction with VDM high-performance nickel alloys, VDM report No. 18 (1991), available by Krupp VDM GmbH, Plettenberger Str. 2, D-58791 Werdohl

[14] R. Kirchheiner, Th. Hoffmann, F. Hofmann: Using Explosion-Clad Plates of Alloy 625 in Flue Gas Desulfirzation Plant, CORROSION 87, Paper No. 257, The National Association of Corrosion Engineers, Houston, Texas, 1987

[15] R. Kirchheiner, F. Stenner: Herstellung und Qualifizierung warmwalzplattierter Verbundwerkstoffe mit Nickelbasisauflagen für den Einsatz in Rauchgasentschwefelungsanlagen, Abschlußbericht zum F&E-Vorhaben BMFT-FB-2.6/7, Krupp VDM GmbH, Altena (1992)

[16] Plattierte Bleche - Die wirtschaftliche Lösung, Druckschrift TR-52/20012-90M-5000d of Voest-Alpine Stahl Linz Ges. m.b.H., Linz/Austria

[17] J. L. McDonald, V. D. Coppolecchia, G. K. Grossmann: Corrosion Resistance and Erection of Nickel Alloy Liners in FGD Systems, Mat. Performance 1987, No. 11, 11 - 16

[18] Texas Utility Goes to Wallpapering to Improve FGD Scrubber Life, Welding Journal, January 1992, Page 76

Index

Contributors

Dr.-Ing. Ulrich Heubner
Krupp VDM GmbH
Werdohl

Dr.-Ing. Ulrich Brill
Krupp VDM GmbH
Werdohl

Schweißfach-Ing. Theo Hoffmann
Krupp VDM GmbH
Werdohl

Dipl.-Ing. Rolf Kirchheiner
Krupp VDM GmbH
Werdohl

Dr.-Ing. Jutta Klöwer
Krupp VDM GmbH
Werdohl

Dipl.-Ing. Reiner Köcher
Düren

Dr. rer. nat. Manfred Rockel
Iserlohn

Dr. Frederick White
Krupp VDM GmbH
Werdohl